BIOGEOGRAPHY
Recent Advances
and future Directions

Biogeography

Recent Advances and Future Directions

Edited by

J. A. Taylor

BARNES & NOBLE BOOKS
Totowa, New Jersey

First published in the USA 1984 by
Barnes & Noble Books
81 Adams Drive, Totowa, New Jersey, 07512

Library of Congress Cataloging in Publication Data
Main entry under title:

Biogeography: recent advances and future directions.

1. Biogeography. I. Taylor, James Allan.
QH84.B55 1984 574.5 84-12428

ISBN 0-389-20507-9

Printed and bound in Great Britain

CONTENTS

LIST OF CONTRIBUTORS

Darch, J.P. Lecturer	School of Environmental Sciences, University of East Anglia, Norwich, U.K.
Flenley, J.R. Senior Lecturer	Geography Department, University of Hull, Hull, Humberside, U.K.
Greenland, D. Associate Professor	Cooperative Institute for Research in the Environmental Sciences, University of Colorado, Boulder, Colorado, U.S.A.
Moss, R.P. Professor of Human Ecology	Geography Department, University of Salford, Salford, U.K.
Putman, R.J. Lecturer	Department of Biology, University of Southampton, Southampton, U.K.
Shimwell, D.W. Senior Lecturer	Geography Department, University of Manchester, Manchester, U.K.
Smith, R.T. Lecturer	Geography Department, University of Leeds, Leeds, U.K.
Stott, P. Lecturer	Geography Department, School of Oriental and African Studies, University of London, London, U.K.
Taylor, J.A. University Reader	Geography Department, University College of Wales, Aberystwyth, Dyfed, UK.
Watts, D. Senior Lecturer	Geography Department, University of Hull, Hull, Humberside, U.K.

PREFACE

Biogeography is currently developing apace but it is not in any sense a new subject. Its modern evolution dates from the last century, but its roots can be traced over two thousand years in the writings of Greek philosophers such as Aristotle (384-322 BC), Theoprastus (c. 370-287 BC) and Virgil (70-19 BC), whose Georgics expounded the art of husbandry and implied many of the basic principles of countryside management which apply today. The foundations of modern biogeography were laid in the nineteenth century, and it is on the shoulders of such major pioneers as Lamarck (1744-1829), Malthus (1766-1834), von Humboldt (1769-1859), de Candolle (1806-1893), Darwin (1809-1882), Wallace (1823-1913), Huxley (1825-1895) and Haeckel (1834-1919) that we now stand. From these sources stemmed the biogeographical ideas that were ancestral to modern theories. Such distinguished but diverse origins were bound to bestow the widest dimensions and the most complex of philosophies to biogeography which in the twentieth century was to undergo separate and sometimes unrelated and variably timed advances, in the context of both the biological and the geographical sciences. The emergence of a double identity and differences in emphasis, methods and objectives have served to confuse the definition and scope of biogeography, and delayed its international establishment as an independent basic science in its own right, aside from its traditional, if varying, role in geographical and biological curricula.

In essence, biogeography is broadly analogous to what biologists traditionally regard as ecology but with the deliberate inclusion, not the specific exclusion, of man both as an animal and as a modifier and replacer of natural ecosystems. Coincidentally, applied ecology has major overlaps with biogeographical interests; so do the environmental sciences as recently regrouped at some British universities. However, biogeographical studies have developed at different rates and at different times in departments of biology and geography so that its current stage and status are difficult to characterise to the satisfaction of all types of practitioner involved.

Notable among the earlier general texts on biogeography is that of Emmanuel de Martonne, whose review, entitled Biogeographie, was published in 1921 as part of the fourth edition of volume III of his Traite de Geographie Physique. It concentrated on explanations of the spatial distributions in plant and animal geography. In 1936, Marion Newbigin's Plant and Animal Geography was published posthumously under the guiding hand of H.J. Fleure. She had been trained initially as a botanist and zoologist and subsequently as a geographer. Part I of the volume presents a broad perspective on 'Life and its environment' and includes two chapters on soils and climate change, written by Margaret Dunlop, which were certainly ahead of their time. Part II focusses on the major plant communities and their animal associates and, again, in Part III plants and animals are discussed together in terms of taxonomic distributions. Part IV deals briefly with factors of plant geography, using global and regional examples. It was Pierre Dansereau's book, however, entitled Biogeography: an Ecological Perspective, published in 1957, which revealed the broader shape of biogeographical studies. The preface opens with the following statement:

> The scope of this book extends across the fields of plant and animal ecology and geography, with many overlaps into genetics, human geography, anthropology and the social sciences.

The chapter sequence includes one on 'Bioclimates' and another on 'Man's impact on the landscape'. The book made the case for the inclusion of aspects of human ecology alongside animal and plant ecology, and anticipated the comprehensive ecosystematic approach adopted by several biogeography texts which appeared subsequently, notably David Watts' Principles of Biogeography: an Introduction into the Functional Mechanisms of Ecosystems, published in 1971, and Ian Simmons' Biogeography: Natural and Cultural, which was published in 1979. Following Dansereau, Watts' approach is systematic and multidisciplinary. Simmons' philosophy is more sharply focussed at the interface between natural ecosystems and their adaptation by man, pointing to fundamental issues about ecological balance in resource use and environmental management.

In reality, biogeography is an assemblage of several discrete but related systematic and integrative studies. Their range and depth lie beyond the expertise of any individual scientist, and that is why this thematic textbook on biogeography has been created on the basis of eleven invited review essays, written by ten different authors who are established specialists in their particular fields of biogeographical research and teaching. The selection of essays is a representative range rather than a comprehensive coverage of the subject. The essays have been pitched at advanced undergraduate level, i.e. the second and third year (usually termed Part II) of a university degree course. However, they are so styled and structured that first year (Part 1) university students, college students and the brighter VIth former or High School student should find them a not inaccessible challenge. Indeed, the mostly jargon-free style, the detailing of principles, the presentation of carefully selected case-studies should also appeal to students who have not received any previous training in biogeography. Again, although the essays form a logical, readable sequence, each one stands on its own theme and may be read separately from the others. The bibliographies following each essay are up-to-date and international, and the 'further reading' lists provide deeper, guided access into the literature and associated research frontiers. A glossary of important terms is included towards the end of the book, along with a detailed subject index. In summary, this book provides rapid, condensed, authoritative penetrations of the literature and prepares the way for further reading and research. The contributors to this volume, taken together, represent varied combinations of expertise in the geographical, biological and environmental sciences, and it follows that students in departments of any one of those three affiliations will find some, if not all, of these essays of relevance and of interest in their courses of study. Biogeography in biology departments is usually concerned with explaining plant and animal distributions as a key to their evolution, expressed in global terms and involving changing land-sea patterns and changing environments, notably of climate. In contrast, biology departments also adopt the micro-scale approach both in the field and the laboratory, currently emphasising autecology rather than synecology. Curiously enough, geography departments tend to follow a meso-scale approach, adopting communities rather than individuals for study, e.g. the analysis of vegetation distributions on a local scale. Again, environmental factors of soil and climate and the impact of man on biota, both now and in the past, are usually given priority. There is thus some welcome, if unrehearsed, complementarity in the current biological and geographical treatments of the subject of biogeography which must work to its long term academic advantage.

Contents

The opening review, Chapter 1, by Philip Stott, traces the history of biogeography from its classical ancestry, through the formative years of the last century to the ups and downs of the present century and its remarkable renaissance in the 1960s. The separate flows of ideas emanating from biological and later geographical intellectual sources are skilfully charted, and their divergent and convergent stages are carefully delineated and timed. Biogeography emerges as a somewhat controversial but challenging science of variable and unpredictable popularity. Sometimes, inadequate basic training in plant and animal taxonomy has proved a deterrent. Again, an insufficient understanding of the physical environment as it affects biological distributions has, from time to time, proved a handicap. It is in joint courses provided by physical geographers and ecologists that such imbalances may be corrected. Some courses in environmental science are designed to do just that.

It is difficult to appraise developments in biogeography during the last twenty-five years or so because, although there has been remarkable growth and, in geography departments, a renaissance, we are still in the midst of the cycle which has engendered as much fierce debate about theory and identity as it has resolved old scores and controversies. The Darwinist/cladist confrontation is discussed by John Flenley in Chapter 3 on 'time-scales' and the Editor, in his final Chapter 11, assesses recent theoretical and methodological advances in the biological sciences in relation to the necessary extension of hypothetico-deductive strategies in physical geography. In Chapter 2, David Watts discusses the spatial dimension in biogeography. Distribution over space is, traditionally, a geographical focus embodying the fundamental concept of location. Yet, the explanation of the distribution of the world's biota has commanded much attention from biologists, in particular zoologists. Many previous attempts to define biogeography have adopted the spatial dimension as a primary base. But location is relative rather than absolute, especially in view of the constantly changing land-sea pattern of the earth's surface over recent, as well as geological time. It is only in relation to operative timescales that the space dimension becomes meaningful. Watts concentrates on species range and density variation, categorisation and type at regional and world scales, and finally deplores the standardising impact of man who, in perpetually mixing the world's biotas together, could create simplistic biogeographical regions and an overall reduction in the earth's gene-pool resources.

Time is important in biogeography for four main reasons, according to Flenley (Chapter 3); these are evolution, migration, environmental change and autogenic succession. Classical Darwinism has been displaced by Neo-Darwinism which accepts the gene-pool as the fundamental unit in evolution and accommodates larger mutations as well as the small heritable variations _a la_ Darwin. Migration of populations has occurred at greatly varying rates and is successful at different distances, but some long distance dispersal to oceanic islands must have occurred. Environmental changes range from continental drift and land-sea level oscillations to climatic variability, especially in recent millenia where soil, vegetation, animals and, not least, man may trigger local environmental changes independently for the general global controls. Autogenic succession incorporates the concept of climatic climax which is becoming less and less acceptable as tropical communities, for example, are being discovered to be quite young, not old, and liable to short-term change and instability.

Rowland Moss in Chapter 4 presents a critical, candid but constructive personal appraisal of the role of the overworked ecosystem concept in both biological and geographical teaching and research with particular reference to ecological territory shared, and sometimes disputed, between departments of geography and biology. Any tendencies to what is internecine strife should be replaced by an awareness of mutual need and reciprocity, as the Editor cogently argues later in Chapter 11. In any event, political in-fighting within the separate biological and geographical camps is as much to blame for intellectual misunderstandings between biologically trained biogeographers on the one hand and geographically trained ones on the other. The challenges and responsibilities of pure and applied research in environmental problems are overwhelmingly too great and associated educational tasks, not least in the area of environmental management, too forbidding for biogeographers of all creeds to be too concerned about academic birthrights and territorialities.

This is not a book about biogeographical techniques as such, and techniques may be regarded as universally available to all sciences rather than belonging to any particular one. Indeed, technology has come of age as a separate science; this has applied to statistics and computing and is now applicable to remote sensing, microcomputing and information technology. Whilst it was deemed appropriate to include a chapter (10) in this book on the special biogeographical application of remote sensing, it was thought that David Shimwell's Chapter 5 on 'Vegetation Analysis' should concentrate on the descriptive classificatory procedures, but included in the 'further reading' list are additional explicit references on the more advanced numerical and statistical methods for analysing biological data. Shimwell outlines some of the basic premises of vegetation description, classification and ordination, and illustrates a selection of descriptive and observational methods using morphology, structure, frequency, etc. For the discipline of adopting the right sampling technique for the field location of quadrats, releves (including their own internal recording) on a grid network or along transect lines, stratified random procedures, adjusted to the spatial realities of the habitat and terrain type-frequency within the prescribed survey area, are normally to be recommended. The bridging of quadrat-based vegetation survey and remotely-sensed vegetation information to meso-scale vegetation maps is discussed in a section in Chapter 11 (p. 359). Shimwell deals finally with several approaches to the classification of world vegetation and points to the biogeographer's future role in the use of rapid vegetation and environmental analysis techniques (e.g. satellite imagery) as a basis for global conservation strategies (see Chapter 10).

In an attempt to encourage the introduction and extension of zoogeographical studies in biogeography curricula as currently taught in many geography departments, an essay (Chapter 6) on the geography of animal communities by Rory Putman has been included herein. He begins with distribution patterns, the limits of tolerance they reveal for specific species and how such limits might be expanded by acclimation. The world's biomes are outlined as animal habitats and the associated network of global zoogeographical realms is presented. The concept of global diversity in animal species is then discussed in the context of the important and much debated theory of 'island biogeography', which postulates relationships between rates of colonisation and distance of the island to the nearest mainland, and between island size and extinction rates. Finally, the structure of animal communities is discussed and it is concluded that 'whilst (animal) population ecology has made notable advances, (animal) community ecology is still in its infancy'.

In Chapter 7, Richard Smith details the role of soils within ecosystems. The soil is conceived as a dynamic system, continuously active and reactive within the ecosystem as a whole and sometimes initiating or directing the nature and rate of change in the ecosystem. Plant-soil relationships are analysed in terms of biogeographical cycling, nutrient budgets and environmental pollution risks. Soils are finally evaluated as environmental, biological and spatial resources which are in grave danger of irrevocable depletion as pressure on land-use increases at accelerated rates the world over.

Because bioclimatology lies precisely in the same sort of intermediate position, i.e. between biology and climatology, as mainstream biogeography, and because climate may be regarded as the independent (initially at least), non-terrestrial, environmental variable and energy source of the biosphere, it seemed logical to include specific treatment of it in a biogeographical context in this volume. Hence Chapter 8 on 'Bioclimates' by David Greenland. He emphasises the intense complexity of the plant climate or the animal climate and the continuing difficulties involved in trying to measure and interpret them at all scales, micro, meso and macro. The initial problem of estimating primary productivity is discussed; further amplification of the point is given in Chapter 9 (p. 256). Using case-studies from the alpine tundra and the tropical rain forest, the difficulties of relating bioclimates to vegetation productivity are demonstrated. What is required is improved instrumentation of greater accuracy, increased knowledge of plant-atmosphere interactions on a species-by-species basis, vigorous research into faunal climates and, finally, the application of bioclimatology to studies of past climate, climate change and contemporary economic and social needs of man.

In Chapter 9, the Editor presents a reappraisal of the man/environment relationship as it operates in both primitive and modern land-use systems. The inherent capacity of natural systems is discussed in relation to energy climate, biological productivity and the broader concept of land productivity, which prepares the way for introducing the management factor and linkages with notions of economic productivity. Possible ways in which the earliest conversions of natural ecosystems by man and related studies in 'cultural biogeography' are detailed, and the continuity of historical choice of the best land resources available to a given settlement is demonstrated. Finally, a range of modern case studies, mostly from Britain, but including some international ones, confirms that ecological constraints still register in the operation and efficiency of some of the most intensively managed modern agricultural systems. The central argument of the essay is that man and environment can be studied, researched and taught together, and their subtle interplays and interdependencies exposed, despite the fact that it is intellectually much easier to study each one separately.

Chapter 10 by Janice Darch is a specifically technical review of the use of remote sensing techniques in biogeography. This rapidly expanding facility has particular relevance in crop-yield recording and predicting, land use survey, disease monitoring and prediction in agriculture and forestry, vegetation mapping, inventories of ground and water surface conditions, e.g. hydrology, temperature fields, ocean current displacement, climatic trends, algal blooms, etc. The continuing need for sampled injections of 'ground truth' will remain, and the frontier of this research area will always relate to the contemporary stage of sophistication of the available satellite technology.

In the final essay (Chapter 11), the Editor endeavours to integrate the arguments portrayed in the ten preceding specialist reviews. Biogeography is redefined as a special interdisciplinary sub-field in terms of its mid-twentieth century renaissance. Its theoretical basis is analysed and the need for hypothetico-deductive models is stressed. The nature and role of applied biogeography is detailed (a) within physical geography, (b) within human geography and (c) as an integrating force between physical and human geography. The essential need for reciprocity between geographers and ecologists is emphasised and areas for extended cooperation are outlined. The conservational commitment of biogeography in a rapidly changing world must adjust to such changes and itself be dynamic and forward looking. New directions for biogeography include (a) innovation in palynological research which has been so repetitive in recent decades, (b) the reconciliation of ecological and economic methods in land use analysis, (c) new biogeographical applications, internationally, in remote sensing and (d) the establishment of public and private biogeographical consultancy following trails already blazed in diverse environments by a distinguished few.

Acknowledgements

The Editor owes much to exchanges of ideas with academic colleagues both in Britain and overseas over the past thirty years, during which period biogeography has come of age. He is especially indebted to the encouragement and support of colleagues at Aberystwyth, to the stimulus of a continuum of research students and, over the past decade, to the enthusiastic 'core' members of the BSG (Biogeography Study Group) which is formally affiliated with the Institute of British Geographers. The Editor is most grateful to Professor C. Kidson and colleagues for allowing him to carry a lighter teaching load than usual in session 1982/83, and to Professor Harold Carter and Professor David Bowen for allowing access to departmental facilities to help to prepare material for this volume. The prompt assistance of the Hugh Owen Library staff at Aberystwyth in facilitating access to references was greatly appreciated. All the contributors have been exemplary in their preparations and submissions and in cooperating with the Editor in what turned out to be a less formidable task than was originally feared. The kind and efficient services of the publishers, through the offices of Peter Sowden and Gill Cox, are finally gratefully acknowledged.

Aberystwyth

J.A. Taylor
Editor

PERMISSIONS

The following institutions, publishers, journals and authors have kindly given permission for copyright material in the Figures and/or Tables specified to be published in this volume.

Academic Press: Table 2.3; Figures 2.8, 3.5.

American Association for the Advancement of Science: Figure 3.1.

Benjamin/Cummings Publishing Company: Figure 6.3.

Biological Review: Table 9.3(a).

Blackwells Scientific Publications: Tables 2.2, 5.6, 5.7, 6.1, 7.2, 11.3; Figures 2.12, 3.6, 3.9.

Blumea (Rijks Herbarium): Figure 3.5.

British Archaeological Reports: Figure 11.1.

British Museum Press, Honolulu: Figure 3.4.

Cambridge University Press: Figure 3.7, Table 4.1.

Centre of Agricultural Geography, Verona, Italy: Figure 11.3.

Council for British Archaeology: Figure 9.4.

Columbia University Press: Figure 2.11.

Ecole Normale, Paris: Figure 2.6.

Edward Arnold: Tables 9.3, 9.4; Figure 11.2.

Elsevier Scientific Publishing Company Ltd: Table 4.3.

Flora Melansiana: Figure 3.4

Geographical Journal: Figure 3.8.

Geographical Review: Figure 9.3.

Harper and Row: Table 4.2.

Harvard University Press: Figures 2.8 (b), 9.2, 11.5.

Heinemann Educational Books Ltd.: Figures 2.2, 6.6

Permissions

Holt-Saunders: Table 4.2.

Institute of British Geographers: Tables 9.6 (a) and (b).

Longman: Figures 2.3, 2.4, 9.1; Table 9.1.

MacMillan Publishing Company: Table 5.5; Figure 6.4.

McGraw-Hill Book Company: Table 4.2.

Memoirs of the Geological Society of America: Figure 3.7.

Methuen: Figures 9.5, 9.6.

Oxford University Press: Figures 3.8, 7.5, 7.11.

Panstwowe Wydawnictwo Naukowe (PWN), Warsaw: Figures 2.8 (a) and (b).

Pergamon Press Ltd.: Table 2.1; Figure 2.4.

Plenum Publishing Company: Figure 11.2 (b).

Proceedings of the Ecological Society of Australia: Figure 11.2 (a).

Quaternary Research: Figure 3.11.

Regional Science Association: Figures 9.11, 11.4.

Revue de Geologie Dynamique et de Geographie Physique: Figure 3.10.

Royal Society: Figure 3.12.

Transactions of the Anglesey Antiquarian Society: Table 9.4.

UNESCO, Paris: Table 9.3 (b).

University of California Press: Figure 2.5.

University of Nottingham (Department of Agricultural Economics): Figures 9.9, 9.10.

Walter de Gruyter (Berlin): Table 5.5 (from Schmithusen, J. (1968), 'Allgemeine Vegetationsgeographie', in Lehrbuch der Allgemeinen Geographie, Vol. 4, 3rd edition (eds. Obst, E., Schmithusen, J.), Walter de Gruyter, Berlin).

Welsh Soils Discussion Group: Table 9.4.

John Wiley & Sons Ltd.: Figures 2.1, 6.5.

J.W. Aitchison: Table 9.5; Figure 9.8.
W.J. Armstrong: Tables 9.6 (a) and (b).
A.R. Aston: Figure 11.2 (a).
D.I. Axelrod: Figure 2.5.
D.F. Ball: Table 9.4.
M.G. Barbour: Figure 6.3.
N.I. Basilevich: Table 9.3 (b).
S.C. Beckett: Figure 3.9.
J.C. Bernabo: Figure 3.11.
J.W. Birch: Figure 11.3.
P.W. Birkeland: Figure 7.5.
K.S. Brown: Figure 2.11.
S.B. Chapman: Table 7.2.
R.H. Collingbourne: Figure 9.2; Table 9.1.
G.R. Coope: Figure 3.10.
J.G. Cruickshank: Tables 9.6 (a) and (b).
P.J. Darlington: Figure 6.5.
R.J. Darlington: Figure 2.8 (e).
M.E. Daw: Figures 9.9, 9.10.
A. Ellison: Figures 9.5, 9.6.
E. Excurra: Figure 2.12.
D.C. Ferns: Figure 10.1.
F.R. Fosberg: Tables 5.6, 5.7.
W. George: Figures 2.2, 6.6.
R. Good: Figures 2.3, 2.8 (c).
D.W. Goodall: Table 4.3.
A.S. Goudie: Figure 3.8.
A.T. Grove: Figure 3.8.
J. Harriss: Figures 9.5, 9.6
D.J.L. Harding: Table 9.2
J.E. Hay: Figure 9.3
T. van der Hammen: Figure 3.6.
F.K. Hare: Figure 9.3.
H. Heatwole: Table 6.1.
A.R. Higgs: Table 11.3.
G.R.J. Jones: Table 9.4.
M.R. Kelly: Figure 3.12.
J. Kornas: Table 2.3.
C.J. Krebs: Table 4.2.
R. Levins: Table 6.1.
G.E. Likens: Figure 7.11.
M.W. McElhinny: Figure 3.1.
K. Newcombe: Figure 11.2 (b).
E.P. Odum: Table 4.2.
N.D. Opdyke: Figure 3.7.
J.R. Packham: Table 9.2.
J.R. Pilcher: Figure 9.4.
E.C. Pielou: Figure 2.1.
E.H. Rapoport: Tables 2.1, 2.2, 2.4; Figures 2.7, 2.9, 2.12.
P.H. Raven: Figure 2.5.
L.E. Rodin: Table 9.3 (b).

J. Schmithusen: Table 5.5.
N. Shackleton: Figure 3.7.
A.G. Smith: Figure 3.4.
E. Soepadno: Figure 3.4.
C.G.G.J. van Steenis: Figure 3.5.
F.A. Street: Figure 3.8.
W. Szafer: Figures 2.4, 2.8 (a) and (b).
J.R. Turner: Figure 2.6.
Thompson Webb III: Figure 3.11.
J. Tivy: Figure 2.4.
J.G. Tyrrell: Figure 9.8.
D. Watts: Table 4.1.
D.F. Westlake: Table 9.3 (a).
R.H. Whittaker: Table 5.5; Figure 6.4.
E.O. Willis: Figure 11.5.
E.O. Wilson: Figure 11.5.

LIST OF FIGURES

List of Figures

List of Figures

LIST OF TABLES

To all students of biogeography,

past, present and future

Chapter 1

HISTORY OF BIOGEOGRAPHY

P. Stott

> '... that complexity which is the charm and at the same time the difficulty of biogeographical study'.
>
> H.J. Fleure in 'A Tribute' to Dr Marion Newbigin, written for the posthumous publication of her book, *Plant and Animal Geography*, 1936.

Both in her person and in her work, Marion Newbigin stands as an exemplar of the tensions that have characterised the history of the complex subject of biogeography. With Emmanuel de Martonne (1927), she was one of the first geographers to write a comprehensive introduction to plant and animal geography. However, as Fleure pointed out in his moving tribute, she was trained as a biologist in the days before the modern geographical movement had got under way. Her original interest was in zoology and as early as 1913 she had written a book entitled *Animal Geography*. But her major work, *Plant and Animal Geography*, which first appeared in 1936, was based on a series of lectures given over a number of years to the geography students of Bedford College in the University of London.

The book has been rightly praised as 'stimulating' by Nicholas Polunin (1960: 21), although it is true to say that the so-called second edition of 1948 was little more than a reprint and failed to remedy some errors of the first, especially, as Polunin notes, in those chapters for which the original author was not wholly responsible.[1] Sadly, the first edition had been published just after Newbigin's death. It is interesting to observe that her main sources of reference were the great ecological classics of Wilhelm Schimper (1903) and Eugenius Warming (1909), both of which will be considered at length later, and the slighter works of Gaussen (1933) and Prenant (1933). She also employed an old 'pot-boiler' by Marcel Hardy called *The Geography of Plants*. This had been issued first in 1920 to replace the author's earlier *A Junior Plant Geography* (1913) and continued to be reprinted until 1952. She also made considerable use, however, of the work of the later contemporary ecologists, including Tansley (1911, 1923), Tansley and Chipp (1926) and Leach (1933), and the

Preface, which was written in fact by her sister, Florence Newbigin, went so far as to claim that biogeography 'corresponds to a large extent to what botanists call Ecology'. This was penned despite the fact that at least one-third of the book deals solely with the taxonomic distribution of plants and animals over the surface of the earth.

Marion Newbigin, therefore, clearly shows us the three major tensions that afflict the biogeographer. First, there is the division between plant geography (phytogeography) and animal geography (zoogeography), with very few practitioners of the art proving equally competent in both. Plant geography has tended to dominate the textbooks and the ecological field of study, although in recent years zoogeography has proved the main stimulus in the generation of theories concerning the world distribution of organisms (e.g. Udvardy 1969). Secondly, there is the fact that biogeography is both taught and researched in two of the broader traditional disciplines, namely biology and geography, which have often developed their own highly distinctive approaches to the subject and between which communication has not always been what it should be. Newbigin herself, of course, like many later biogeographers, crossed the divide, in her case from zoology to geography. Thirdly, and perhaps most seriously, there is the problem of defining the 'geography' element in the subject.

Generally speaking, and notwithstanding the difficult division between plants and animals, the bio aspect of the subject has posed few problems. All approaches are concerned in some way, however tenuous, with living organisms. But the 'geography' element has constantly created tension, well exemplified by the uneasy division between taxonomic geography and ecology in Newbigin's own book. This same problem was encountered by another respected geographer, Dudley Stamp, when he was asked in 1962 to provide a geographer's postscript to the Systematic Association's volume on Taxonomy and Geography, a publication dominated by the work of biologists. Stamp was forced to observe that:

> 'Although the title of this publication is Taxonomy and Geography, and although the words 'geography', 'geographical', and 'geographic' appear in the titles of four of the contributions and are used repeatedly in all, it is by no means clear whether any consideration has been given to their precise meaning, or whether the separate authors mean the same thing in using them'.

The divide is between a purely spatial view of geography, in which the biogeographer is primarily concerned to explain the distributions of certain biologically defined phenomena, such as species, genera and families, and a more ecological approach, particularly when this is taken to embrace the relationships of man with his environment. It is especially interesting to note that the Preface to Newbigin's book totally dismisses this latter problem with the peremptory statement, 'the complex subject of man is of necessity omitted'. No modern geographer could escape with such a clause, for, as Chorley has written (1973: 157) 'even the Garden of Eden had its entrepreneur' and 'man's relation to nature is increasingly one of dominance and control, however lovers of nature may deplore it.' Indeed, since the important address of Barrows to the American Association of Geographers in 1923, human ecology has inevitably become a central tenet in any definition of geography, even when it carries the prefix, bio. This wider definition has, moreover, frequently taken the subject of biogeography out of its traditional home in departments of biology and geography to place it in the wider context of environmental science in general, with its inevitable bias towards the applied aspects of the subject, both social and political. Hence, biogeography has also come to embrace the great themes of conservation and environmental management.

Thus, in any history of biogeography written by a geographer, it is essential to recognise at least three main traditions in the subject. First, there is the study of ecology. Secondly, there is the study of the role of man in nature. Thirdly, there is the traditional core of the subject which deals with the spatial distribution of living organisms and structures over the earth's surface. It is a great pity that many geographers have come to dismiss this last tradition with the rather snide phrase that it is 'geography for the biologists'. As has already been noted elsewhere (Stott, 1981), 'we must retain a comprehensive view of the science', if we are to maintain the 'charm' of this complex subject referred to by Fleure.

The Ecological Tradition

It appears that the first writer to use the word 'ecology' which, like 'economy', is derived from the Greek oikos for 'home' or 'household', was the American essayist and nature-lover, Henry David Thoreau (1817-62), who introduced the term into a letter in 1858. However, the earliest scientist to put meaning into the word was a German correspondent of Charles Darwin, Ernst Haeckel (1834-1919), a biologist and popularist philosopher, who defined it in 1866 as 'the total science of the relationships of the organism to the surrounding environment, within which we can include in a further sense all conditions for life'. This is a distinctly autecological circumscription but since then the term has taken on an increasingly synecological character. By 1876, in The History of Creation, Haeckel himself had redefined the subject as the study of 'the correlations between all organisms living together in one and the same locality and their adaptation to their surroundings.' During the twentieth century, there have been many attempts to improve on this definition, from that of Charles Elton (1927) in which ecology is seen as 'scientific natural history', through Eugene Odum's 'the structure and function of nature' to Charles Krebs' (1978) 'the scientific study of the interactions that determine the distribution and abundance of organisms'. All agree, however, that ecology is essentially concerned with the relationships of the organism or organisms and the environment or 'home'.

Although the term first made an appearance in the latter part of the nineteenth century, 'the roots of ecology' as Krebs (1978: 4) puts it, are as old as man himself, for they 'lie in natural history'. As early as the fourth century B.C., in Historia Animalium, Aristotle was expounding sound ecological principles about plagues of locusts and field mice, and Egerton (1968) has demonstrated a clear link between the Greek view of the harmony of nature and the more modern concept of the balance of nature.

However, the analytical framework of the subject was to take a long time to develop and eventually depended on the work of the pioneer human demographers, such as Graunt (1662) and Malthus (1798), and then on the evolutionary understanding of the scientific giants of the nineteenth century, such as the geologist, Lyell, and the naturalists, Darwin, Wallace, Huxley and Hooker, and, among many others, Lamarck. It is surely not without considerable significance that the word 'ecology' was coined just one year before the publication of On the Origin of Species by Means of Natural Selection (1859), and defined but seven years later. It is worth noting, however, that as early as 1844, Edward Forbes was describing the distribution of animals in the coastal waters of Britain and part of the Mediterranean, demonstrating how different associations were characteristic of zones of differing depths.

As Stoddart (1966) has rightly demonstrated, Darwin and the subsequent Darwinists have influenced the development of geography in at least four ways. Obviously, they helped to establish in the subject the concept of change through time or evolution. Similarly, the ideas of struggle or natural selection and of randomness or chance variations in nature have attracted interest. But lastly, they also developed the crucial concepts of association and organisation in nature, with man as a part of this organisation. The very discussion on evolution brought into closer focus the relationships of organisms with their environment, but the earliest view of these relationships was, as we have seen, autecological in character. However, this was soon modified to take into account the relationships of groups of organisms, thus establishing the synecological tradition with its emphasis on association and organisation in nature. An early indication of this more holistic approach is to be found in the classic work of Karl Möbius (1877) on the oyster-bed community, for which he coined the word biocoenosis to describe the internal relationships of living communities. Much later, in the 1940s, a Russian forester, Vladimir Sukachev, was to develop this term further, introducing the expanded word biogeocoenosis to describe the sum total of ecological niches (both plant and animal) with their environment (Sukachev and Dylis, 1968). Another significant study in the new mould was that of S.A. Forbes (1887) on the lake as a 'microcosm', in which he argued that the species assemblage of a lake was an organic complex in which a change in one species would affect the whole. Here we have a clear precursor of the later concept of the ecosystem which has become so fashionable in recent times.

By the end of the nineteenth century, the synecological approach had seen the publication of its first two major textbooks, both of which emphasised the study of whole vegetation formations.

The earlier of these was by a Danish Professor of Botany at the University of Copenhagen, Eugenius Warming (1895), and was entitled Plantesamfund. Grundträk af den ökologiska Plantegeografi. A German edition, edited by E. Knoblauch, appeared the following year, but the English translation by P. Groom and I.B. Balfour, called Oecology of Plants, was not available until 1909. The second was by a German professor at the University of Bonn, Wilhelm Schimper, which was published in German in 1898. The English version of this now classic work was a translation by W.R. Fisher entitled Plant Geography upon a Physiological Basis. It appeared in 1903, six years before the English edition of Warming's book, and has inspired and influenced Anglo-American biogeographers ever since.

In the Preface to his book, Warming boldly claims that his 'was the first attempt to write a work in Oecological Plant-geography', which, as Seddon (1974) has rightly intimated, is something of an overstatement. Elements of ecology, though not named as such, exist in many previous works, and not least in the Historia Plantarum of Theophrastus (c.370-287 B.C.; 1754). However, Warming (1909: 41) did admit that 'the ecology of plants is still in its infancy'. Yet it is undeniable that the achievements in synthesis of both Warming and Schimper were quite remarkable and the two books dominated thought in ecological biogeography until the concept of the ecosystem began to influence the structure of textbooks from the 1950s onwards.

Ecological Units. At the heart of both these great works were two major objectives. The first was the description and classification of the world's vegetation formations. The leader in this field was, without question, Schimper and many of his terms are still standard. For example, it was Schimper who first introduced the term 'tropical rain forest' (tropische Regenwald) and put flesh on the bare bones of its meaning by a world-wide survey of the structure and floristics of the formation. Today, the description and classification of vegetation is a very advanced science, often marred by rather fruitless methodological controversy, and dealing with the subject at all levels of enquiry, from the world surveys of Schimper to the study of the variations of pattern encountered in a single field. Recent reviews are provided by Shimwell (1971), Harrison (1971, 1975) and Greig-Smith (1983), among many others.

The second major objective, however, was the discovery of the 'ecological units' of nature, in contradistinction to the systematic or taxonomic units of the taxonomist. It was Warming who was particularly active in this search, which arose from an early recognition of the fact that certain groups or organisms, although quite unrelated taxonomically and by ancestry, might still possess very similar ecological characteristics. This understanding of the ecological importance of convergent evolution led to the development of the concept of the growth form, a type of common attribute in groups of different plants indicating adaptive significance. The approach was to a large extent autecological in nature, in that each organism was classified according to its key environmental or ecological relationship.

The products and terms of this era of ecology now litter the literature of the subject, but are not always to its ultimate benefit or elucidation. Classic examples of growth-form terms still in common usage are 'xerophyte', 'mesophyte' and 'hydrophyte', as well as more specialist terms like 'psammophyte' and 'sclerophyllous'. The confusion in understanding brought about by some of these rather vaguely defined 'catch-all' terms has been well reviewed by Seddon (1974). Later research has so often demonstrated that plants which appear to possess, say, xeromorphic characters do not always actually function as xerophytes when put to the experimental test. The most enduring growth-form system is undoubtedly that of the Danish botanist, Raunkiaer (1907, 1934), who devised a life-form system, still of value in modern studies, which classified plants according to the manner in which they survived the season of adverse conditions. He did this by grouping plants in relation to the position and character of their key perennating organs, the bud and the seed.

A Functional Approach. This early phase of ecology was, therefore, essentially descriptive in character and it was perhaps inevitable that a more functional view would eventually succeed it as the dominant approach in the science. The first hint of this change came in the study of the dynamics of vegetation succession, with North American ecologists leading the way, such as H.C. Cowles (1899), who analysed plant succession on the sand dunes at the end of Lake Michigan. The earliest major text on the subject, however, was the stimulating work of Frederic Edward Clements (1916; see also 1936, 1949) entitled Plant Succession: an Analysis of the Development of Vegetation, published by the Carnegie Institution of Washington DC, USA. This was, and still is, a controversial book. It perhaps symbolises more than any other work the approach of those ecologists who treat the community as a 'superorganism'.

Clements argued that the community was an 'organism' which grew and matured over time, ultimately achieving a relatively stable stage in balance with the prevailing climate of the area and the era: he called this final stage the climatic climax. This view has come to be known as Clements' monoclimax theory, the 'mono' referring to the fact that, for Clements, climate was the one overriding factor governing the ultimate development of vegetation. In Britain, however, the great ecologist, Arthur Tansley, although he too took an organismic view of the community, disagreed with Clements about the dominant role of climate, and he developed instead a polyclimax theory. In this approach, a wide range of environmental factors, such as soil (the edaphic climax) and man (the anthropogenic climax), might ultimately control the composition of the final stage of the climax. The finest monument to this polyclimax interpretation is Tansley's own masterpiece, The British Islands and their Vegetation (1939).

Gleason (1926), on the other hand, disagreed with both Clements and Tansley, and, more fundamentally, with the organismic view itself. He concluded that no plant assemblage could persist in a stable state over long periods of time. For Gleason, and for many later ecologists, a community is little more than an ecological abstraction, being in reality a chance collection of populations with the same environmental requirements. The convergence of succession on a wide scale is regarded as a highly improbable event (Whittaker 1953, 1962, 1970) and some recent studies have even indicated that successional divergence might be the norm (Matthews, 1979). Each species is seen as possessing its own specific range, and associations are thought of as mere assemblages of wandering populations. Whittaker (1953, 1970) thus defined climax vegetation as a pattern of species of abundances, which, while locally constant, varies from place to place in a continuous fashion.

The ecosystem. Tansley, however, is celebrated for more than his polyclimax theory, and his most lasting contribution to the science will inevitably be that now commonplace and much overworked word he coined in 1935, the 'ecosystem'. Ecosystems, of course, do not exist in reality, although one is hard put to grasp this simple fact when reading some textbooks of ecology. The ecosystem is a concept in which the living (biotic) community and the non-living (abiotic) environment are viewed as a functioning integrated system. The approach necessitates, therefore, the application of general systems theory to the study of ecology, again a truth not always wholly appreciated (Jeffers, 1978). The approach is essentially synecological in character and by the mid-1950s, following the pioneering work of Lindeman (1942) on Cedar Lake, had already become the major centre of attention in ecology and biogeography. This was, indeed, so much so that, by 1956, Evans could write, 'the ecosystem thus stands as a basic unit of ecology, a unit that is as important to this field of natural science as the species is to taxonomy and systematics.' It is especially interesting to observe the shift in emphasis indicated by this statement, from the earlier search for the autecological growth form as the basic unit of ecology to the new synecological unit of the ecosystem. Moreover, the approach is now far less descriptive and much more functional and dynamic, focussing as it does on the movement of matter and energy through defined systems.

Stoddart (1967) argues that the ecosystem concept appeals particularly to geographers because of four main basic characteristics in its make-up. First, it is monistic and embraces within the single concept the worlds of plants and animals and of man, thus permitting the ready study of the interactions between man and the living environment, often by means of quantitative methods and models. It thus bypasses neatly the age-old confrontation between determinism and possibilism. Secondly, the approach is essentially functional and looks at nature dynamically. Thirdly, it is a structured approach and organises nature in an orderly fashion. Lastly, it is a type of general system which, as we have already noted, can make use of the general theories of systems analysis. However, Chorley (1973) has had doubts about the value of the ecosystem concept for geographers as a whole and he has argued that it will be of true geographical significance only in so far as men can be said to function in the same way as other species. 'The ecological model' he predicts 'may fail as a supposed key to the general understanding of relations between modern society and nature, and therefore as a basis for contemporary geographical studies, because it casts social man in too subordinate and ineffectual a role.'

It is interesting to note, however, that the ecosystem concept recalls in many aspects the ideas of a number of earlier geographers, not the least of which is A.J. Herbertson (1865-1915), who devised the term 'macro-organism' to describe the 'complex entity' of the physical and organic elements of the earth's surface. In its turn, this work harks back to the thinking of the great German geographers, such as Friedrich Ratzel (1844-1904) and his concept of the state 'as an organism attached to the land', a view he propounded in Politische Geographie (1897). Similarly, the ecosystem closely parallels the independent work and ideas of the German geographer, Carl Troll (1899-1975), on landschaft ecology (Lauer and Klink, 1978) and of the inspiring Berkeley geographer, Carl Sauer (1889-1975), on landscape morphology (1925, 1963). It is not surprising, therefore, that geographers were so quick to appreciate the merits of the ecosystem concept, and in many ways it is little more than a functional approach to regional geography, though many would be somewhat aghast to recognise it as such. Whatever its merits and demerits, however, the concept is now clearly dominant in the biogeography textbooks of both geographers and ecologists, along with a number of more specialist approaches, based on population ecology (e.g. Kellman, 1975), conventional mathematics (Pielou, 1977, 1979), autecology (Daubenmire, 1974; Vermeij, 1978), and physiographic biogeography (Zimmermann and Thom, 1982). No one textbook, however, can today do justice to the multifarious complexity of the ecological branch of biogeography, particularly when the role of man in nature is also taken into due consideration.

The Role of Man

The ecological approach has also proved basic to the biogeographer's understanding of the role of man in nature. As Carl Sauer (1956) so succinctly expressed it, the central theme here 'is the capacity of man to alter his natural environment, the manner of his doing so, and the virtue of his actions'. For the biogeographer, the concern has been the cumulative effect of the biological processes that man has set in motion, or inhibited or deflected. The ancestry of this concern is long, but perhaps finds its fullest early expression in the pioneer works of the statesman-scholar of Vermont, George Perkins Marsh (Mumford, 1931; Lowenthal, 1958), which were suitably entitled Man and Nature (1864) and The Earth as Modified by Human Action (1885). Since then, many biogeographers and ecologists (e.g. Fosberg, 1963; Glacken, 1967) have focussed their attention on the delicate relationships between man and his environment.

First, man has obviously had a significant role to play in the domestication of both plants and animals for his own use. The study of this process constitutes what has come to be known as cultural biogeography (Simmons, 1980), although the terms ethnoecological, ethnobotanical and ethnozoological are increasingly in vogue (Carter, 1950). But as Harris (1982) has wisely pointed out, the latter terms tend to mean different things to different scholars. Many take ethnobotany, for example, as an all-inclusive designation to denote any study of the relationships between plants and people. Certain anthropologists, on the other hand, restrict its meaning to the study of indigenous peoples' understanding of their own plant worlds, a development closely related to the current growth of curiosity in perception geography. A classic example of work in this latter mould is provided by the recent study of Berlin et al (1974) into the plant world of the Tzeltal Indians of southern Mexico. Interestingly, this was a collaborative exercise by one anthropologist and two botanists.

The narrower topic of research into the origins, evolution and dispersal of crops is, however, already more than a century old. Although 'it did not escape the attention', to use Harris' (1982) words, of the great German polymath, Alexander von Humboldt (1805), it was really the Swiss botanist, Alphonse de Candolle, who first attempted a systematic treatment of the subject. This was his Origine des Plantes Cultivées of 1882. The Origine was, to quote Harris, 'remarkable in its comprehensiveness, its use of archaeological, historical, and linguistic, as well as botanical evidence, and in the certitude of its conclusions'. It also contains what appears to be the earliest reference to the biogeographical concept of the centres of origin of cultivated plants, a subject which was much developed by the later work of the Russian botanist and plant breeder, Nicolai Vavilov (1926, 1951). Since the Second World War there has been an explosion of interest in plant and animal domestication and early agriculture among many scientists, including geographers, beginning with the stimulating contribution of Carl Sauer in his Agricultural Origins and Dispersals (1952). In the context of biogeography, however, the contribution of David Harris, a geographer who now holds a Chair in the University of London in the field of archaeology, has been especially noteworthy, because he has come to stress the need to study the ecosystems within which the early domesticates were embedded (Harris, 1969 and 1973). Once again, we see the move from an autecological approach to the synecological.

The second major area of study has been in the applied aspects of ecology. This has proved an area of enormous importance in the last twenty years or so, with biogeographers and ecologists considering such crucial themes as land-use ecology, the ecology of resources, recreation ecology, the management of biotic communities and impact studies. These developments have been well reviewed in the work of Simmons (1974, 1979 and 1980) and they are also considered at length in later chapters of the present book, particularly in the contributions by Moss (Chapter 4) and Taylor (Chapter 11). In terms of the history of the subject, they are the natural outcome, especially for geographers, of an increased concentration on the relationships between man and the environment, which was perhaps inevitable with the pre-eminent growth of the synecological tradition within biogeography.

The Spatial Tradition

Whilst most geographers have thus busied themselves with the integrating concept of the ecosystem and with the applied aspects of ecology, it is interesting to find that many botanists and zoologists have been developing a new spatial focus in biogeography (Stoddart, 1977). They have been concentrating their research efforts on the geographical distribution of plant and animal taxa over the surface of the earth, and on 'geography' in the sense normally understood by the proverbial 'man-in-the-street', as one of my botanical friends pithily commented. Undoubtedly, this return to the spatial tradition is timely. First, it is clearly the oldest tradition in what we may term 'classical' biogeography and should not be neglected if biogeographers are to be true to their origins and their forebears. Secondly, however, its revival comes at a time when advances in our understanding of the history of the earth's surface, through the models of plate tectonics and sea-floor spreading, and when a wide-ranging controversy about approaches to phylogenetics, focussing on the views of the cladists and transformed cladists, both mean that the subject is experiencing an exciting and stimulating new lease of life (Nelson and Rosen, 1981; Nelson and Platnick, 1981; Patterson, 1981; Stoddart, 1982). The old tradition is therefore still a very vital element of work in biogeography.

It is also worth noting that, in the late 1960s and 1970s, the subject also received another significant stimulus with the publication of MacArthur and Wilson's The Theory of Island Biogeography (1967). They argued that the biota of any island must be in a dynamic equilibrium between the immigration of species to the island (which is a function of distance from source) and the extinction of species already present (which is a function of island area). Support for this thesis has come mainly from zoogeography, with work such as Hellier's (1976) on the snails of the Aegean islands, the study of mammals on mountain tops (Brown, 1971), the ideas of Diamond (1973 and 1975) derived from his work on the birds of the New Guinea region, and the study of Terborgh and Faaborg (1973) on the bird populations of the West Indies. However, many plant geographers have taken a rather more critical stance in relation to the theory, as exemplified by the scathing review of Gilbert (1980). One recent study of especial interest for plant geographers concerned with the theory is that of Flenley and Richards (1982) on the latest survey of the flora and ecology of that most 'experimental' of tropical islands, the caldera of Krakatau (Krakatoa), in Indonesia. All this focus of attention on islands has, of course, a long pedigree and recalls the famous work of Darwin on the Galápagos islands and Wallace's classic book, Island Life (1880).

Clearly, man has always had some interest in the geographical 'where' of plants and animals, and particularly of those organisms that have proved of direct value to him. Even in the tropics, it is salutary to note that the earliest flora goes back to the fourth century AD, with the Nan-fang Ts'ao-mu Chuang (Plants of the Southern Regions) of Chi Han, the usually accredited author. This covers the region of the world represented by present-day Kwangtung and Kwangsi provinces in southern China and the central and northern parts of Viet-Nam (Hui-Lin Li, 1979). It was probably compiled in AD 304.

The true analytical basis of the spatial tradition, however, begins with the work of de Tournefort in the seventeenth century and is thus distinctly pre-Darwinian in its ancestry, a fact rather overplayed by the somewhat stridently evangelical modern protagonists of the 'new biogeography' of cladistics and vicariance, who generally regard Darwin as someone who has subverted the real subject for a century or so. *On the Origin of Species* is dismissed merely as 'piffle' (Croizat, 1981). A brief history of biogeography is not the place to enter this often needlessly acrimonious and frequently *ad hominem* debate, but it is necessary to note that the vicariance biogeographers would see their tradition stemming from the pioneer writings of the genuinely great nineteenth century plant geographer, de Candolle (Nelson, 1978), particularly his masterpiece, *Géographie Botanique Raisonnée* (1855). The ancestry is then traced through the phylogenetic systematics (now normally referred to as cladistics) of Hennig (1966). Cladistics is simply a method of classification, giving rise to graphs of relative affinity, termed cladograms. These make no *a priori* assumptions about the nature of the relationships involved. On replacing the taxa represented by cladograms with the localities they inhabit, cladograms of affinities of areas then result. Not all biologists are convinced of their value, however, and Mccuoc (1981) recently suggested that the main use of dendrograms might be to 'frame them to hang them on the wall'. Admittedly, in the original, this quotation carried a question mark and was a plea for cladists to explain exactly why their products were 'really worth the consumption of so much research time'. The cladists, of course, reject utterly the evolutionary understanding of Darwin, Wallace and Huxley, with Darwin as their especial *bête noir*, because of his 'tragic' interest in dispersal as distinct from vicariance as an explanation of disjunct and endemic distributions (Croizat *et al*, 1974, 273-77). As Stoddart (1981) rightly says, this is a totally false and naive opposition which unnecessarily polarises the arguments. But then, Croizat is rather prone to such polarisations, having declared 'either the Darwinians bury me, or I them' (Croizat 1981, 517).

However, well before both Darwin *and* Croizat, the earliest major compendium in plant geography was being written by the German polymath, Alexander von Humboldt (1769-1859), who has already been mentioned in respect of the study of domesticated plants and animals. His now famous research expeditions, particularly to South America, were to lay the foundations of much of physical geography. For both the spatial and ecological traditions, his masterpiece, entitled *Essai sur la Géographie des Plantes* (Humboldt and Bonpland 1805, German edition 1807), is so often the true starting point of the subject. This was, indeed, then furthered by the work of de Candolle, among others, until, as we have seen, the evolutionary viewpoint of Darwin, Wallace, Huxley and Hooker, and many others, such as Lamarck, gave the subject its modern theoretical dimensions, a fact lamented and now challenged by the cladists.

In the case of zoogeography, an interest in the distribution of animals is clearly traceable to Aristotle (if not to Noah!), but became especially obvious after the discovery of the Americas, so that Georges L.L. Buffon (1707-88) in his Histoire Naturelle (1749-1804) was one of the first to perceive clearly that the faunas of the Old and New Worlds were entirely different. The identification of the main zoogeographical regions of the world was then furthered in the nineteenth century by the work of Sclater (1858), Huxley (1868) and, of course, Wallace (1860, 1878), who worked mainly in South-East Asia. For plants, the post-Darwinian theory of the subject was taken forward by many workers, including Grisebach, Engler, Drude, de Vries, Diels, Guppy, Ridley and Willis. Willis developed a stimulating, but eventually doomed, theory of 'Age and Area', which attempted to lay down laws for the relationship between distribution and time (Willis, 1922, 1940 and 1949; see also Good, 1974, and Stott 1981: 68-70). This involved a mathematical treatment of the concept using a kind of graph known as 'the hollow curve'. Unfortunately, this did not work, and it certainly alienated many botanists from his ideas (Nicholson, 1951).

The earth's moving surface. The next major theoretical breakthrough, however, came with the theory of moving continents or continental drift. The concept is, in fact, a very old one and, as early as 1620, in Novanum Organum, Francis Bacon was commenting that the obvious correspondence between the coastlines of Africa and South America could hardly be accidental, whilst in 1658 the Frenchman, Francois Placet, was writing a memoir in which he suggested that the Old and New Worlds might have become separated following the Biblical Flood. It is particularly interesting to note that, in 1800, one of the founders of biogeography, no less a person than von Humboldt, was arguing that the Atlantic Ocean was essentially a hugh river valley, the sides of which had become separated by a great volume of water across which Noah's ark had sailed. In 1858, Antonio Snider-Pelligrini produced the first actual diagram to fit together the continents bordering the Atlantic and, in 1861, Pepper invoked the same concept to explain the occurrence of identical fossil plants in the coal deposits of both Europe and North America. Then, early in the twentieth century, two Americans, Taylor and Baker, gave further and intelligent expressions of the theory.

However, it was with the publication of a book called Die Enstehung der Kontinente und Ozeane by the German geophysicist, astronomer and meteorologist, Alfred Wegener, that the modern and biogeographically significant theory of continental drift was truly launched. Although the earliest statement of Wegener's hypothesis was in 1912, and the first German edition of his book appeared in 1915, it was the 1924 English translation by J.G.A. Skerl of the third German edition of 1922 which really started the debate that was to engage so many disciplines until the final vindication of the now much wider and advanced principles of plate tectonics and sea-floor spreading in the 1960s.

For forty years, therefore, the idea that Wegener started was to divide biogeographers, many of whom remained bitterly opposed to the concept of moving continents. Others, however, such as the able Russian plant geographer, E.V. Wulff (1943, 1944), warmly embraced the new framework provided by the theory to explain the world patterns of distribution in animals and plants. It is fascinating to see how many authors of textbooks avoided or skirted around the subject, or failed to cope with it at all. Even Newbigin (1936), in her otherwise excellent book, was to hide behind an enigmatic phrase. Having dealt with the distribution of lemurs (p. 221), she adds, and it is one of the only two references to Wegener in her book, 'Alfred Wegener's theories of continental drift offer a hypothesis alternative to that of a lost land-bridge, Lemuria, in the Indian Ocean'. It should be noted that land-bridges are still in vogue with a number of writers, not least van Steenis (1962). The dilemma is also well documented by the way Ronald Good has dealt with the changing position of the theory in the four editions of his classic work, <u>The Geography of the Flowering Plants</u>, which span the crucial period from 1947 to 1974. Each edition has an added section on the subject, each clearly moving towards a more general acceptance of the theory and, in the last edition of 1974, Good is forced to comment that, 'in the last ten years there has been a dramatic and astonishing change in attitude towards the theory'. It is now an undoubted principle of earth science (Hallam, 1973) and its importance in biogeography is greatly enhanced by the key fact that it is attested to by many lines of independent geophysical evidence which have nothing whatsoever to do with the study of plant and animal distributions. The dangers of a biogeographically circular argument are thus circumvented and the theory of movement can be rightly used as an independent tool of interpretation in world-scale spatial biogeography.

Moreover, plate tectonics and sea-floor spreading, the current expressions of the theory, are crucial in the recent rise of the rather prematurely and self-styled 'new' biogeography based on cladistics and vicariance. In fact, one of the major limitations of this new school is that its approach is valid only at the world scale and it ignores the many other scales at which biogeographers work. Plate tectonics and sea-floor spreading have also forced us to look again at some of the great biogeographical problems studied by our forebears in the nineteenth century and are casting new light on such long-standing enigmas as the zoogeographical status of Wallace's line, Wallacea and Celebes (Sulawesi) (Whitmore, 1981). It is fascinating to recall that Wallace himself wrote, as early as 1880 in <u>Island Life</u>, that 'the question at issue can only be finally determined by geological investigations'. We now know, through our understanding of plate tectonics, that his famous anomalous island of Celebes, which refused to fit nicely into any system, is actually composed of rocks of both Gondwanaland and Laurasia, a collision having taken place there some 15 million years ago. No wonder Wallace had problems in analysing the faunal distribution of the island and the region.

All this underlines the simple fact that the interpretation of the spatial expression of plant and animal distributions demands a highly complex synthesis of many more specific disciplines that are concerned with the geology, geography and biology of the earth. Hence again, the 'charm' that Fleure found in the difficulties of the subject.

Biogeographies of the Past

One especially interesting characteristic of the recent contributions made by geographers to biogeography has been the, perhaps, unexpected concentration of work on what might be termed historical biogeography, and especially on Quaternary studies (Edwards, 1982). In his analysis of the fields of research occupying British biogeographers in the 1970s, Simmons (1980) discovered that there were as many scholars concerned with this topic as with the study of ecosystem description and dynamics. These two fields of activity were clearly joint-top in the table of biogeographical preferences, well above most other subjects, except for the applied topic of land-use ecology, which came a close second to the two leaders. Historical biogeography is essentially involved with attempts to reconstruct 'past biogeographies'. The techniques used are primarily geological, archaeological and biological, and include pollen and diatom analysis, macrofossil studies and stratigraphical work, bolstered by the evidence from radiocarbon dating, archaeological surveys and historical sources, where appropriate. Particular interest has been shown in the ecological impact of human societies during the prehistoric period.

But all this is in marked contrast with the sad fact that there has been remarkably little interest in the biogeography of the past, that is, in the history of the subject itself, in which so many people of so many different persuasions now claim that they work. Scholars have drifted into the realm of biogeography from their various fields, often quite unconsciously and inadvertently, so that the left hand of the subject has hardly known what the right is doing. Moreover, and perhaps in tenor with the times, much of modern biogeography has proved a little strident and perhaps a fraction arrogant in its assumptions that the approaches being newly adopted are the only valid ones. So often these approaches prove to be far from new and have been clearly anticipated in much earlier work. Workers have also frequently failed to see that their approach to the subject is but one of many and that they are only highlighting one facet of a very complex whole.

Without question, biogeography is in a new phase of exciting growth, with crucial debates taking place on the relative merits of descriptive, narrative and analytical approaches to the subject (Ball, 1975). It is, therefore, all the more important that we do not forget or neglect our great heritage. Each generation of biogeographers is, of course, moulded by the concerns and preoccupations of their own period, particularly today when academic lives and jobs so often depend on being in the forefront and right up-to-date. But at no one time are all the subject's many facets fully appreciated by all its practitioners. Moreover, there can be no such thing as the 'definitive' approach or the 'final' biogeography. For any biogeographer, in any period of the subject's development, the compliment paid by H.J. Fleure to Marion Newbigin should be sufficient, that they have proved 'so valuable a contributor to the progress of knowledge and still more of understanding'.

The basic aim of this current book is to survey the understanding of the present generation of biogeographers who are first and foremost geographers. There can be few more demanding tasks when one remembers the description of biogeography presented by Charles Darwin to Joseph Dalton Hooker in a letter of 1845: '... that grand subject, that almost keystone of the laws of creation, Geographical Distribution'.

NOTES

1. The book was also re-issued in 1968, with an additional Preface by Monica Cole.

REFERENCES

Ball, I.R. (1975) 'Nature and formulation of biogeographical hypotheses', Systematic Zoology, **24**, 407-30.

Barrows, H.H. (1923) 'Geography as human ecology', Annals of the Association of American Geographers, **13**, 1-14.

Berlin, B., Breedlove, D.E. and Raven, P.H. (1974) Principles of Tzeltal Plant Classification: an Introduction to the Botanical Ethnography of a Mayan-speaking People of Highland Chiapas, Academic Press, New York and London.

Brown, J.H. (1971) 'Mammals on mountain tops: non equilibrium insular biogeography', American Naturalist, **105**, 467-78.

Buffon, G.L.L. Comte de (1749-1804) Histoire Naturelle, Générale et Particulière, 44 vols., L'Imprimerie royale, Paris (see recent edition by Jean Piveteau, Paris, 1954).

Candolle, A. de (1855) Géographie botanique raisonnée ou exposition des faits principaux et des lois concernant la distribution géographique des plantes de l'époque actuelle, vols. I and II, Masson, Paris and J. Kessman, Geneva.

Candolle, A. de (1882) Origine des Plantes Cultivées, Germer Baillière, Paris.

Carter, G.F. (1950) 'Ecology-geography-ethnobotany', Scientific Monthly, **72**, 73-80.

Chorley, R.J. (1973) 'Geography as human ecology' in Chorley, R.J. (ed.), Directions in Geography, Methuen, London, pp. 155-69.

Clements, F.E. (1916) Plant succession: an Analysis of the Development of Vegetation, Carnegie Institution, Washington DC, USA.

Clements, F.E. (1936) 'Nature and structure of the climax', Journal of Ecology, **24**, 252-84.

Clements, F.E. (1949) Dynamics of Vegetation, Hafner, New York, USA.

Cowles, H.C. (1899) 'The ecological relations of the vegetation on the sand dunes of Lake Michigan', Botanical Gazette, **27**, 95-117, 167-202, 281-308, 361-91.

Croizat, L. (1981) 'Biogeography: past, present, and future', in Nelson, G. and Rosen, D.E. (eds.), Vicariance Biogeography: a Critique, Columbia University Press, New York, USA, pp. 501-23.

Croizat, L., Nelson, G. and Rosen, D.E. (1974) 'Centres of origin and related concepts', Systematic Zoology, **23**, 265-87.

Darwin, C. (1859) On the Origin of Species by Means of Natural Selection, John Murray, London (the 6th edition of 1872 is the edition most frequently cited).

Daubenmire, R.F. (1974) Plants and Environment. A Textbook of Autecology, 3rd edition, John Wiley, New York, USA.

Diamond, J.M. (1973) 'Distributional ecology of New Guinea birds', Science, NY, **179**, 759-69.

Diamond, J.M. (1975) 'The island dilemma: lessons of modern biogeographic studies for the design of natural reserves', Biological Conservation, **7**, 129-46.

Edwards, J.K. (1982) 'Palynology and biogeography', Area, **14**, 241-48.

Egerton, F.N., III (1968) 'Ancient sources for animal demography', Isis, **59**, 175-89.

Elton, C. (1927) Animal Ecology, Sidgwick and Jackson, London.

Evans, F.C. (1956) 'Ecosystem as the basic unit in ecology', Science, NY, **123**, 1127-8.

Flenley, J.R. and Richards, K. (eds.) (1982) The Krakatoa Centenary Expedition: Final Report, Department of Geography, University of Hull (Miscellaneous Series No. 25), Hull, UK.

Fleure, H.J. (1936) 'Dr Marion Newbigin: a tribute' in Newbigin, M.I., Plant and Animal Geography, Methuen, London, pp. vii-viii.

Forbes, E. (1844) 'Report on the molluscs and radiata of the Aegean Sea, and on their distribution considered as bearing on geology, Report of the British Association for the Advancement of Science, **13**, 130-93.

Fosberg, P.R. (1963) Man's Place in the Island Ecosystem (Symposium of Tenth Pacific Science Congress), Bishop Museum Press, Bishop, Honolulu.

Gaussen, H. (1933) Géographie des Plantes, Armand Colin, Paris, France.

Gilbert, F.S. (1980) 'The equilibrium theory of island biogeography: fact or fiction?', Journal of Biogeography, **7**, 209-35.

Glacken, C.J. (1967) Traces on the Rhodian Shore, University of California Press, USA.

Gleason, H.A. (1926) 'The individualistic concept of the plant association', Bulletin of the Torrey Botanical Club, **53**, 7-26.

Good, R. (1974) The Geography of the Flowering Plants, 4th edition, Longman, London.

Graunt, J. (1662) Natural and Political Observations mentioned in a following Index, and made upon the Bills of Mortality, Roycroft, London.

Greig-Smith, P. (1983) Quantitative Plant Ecology, 3rd edition, Blackwell Scientific, Oxford, UK.

Haeckel, E. (1866) Generelle Morphologie der Organismen, Reimer, Berlin.

Haeckel, E. (1876) The History of Creation, New York, USA.

Hallam, A. (1973) A Revolution in the Earth Sciences: from Continental Drift to Plate Tectonics, Clarendon Press, Oxford, UK.

Hardy, M.E. (1913) A Junior Plant Geography, Clarendon Press, Oxford, UK.

Hardy, M.E. (1920) The Geography of Plants, Clarendon Press, Oxford, UK.

Harris, D.R. (1969) 'Agricultural systems, ecosystems and the origins of agriculture' in Ucko, P.J. and Dimbleby, G.W. (eds.), The Domestication and Exploitation of Plants and Animals, Duckworth, London, pp. 3 16.

Harris, D.R. (1973) 'The prehistory of tropical agriculture: an ethnoecological model' in Renfrew, C. (ed.), The Explanation of Culture Change. Models in Prehistory, Duckworth, London, pp. 391-417.

Harris, D.R. (1982) 'A tropical view of ethnobotany' in Whitmore, T.C., Flenley, J.R. and Harris, D.R., 'The tropics as the norm in biogeography?', Geographical Journal, **148**, 8-21.

Harrison, C.M. (1971) 'Recent approaches to the description and classification of vegetation', Transactions of the Institute of British Geographers, **52**, 113-27.

Harrison, C.M. (1975) 'The description and analysis of vegetation' in Chapman, S.B. (ed.), Methods in Plant Ecology, Blackwell, Oxford, pp. 85-155.

Hellier, J. (1976) 'The biogeography of Enid landsnails on the Aegean Islands', Journal of Biogeography, **3**, 281-92.

Hennig, W. (1966) Phylogenetic Systematics, University of Illinois Press, Urbana, USA.

Humboldt, F.H.A. von and Bonpland, A.J.A. (1805) Essai sur la Géographie des Plantes: accompagné d'un Tableau Physique des Régions Equinoxiales, Levrault, Schoell & Compagnie, Paris (German edition, 1807).

Huxley, T.H. (1868) 'On the classification and distribution of the Alectoromorphae and Heteromorphae', Proceedings of the Zoological Society of London, **1868**, 294-319.

Jeffers, J.N.R. (1978) An Introduction to Systems Analysis: with Ecological Applications, Edward Arnold, London.

Kellman, K.C. (1975) Plant Geography, Methuen, London (2nd edition 1980).

Krebs, C.J. (1978) Ecology: the Experimental Analysis of Distribution and Abundance, 2nd edition, Harper and Row, New York, USA.

Lauer, W. and Klink, H.J. (1978) Pflanzengeographie, Wissenschaftlich Buchgesellschaft, Darmstadt.

Leach, W. (1933) Plant Ecology, Methuen, London.

Li, Hui-Lin (1979) Nan-fang Ts'ao-mu Chuang: a Fourth Century Flora of Southeast Asia, Introduction, Translation, Commentaries, The Chinese University Press, Hong Kong.

Lindeman, R.L. (1942) 'The trophic-dynamic aspect of ecology', Ecology, **23**, 399-418.

Lowenthal, D. (1958) George Perkins Marsh: Versatile Vermonter, Columbia University Press, New York, USA.

Malthus, T. (1798) An Essay on the Principle of Population, reprinted by Macmillan, New York, USA.

MacArthur, R.H. and Wilson, E.O. (1967), The Theory of Island Biogeography, Princeton University Press, Princeton, NJ, USA.

Marsh, G.P. (1864) Man and Nature or Physical Geography as Modified by Human Action, Charles Scribner, New York, USA.

Marsh, G.P. (1885) The Earth as Modified by Human Action, 2nd edition, Scribner's Sons, New York, USA.

Martonne, E. de (1921) Traité de Géographie Physique, 4th edition, vol. III, Biogéographie, Paris.

Matthews, J.A. (1979) 'Refutation of convergence in a vegetation succession, Naturwissenschaften, **66**, (1), 47-9, Springer Verlag, Berlin.

Meeuse, A.D.J. (1981) 'Again, cladistics in botany', Taxon, 30, 642-44.

Möbius, K. (1877) Die Auster und die Austernwirtschaft, Wiegundt, Hempel and Parey, Berlin.

Moore, D.M. (ed.) (1982) Green Planet: the Story of Plant Life on Earth, CUP, Cambridge, UK.

Mumford, L. (1931) The Brown Decades: a Study of the Arts in America 1865-1895, Dover, New York, USA.

Nelson, G. (1978) 'From Candolle to Croizat: comments on the history of biogeography', Journal of the History of Biology, **11**, 269-305.

Nelson, G. and Platnick, N.I. (1981) Systematics and Biogeography: Cladistics and Vicariance, Columbia University Press, New York, USA.

Nelson, G. and Rosen, D.E. (eds.) (1981) Vicariance Biogeography: a Critique, Columbia University Press, New York, USA.

Newbigin, M.I. (1913) Animal Geography, Clarendon Press, Oxford.

Newbigin, M.I. (1936) Plant and Animal Geography (2nd edition 1948), Methuen, London.

Nicholson, R.J. (1951) 'A note on hollow curves', New Phytologist, **50**, 138-9.

Patterson, C. (1981) 'Biogeography: in search of principles', Nature, **292**, 612-13.

Pielou, E.C. (1977) Mathematical Ecology, Wiley, New York, USA.

Pielou, E.C. (1979) Biogeography, John Wiley, New York, USA.

Polunin, N. (1960) Introduction to Plant Geography and some Related Sciences, Longman, London.

Prenant, M. (1933) Géographie des Animaux, Armand Colin, Paris.

Ratzel, F. (1897) Politische Geographie, Oldenburg, Munich.

Raunkiaer, C. (1907) Planterigets Livsformer og deres Betydning for Geografien, Copenhagen.

Raunkiaer, C. (ed. Tansley, A.G.) (1934) The Life Forms of Plants and Statistical Plant Geography, Clarendon Press, Oxford.

Sauer, C.O. (1925) 'The Morphology of Landscape', University of California Publications in Geography, **2**, 19-35.

Sauer, C.O. (1956) 'The agency of man on the Earth' in Thomas, W.L. Jr. (ed.), Man's Role in Changing the Face of the Earth, University of Chicago Press, Chicago, USA, pp. 49-60.

Sauer, C.O. (1963) Land and Life, ed. Leighley, J.B., Berkeley, USA.

Schimper, A.F.W. (1898) Pflanzengeographie auf Physiologischen Grundlage, Fischer, Jena, West Germany.

Schimper, A.F.W. (1903) Plant Geography upon a Physiological Basis, trans. Fisher, W.R., Clarendon Press, Oxford, UK.

Sclater, P.L. (1858) 'On the general geographic distribution of the members of the class Aves', Journal of the Linnean Society of London, **2**, 130-45.

Seddon, G. (1974) 'Xerophytes, xeromorphs and sclerophylls: the history of some concepts in ecology', Biological Journal of the Linnean Society, **6**, 65-87.

Shimwell, D.W. (1971) The Description and Classification of Vegetation, Sidgwick and Jackson, London.

Simmons, I.G. (1974) The Ecology of Natural Resources, Edward Arnold, London.

Simmons, I.G. (1979) Biogeography: Natural and Cultural, Edward Arnold, London.

Simmons, I.G. (1980) 'Biogeography' in Brown, E.H. (ed.), Geography Yesterday and Tomorrow, Oxford University Press, Oxford, UK, pp. 146-66.

Stoddart, D.R. (1966) 'Darwin's impact on geography', Annals of the Association of American Geographers, **56**, 683-98.

Stoddart, D.R. (1967) 'Organism and ecosystem as geographical models', in Chorley, R.J. and Haggett, P. (eds.), Models in Geography, Methuen, London, pp. 511-48.

Stoddart, D.R. (1977) 'Progress report: Biogeography', Progress in Physical Geography, **1**, 537-43.

Stoddart, D.R. (1978) 'Progress report: Biogeography', Progress in Physical Geography, **2**, 514-28.

Stoddart, D.R. (1981) 'Progress report: Biogeography: dispersal and drift', Progress in Physical Geography, **5**, 575-90.

Stamp, L.D. (1962) 'A geographer's postscript', in Taxonomy and Geography, Systematics Association Publication, No. **4**, London, pp. 153-8.

Steenis, C.G.G.J. van (1962) 'The land-bridge theory in botany', Blumea, **11**, 235-542.

Stott, P.A. (1981) Historical Plant Geography, George Allen and Unwin, London.

Sukachev, V. and Dylis, N. (1968) Fundamentals of Forest Biogeocoenology (English trans. from Russian by MacLennan, J.M.), Oliver and Boyd, Edinburgh and London.

Tansley, A.G. (1911) Types of British Vegetation, Cambridge University Press, Cambridge, UK.

Tansley, A.G. (1923) Practical Plant Ecology, George Allen and Unwin, London.

Tansley, A.G. (1935) 'The use and abuse of vegetational concepts and terms', Ecology, **16**, 284-307.

Tansley, A.G. (1939) The British Islands and Their Vegetation, Cambridge University Press, Cambridge, UK.

Tansley, A.G. and Chipp, T.F. (eds.) (1926) Aims and Methods in the Study of Vegetation, Crown Agents for the Colonies, London.

Terborgh, J.W. and Faaborg, J. (1973) 'Turnover and ecological release in avifauna of Mona Island, Puerto Rico', Auk, **90**, 759-79.

Theophrastus (1754) Historia Plantarum, Andreas Hedenberg, respondent.

Udvardy, M.D.F. (1969), Dynamic Zoogeography, Van Nostrand Reinhold, New York, USA.

Vavilov, N. (1926) 'Studies on the origin of cultivated plants', Bulletin of Applied Botany, **16**, 139-248.

Vavilov, N. (1951) The Origin, Variation, Immunity and Breeding of Cultivated Plants, trans. by Chester, K.S., Chronica Botanica, Waltham, Mass., USA.

Vermeij, G.J. (1978) Biogeography and Adaptation: Patterns of Marine Life, Harvard University Press, Bridge, Mass., USA.

Wallace, A.R. (1860) 'On the zoological geography of the Malay archipelago', Journal of the Linnean Society of London, 14, 172-84.

Wallace, A.R. (1876) The Geographical Distribution of Animals, 2 vols., Macmillan, London.

Wallace, A.R. (1880) Island Life (2nd edition 1882), Macmillan, London.

Warming, E. (1895) Plantesamfund. Grundtrak af den ökologiscka Plantegeografi (German edition, 1896, Knoblauch, E. (ed.), Borntrager, Berlin), Philipsen, Copenhagen.

Warming, E. (1909) Oecology of Plants. An Introduction to the Study of Plant Communities, trans. Groom, P. and Balfour, I.B., Clarendon Press, Oxford, UK.

Wegener, A. (1915) Die Enstehung der Kontinente und Ozeane (2nd edition 1920, 3rd edition 1922, 4th edition 1929), F. Vieweg und Sohn, Braunschweig.

Wegener, A. (1924) The Origin of Continents and Oceans (trans. of 3rd German edition by Skerl, J.G.A.; trans. of 4th German edition by Biram, J., 1966, Dover, New York), Methuen, London.

Whitmore, T.C. (ed.) (1981) Wallace's Line and Plate Tectonics, Clarendon Press, Oxford, UK.

Whittaker, R.H. (1953) 'A consideration of climax theory: the climax as a population and pattern', Ecological Monographs, **23**, 41-78.

Whittaker, R.H. (1962) 'Classification of natural communities', Botanical Review, **28**, 1-239.

Whittaker, R.H. (1970) Communities and Ecosystems, Macmillan, New York, USA.

Willis, J.C. (1922) Age and Area. A Study in Geographical Distribution and Origin of Species, Cambridge University Press, Cambridge, UK.

Willis, J.C. (1940) The Course of Evolution by Differentiation or Divergent Mutation rather than by Selection, Cambridge University Press, Cambridge, UK.

Willis, J.C. (1949) The Birth and Spread of Plants, Chronica Botanica, Waltham, Mass., USA.

Wulff, E.V. (1943) An Introduction to Historical Plant Geography (trans. from Russian by Brissenden, E.), Chronica Botanica, Waltham, Mass., USA.

Wulff, E.V. (1944) Historical Plant Geography: History of the World, Akademiya Nauk SSSR (in Russian).

Zimmermann, R.C. and Thom, B.G. (1982) 'Physiographic plant geography', Progress in Physical Geography, **6**, 45-59.

FURTHER READING

Ball, I.R. (1975) 'Nature and formulation of biogeographical hypotheses', Systematic Zoology, **24**, 407-30.

Browne, J. (1983) The Secular Ark: Studies in the History of Biogeography, Yale University Press, New Haven and London.

Chorley, R.J. (1973) 'Geography as human ecology' in Chorley, R.J. (ed.), Directions in Geography, Methuen, London, pp. 155-69.

Compte rendu des séances de la Société de Biogéographie, published by the Société de Biogéographie, 57, rue Cuvier, Paris V^{e}, France.

Harrison, C.M. (1979) 'Ecosystems and communities: patterns and processes' in Gregory, K.H. and Walling, D.E. (eds.), Man and Environmental Processes, Dawson, Folkestone and Westview Press, Boulder, Colorado, USA, pp. 225-40.

Holt-Jensen, A. (1980) Geography: its History and Concepts, Harper and Row, London, see pp. 9-36.

Nelson, G. (1978) 'From Candolle to Croizat: comments on the history of biogeography', Journal of the History of Biology, **11**, 269-305.

Nelson, G. and Rosen, D.E. (eds.) (1981) Vicariance Biogeography: a Critique, Columbia University Press, New York, USA.

Polunin, N. (1960) Introduction to Plant Geography and Some Related Sciences, Longmans, London, see pp. 1-23.

Ridley, M. (1983) 'Can classification do without evolution?', New Scientist, 1 December, 647-51.

Seddon, G. (1974) 'Xerophytes, xeromorphs and sclerophylls: the history of some concepts in ecology', Biological Journal of the Linnean Society, **6**, 65-87.

Simmons, I.G. (1980) 'Biogeography' in Brown, E.H. (ed.), Geography Yesterday and Tomorrow, Oxford University Press, Oxford, UK, pp. 146-66.

Stoddart, D.R. (1966) 'Darwin's impact on geography', Annals of the Association of American Geographers, **56**, 683-98.

Stoddart, D.R. (1967) 'Organism and ecosystem as geographical models' in Chorley, R.J. and Haggett, P. (eds.), Models in Geography, Methuen, London, pp. 511-48.

Stoddart, D.R. (1977, 1978, 1981) 'Biogeography', Progress in Physical Geography, **1**, 537-43; **2**, 514-28; **5**, 575-90.

Whitmore, T.C. (ed.) (1981) Wallace's Line and Plate Tectonics, Clarendon Press, Oxford, UK (see pp. 3-8, 24-35).

Whitmore, T.C., Flenley, J.R. and Harris, D.R. (1982) 'The tropics as the norm in biogeography', The Geographical Journal, **148**, 8-21.

Williamson, M. (1981) Island Populations, Oxford University Press, Oxford, UK.

Chapter 2

THE SPATIAL DIMENSION IN BIOGEOGRAPHY

D. Watts

Concepts of space, and spatial distributions and interlinkages are central to most biogeographic research and writing: as Pielou (1979) has put it, 'the work of the biogeographer consists in observing, recording and explaining the geographic ranges of all living things'. If biogeography is to mean anything at all, either in theory or in application, one needs to know why individual organisms, or populations, at species, generic or supra-generic levels, are located where they are, and how they came to be there. But despite the detailed advocacy of Cain (1944), Polunin (1950) and Udvardy (1969), studies of the spatial aspects of this field have been unaccountably de-emphasised during the last decade. In attempting to redress the balance to some extent, it is the aim of this paper to direct attention to the more salient features of range biogeography, as exemplified by both past and recent work.

It was Cain (1944) who first suggested that ranges of living organisms might be categorised according to their shape or form. He noted that ranges tended towards circularity, because of the generally random dissemination patterns of diaspores; and that the most common deformation was towards an ovate, east-west dominant form, reflecting climatic controls, and the fact that climate is more responsive to variations in latitude than to those in longitude. Other deformations might arise from conditions of topography or soil. Moreover, the ranges of most species are normally wider than those of the communities of plants or animals to which they belong. To these comments one may add a few other general points. Ranges may be large or small: they are constantly changing, expanding, contracting or being modified in detail through the agencies of environmental or biological (competitive) amelioration or deterioration, and the mechanisms of plant and animal adaptation and evolution. They may fragment, reconsolidate or, in extreme cases, move entirely from one part of the world to another.

Evidence for these situations, presented herein, is taken largely at species, or taxon level, the term taxon referring to a set of closely-related species.

Range Variation, Categorisation and Type, at Regional and World Scale

Ranges of earth organisms vary enormously in size. Some indication of this, for one particular animal group (the mammals), is given in Table 2.1.

Table 2.1: Mean range size of species in selected orders of mammals in Central and North America (n = 697 species)

Order	Area ($10^4 km^2$)	S.D.
Carnivora	617.4	613.1
Artiodactyla	507.2	440.0
Lagomorpha	192.6	275.7
Chiroptera	148.7	244.8
Marsupialia	113.0	204.9
Insectivora	117.7	250.0
Edentata	88.9	99.8
Rodentia	76.4	165.0
Primates	24.9	34.8
Mammals (mean)	157.2	321.8

Data from Rapoport (1975).

At the species level, a range may be taken to be the area normally occupied by a breeding, reproductive population. Such actual ranges may be very different in scale to potential ranges, which include the whole gamut of environments a species is technically capable of occupying. Actual ranges are limited by ecological or biological barriers, which restrict the further dispersal of species at any given point in time: such barriers are rarely permanent, except for those arising from the broad division of the world into land or water environments, and even these are breached by amphibia, and certain other groups. For land-based organisms, conditions of climate, topography, and soil may all, individually or collectively, form habitat barriers at range limits, and these may also be created biologically by patterns of predation, competition, the activities of parasites, and other similar influences. Historical events, and human activity, can further modify range extent. Barriers to the spread of water-based organisms, on the other hand, are more clearly associated with differences in water chemistry, temperature, light penetration, and the general condition of the food-chain web within the water body.

On land, a very few species are essentially worldwide in their distribution, being present on all continents except Antarctica, which currently is too cold to support most forms of organic life. These are mainly plant species which have few limiting factors to their successful dispersal over long distances, and to their establishment and survival in a wide range of habitats. Plants which fall into this category are termed cosmopolitan: more often than not, they are low-growing, heliophytic forms which prefer open, and often man-disturbed ground. Ecologically, they may be termed eurytopic, or tolerant, species, and genetically they are usually very flexible. Thus, the complex of largely self-pollinating forms of procumbent yellow sorrel (Oxalis corniculata) is a good example of a cosmopolitan plant which has reached not only all continents except Antarctica, but also many remote islands, including Tristan da Cunha (Baker, 1972). Other very wide-ranging plants include the broad-leaved plantain (Plantago major), a species of European origin which tolerates both acidic and alkaline environments, and all but the coldest and driest parts of the world (Sagar and Harper, 1964); and bracken (Pteridium aquilinum), which is equally at home in tropical and cooler regions. No animal species is truly cosmopolitan in its natural distribution, although some animal families (e.g. the bat family) effectively have a world-wide range.

Clearly, the vast majority of plant and animal species have much more restricted distribution patterns: they are stenotopic, or less tolerant organisms, confined to particular environments or regions, or suites of environments or regions. How may one best categorise these, bearing in mind the multiplicity of range size differences which may be expected among the six million animal species, and the two to three million plant species which inhabit the face of the earth? Are they determined by the size of organism in any way, or by their location? What other factors may influence range size? Do range sizes have any clear, repetitive spatial patterns, or are they largely random in their disposition? All these are questions which should concern the biogeographer.

In respect of range size, one useful, new attempt to move away from descriptive analysis has been put forward by Rapoport et al (1976), who have proposed a scheme whereby the degree of cosmopolitanism of any species or taxon may be measured according to how far its range impinges into each of the major biogeographic regions of the world. The term major biogeographic region perhaps requires further clarification here. It refers to a concept first used widely by Wallace (1876), who divided the world into six large-scale regions, based on their overall broad faunal similarities (Fig. 2.1). As knowledge of species distribution patterns subsequently increased, Wallace's work has come to be challenged by others, such as Schmidt (1954) who, on the basis of zoogeography, has delimited five major world biogeographic regions; and Pielou (1979), who produced eight distinguishable regions, using the scatter of both plant and animal ranges. Nevertheless, Wallace's faunal regions retain a broad utility as a datum for the objective consideration of biogeographical range analysis. Rapoport et al (1976) suggest that species or taxa may be classified as endemic, characteristic, or semi-cosmopolitan, dependent on whether their ranges extend into 1, 2 or 3-4 of Wallace's regions, respectively: and cosmopolitan species should occur in at least five of the regions. The degree of cosmopolitanism

Figure 2.1: The major biogeographic regions of the world, as delimited (a) by faunal groups, after Wallace (1876); and (b) by faunal and vegetation groups together, after Pielou (1979).

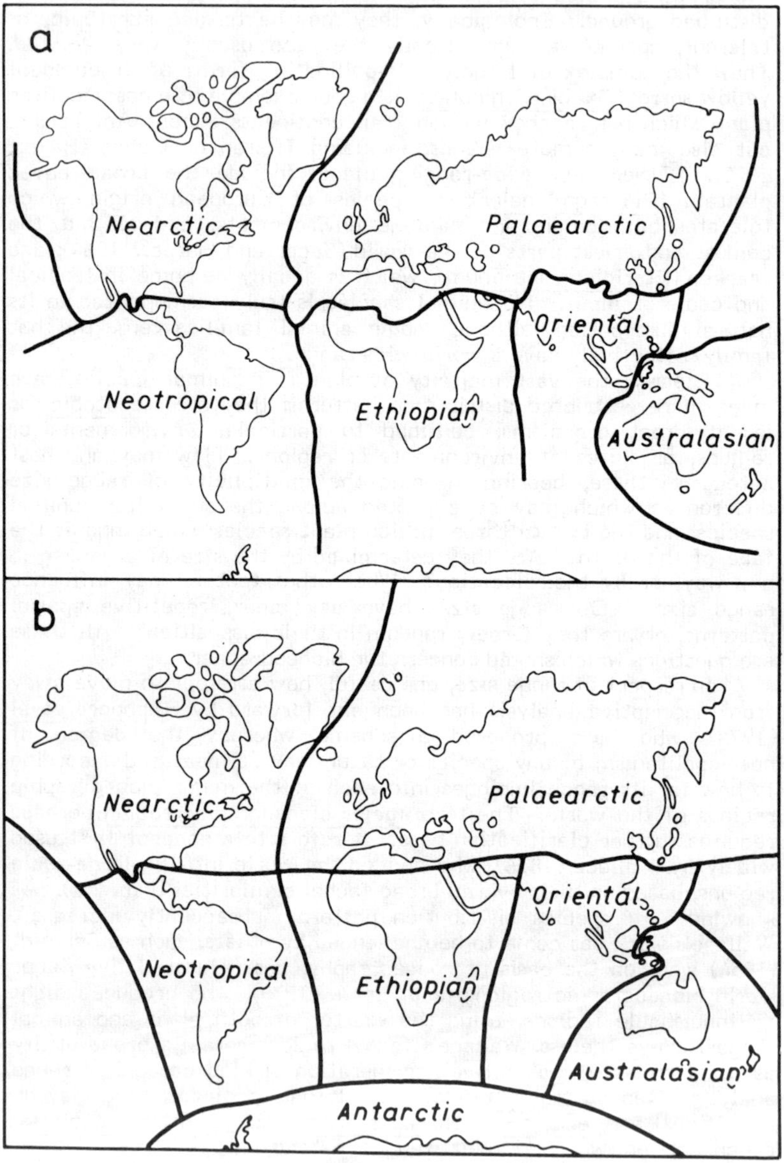

of species or taxa may also be determined numerically. If r is the number of regions occupied by each species or taxon, Yr the number of species or taxa that occupy r regions, n the total number of species or taxa, and r_{max} = 6 regions, then the degree of cosmopolitanism (C) may be expressed by:

$$C = \frac{\left(\sum_{r=1}^{r_{max}} r\,Yr/n \right) - 1}{r_{max} - 1}$$

Values of C will range from a minimum of 0 to a maximum of 1, being $\sum_{r=1}^{r_{max}} r\,Yr/n$, the mean occupation ($\bar{x}$). Table 2.2 provides an illustration of how the scheme may work.

Table 2.2: The degree of cosmopolitanism of selected taxa

No. of regions occupied by taxa	1 Tinamid birds	2 European -based birds	3 Insect pests	4 Phytopathogens (viruses, bacteria, fungi)
1	40	204	74	45
2	-	45	43	31
3	-	25	24	38
4	-	12	28	41
5	-	8	51	67
6	-	0	0	0
n	40	294	220	222
$\bar{x}$	1.00	1.55	2.72	3.24
C	0	0.110	0.344	0.448

Data from Voous, 1962 and Rapoport et al (1976).

In this, tinamid birds are accorded a minimum C value, being restricted to one major biogeographic region alone. Among more widely-distributed groups, birds which are predominantly based in Europe, but which are also found elsewhere, maintain very low C values, as also do groups of insect pests and phytopathogens, even though the ranges of these have been greatly extended recently by man throughout the world. The data display how exceptional the dispersal, establishment and survival qualities of plant and animal species and taxa must be if they are to achieve cosmopolitan status with high C values.

In real life terms, semi-cosmopolitan species most commonly are those which are widely distributed in either tropical or temperate regions of the world, but not in both. They include many temperate-latitude weeds, such as the dandelion (Taraxacum vulgare), the English plantain (Plantago lanceolata) and wild oats (Avena fatua), the latter now being as well adapted to the dry hills of central California and the State of South Australia as it is in its native Mediterranean. Most of these weeds are annuals or perennials, the former reproducing largely by seed, and the perennials (e.g. couch grass, Agropyron repens) vegetatively (Salisbury, 1961). The vast majority of annual weeds have a very high output of seeds per plant (up to 230,000 in the case of particular species: Hill, 1977), a prolonged within-soil seed survival rate (Major and Pyott, 1966; Sarukhan, 1974; Kellman, 1974, 1978), rapid seedling growth once germination has taken place, and an especially wide tolerance of different habitat conditions. They compete well not only through the production of large numbers of propagules, but also by means of a rapid progression towards maturity, and further seed production, growth strategies which have been referred to as r-selection in their application (Pianka, 1970). Many temperate latitude weeds are polyploid, and of fairly recent (possibly late-Pleistocene) origin.

Semi-cosmopolitan weeds are also dispersed widely in the tropics: according to May (1981), all of the world's ten worst weeds are located here. These are the purple nutsedge (Cyperus rotundus, L.), Bermuda grass (Cynodon dactylon (L.), Pers.), barnyard grass (Euchinochloa crusgalli, L.), jungle rice (Euchinochloa colonum, L.), goosegrass (Eleusine indica (L.), Gaertn.), Johnson grass (Sorghum halepense (L.), Pers.), Guinea grass (Panicum maximum, Jacq.), water hyacinth (Eichornia crassipes (Mart.), Solms.), cogon grass (Imperata cylindrica (L.), Beauv.) and West Indian sage (Lantana camara, L.), of which eight are grasses and sedges, five are perennial grasses, and West Indian sage is a dry-land shrub. The range of most of these species has been inadvertently extended between continents by man within the last two centuries. As an example, West Indian sage is now thoroughly naturalised along the eastern and north-eastern coasts of Australia, to which its seeds were carried accidentally by sailing ships, in the nineteenth century, on the England-Australia run. However, most such transfers, whether in the tropics or in temperate latitudes, have been not in this direction, but from the Old World to the New. Thus Parsons (1972) has described in detail the take-over by African pasture grasses of large areas of tropical South American grazing land; and grasses of the genera Avena and Bromus, of Mediterranean origin, now comprise much of the ground cover of open land in California, below 2,000 m elevation.

Among the larger animals, some semi-cosmopolitan distributions may also be found, especially in bird and fish taxa. Orioles, for example, have a range which is predominantly eastern-hemisphere, including the warmer parts of Asia, Africa and Australia, and extending into western Europe as well: but they are absent from the New World. The carp family is dispersed widely, although their overall distribution pattern is unusual (Fig. 2.2):

Figure 2.2: Semi-cosmopolitan distributions: the carp family (after George, 1962).

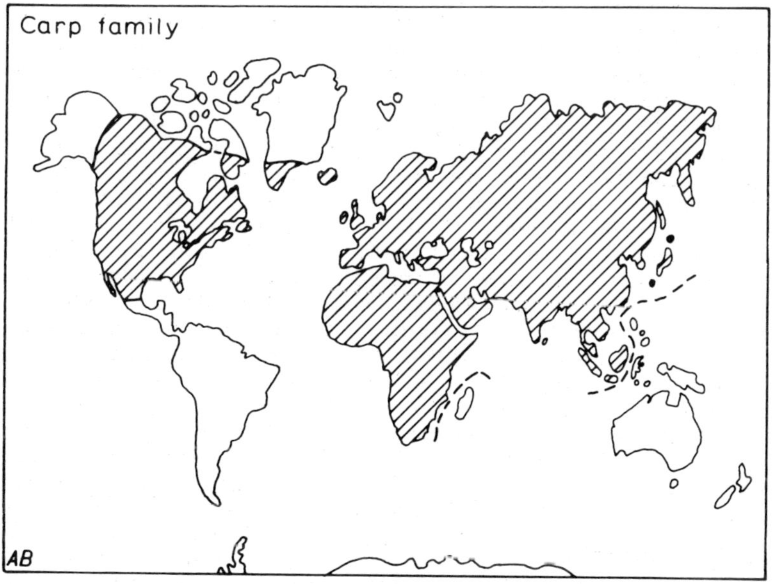

They exist in the northern hemisphere across all continents, and in Africa too, but they have not colonised either South America or Australia (George, 1962). Semi-cosmopolitan distributions are much more common among some of the smaller animals, especially among the viruses and bacteria.

Characteristic species, which are present in two major biogeographic regions, are perhaps best exemplified in those parts of the northern hemisphere which have been recolonised by plants and animals over the last 9,000 to 10,000 years, following the decline of the last, major, Würm-Wisconsin glaciation. Here, within tundra and taiga vegetation, suites of similar or identical species occur today in circumpolar ranges which extend throughout both North America and Eurasia; and somewhat farther south, where nemoral, broad-leaved deciduous forests are established, corresponding pairs of closely-related taxa (trees, shrubs, and herbs) are located on the two continents (Table 2.3).

Table 2.3: Selected corresponding taxa in nemoral, broad-leaved forests of south-eastern Canada and Poland (after Kornas, 1972)

	S.E. Canada	Poland
Trees		
Hornbeam	Carpinus caroliniana, Walt.	C. betulus, L.
Beech	Fagus grandifolia, Ehrh.	F. sylvatica, L.
Ash	Fraxinus nigra, Marsh	F. excelsior, L.
Wild Cherry	Padus serotina (Ehrh.) A gardh.	P. avium, Mill.
Oak	Quercus alba, L.	Q. robur, L.
Lime	Tilia americana, L.	T. cordata, Mill.
Elm	Ulmus americana, L.	U. laevis, Pall.
Elm	Ulmus rubra, Muhl.	U. carpinifolia, Gled.
Shrubs		
Elder	Sambucus racemosa ssp. pubens (Michx) House	S. racemosa ssp. racemosa
Viburnum	Viburnum trilobum, Marsh	V. opulus, L.

Other groups of plants are common to the Himalayan region, China, Japan and North America (Hara, 1972), including Phryma leptostachya, L., Monotropa uniflora, L. and Symplocarpus foetidus (L.), Nuttall. Further instances of plants with characteristic ranges may be found along the littoral of the opposing sides of some major oceans, as in the case of the four mangrove species Avicennia germinana (L.), L., Laguncularia racemosa (L.), Gaertn., Rhizophora mangle, L. and Conocarpus erecta, L., all of which are widely dispersed both in the West Indies and in West Africa (Exell and Stace, 1972).

In biogeographical and ecological literature, the term endemic has been accorded various meanings. In the sense in which it is used by Rapoport et al (1976), it serves almost as an antonym for the term pandemic, this latter referring to species that exist in more than one biogeographic region. It would include, therefore, both broad endemic species, which have a large range within one continent, and those narrow endemics which have a much more restricted distribution and which, in Tivy's (1982) words, are 'peculiar or exclusive to a particular area'. Endemic species with small ranges may either have evolved very recently, or they may have been present on a long-term basis, but prevented by environmental or biological factors from expanding out of their immediate locality; they may, accordingly, be clearly subdivided into those which are 'new', and those which are 'relict'. These two interpretations of endemism surely refer to very different biogeographical situations, each of which in turn is discussed below.

Most of our knowledge of the nature of broad endemic species was first compiled and assorted by Good (1947, 1974), using vegetation data. It was he who observed that the ranges of many broad endemics roughly coincided, so that floristic regions, each containing unique species groups, might be defined. On a world scale, he identified 37 major floristic regions, the aim having been to:

> divide the land surfaces of the world into a convenient but not too large number of regions, each of which may be regarded as supporting a flora of its own, that is to say a flora which is characteristic of the region, which ... has largely developed within the region, and which has, to a like extent, been conditioned by the history and circumstances of the region.

A strong association between the distribution patterns of broad endemics and the environments in which they are located is inferred. The floristic regions which he delimited range in size from the very large Euro-Siberian to the very small Hawaiian (Fig. 2.3). Within all of them, broad endemic plants tend to live in a variety of established communities, in which selection for stability, with the production of small numbers of propagules, and often a slow pattern of development towards maturity (K-selection) is appropriate (Grime, 1979). Though of necessity, such species are usually very well adapted to the physical and biological habitats in which they live, sensitivity to stress - both physical and biological - at different levels of plant activity (germination, seedling growth, flowering, seed production, etc.) will ultimately act as a limiting factor to range extension. Thus the response to one type of stress, winter cold, by the English yew (Taxus baccata, L.) has restricted the distribution of this species in the north-east of England compared with other parts of the country (Melzack and Watts, 1982). On the other hand, the genetic flexibility normally found in broad endemics means that they are well capable of adopting different strategies for survival in different parts of their range. The Scots pine (Pinus sylvestris, L.), for example, which has an enormous range in Europe and northern Asia (Fig. 2.4: Sfazer, 1966), is a xerophyte towards its northern limits, but is mesophytic further south (Mirov, 1967); the Canadian black spruce (Picea mariana) is a wet-land species towards the southern end of its range in Minnesota and southern Ontario, but prefers dry, stony soil north of Lake Superior (Hosie, 1969); and the European beech (Fagus sylvatica) is a mountain species which occurs close to the tree line in central Corsica, despite having a lowland distribution in the British Isles. But not all broad endemics have ranges as wide as those described above. Some have much narrower habitat tolerances, as does the bog myrtle (Myrica gale), which grows well only in moist (and especially wet-ground) environments which lie close to the Atlantic coast of Europe.

In contrast to the above, narrow endemics are present most frequently in environments which are, or have been in the past, characterised by a good deal of isolation. Most are found in oceanic islands (i.e. those which are far removed from, and have no direct geological connections with, the nearest continents), or on the tops

Figure 2.3: Floristic regions of the world, after Good (1974).

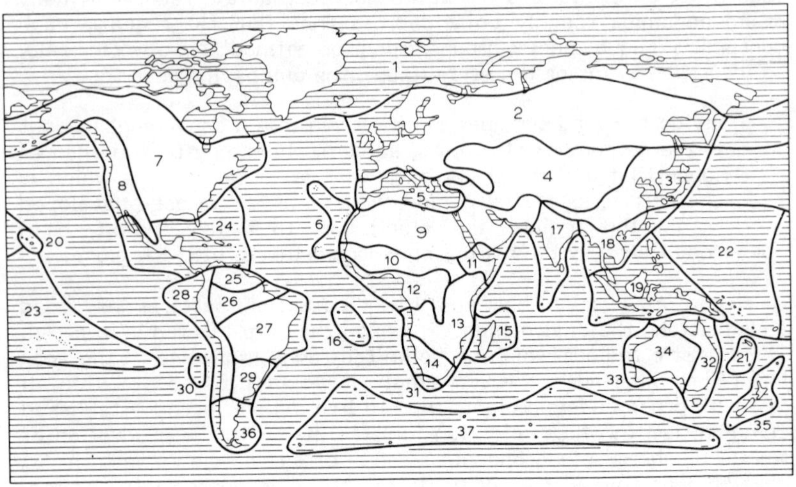

1 Arctic and sub-Arctic; 2 Euro-Siberian; 3 Sino-Japanese; 4 Western and Central Asiatic; 5 Mediterranean; 6 Macaronesian; 7 Atlantic North American; 8 Pacific North American; 9 African-Indian desert; 10 Sudanese park-steppe; 11 North-east African highland; 12 West African rainforest; 13 East African steppe; 14 South African; 15 Madagascan; 16 Ascension and St Helena; 17 Indian; 18 Continental south-east Asiatic; 19 Malaysian; 20 Hawaiian; 21 New Caledonian; 22 Melanesian and Micronesian; 23 Polynesian; 24 Caribbean; 25 Venezuelan and Guyanan; 26 Amazonian; 27 South Brazilian; 28 Andean; 29 Pampas; 30 Juan Fernandez; 31 Cape of Good Hope; 32 North and East Australian; 33 South-west Australian; 34 Central Australian; 35 New Zealand; 36 Patagonian; 37 South temperate oceanic islands.

of large mountain masses. The vast majority of narrow endemics have evolved relatively recently. The speed at which such evolution takes place varies with the type of organism, but it can be exceptionally rapid. Thus, in the American south-west, creosote bush (Larrea divaricata) communities have emerged only within the post-Pleistocene (Holocene) period, over the last 10,000 to 12,000 years, yet there is a highly numerous and specialised insect fauna therein, which has very few close relatives anywhere else (Endler, 1982). Some fruit flies appear to have been capable of speciation within historic time (Bush, 1974), and certain Hedylepta moths in Hawaii within the last 1,000 years (Zimmerman, 1960). Larger

Figure 2.4: Contrasting broad-endemic ranges in Eurasia: the Scots pine (Pinus sylvestris) and the bog myrtle (Myrica gale) (after Szafer, 1970 and Tivy, 1982).

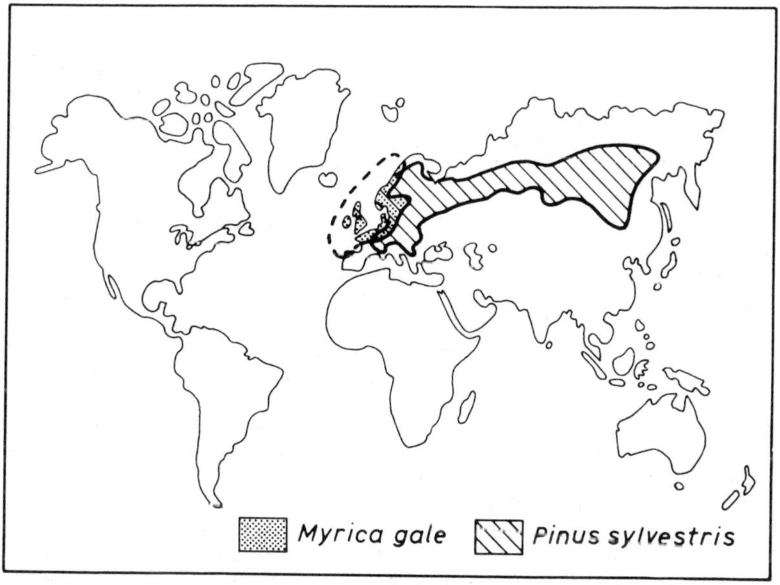

animals in general may take longer to produce new forms, but even so Haffer (1969, 1974) has speculated that new bird species have become differentiated in the Amazon basin over the last 30,000 years. The evidence for mean rates of plant speciation is less certain: however, many of the more complex plant forms may require up to 100,000 years to evolve at the species level.

The scale of narrow endemism in particular localities thus may be expected to vary not only with the degree of isolation and the characteristics of environment, but also with the types of organisms which are present. In the Hawaiian group of islands, the nearest continental neighbour of which (North America) lies some 4,000 km distant, c. 16.5 per cent of all native plants are endemic, but 20 per cent of all native flowering plants; and for alpine areas therein, which may be regarded as 'islands within islands', and so doubly isolated, the percentage of endemic plant species rises to 91 (Stone, 1967; Amerson, 1975). Most endemic plants in the Hawaiian group have emerged through recent adaptive radiation. Lobeliads, for example, developing in habitats in which their normal competitors (orchids) are rare, have radiated into six new endemic genera, and 150 new endemic species and varieties, and 60 of the latter are placed in one endemic genus (Cyanea) alone. In the absence of suitable insect pollinators, species expansion in the lobeliads has been accompanied by the parallel evolution of nectar-feeding birds, namely honey-creepers. These form an endemic family, the Drepanididae, which now contains 11 endemic genera, and 23 endemic species, the original ancestors of which were probably finch-like creatures, blown a long way off course from their former broad endemic ranges in North America (Amadon, 1950; Raikow, 1976; Juvik and Austring, 1979). Over one-third of the 1,200 known species of fruit flies (Drosophila and Scaptomyza) are present as endemics in the Hawaiian chain, and in this instance endemism has been encouraged by spasmodic but repetitive lava flow inception, which has created major barriers for fruit fly dispersal, and which has accordingly increased substantially the extent of their local isolation from time to time. The Hawaiian chain also has many endemic species of crickets and carabid beetles (Zimmerman, 1970).

Narrow endemics are indeed found on all but the smallest volcanic and coral islands in open oceans, and on most summit cones of major mountains; and in general the degree of endemism is greater in the former than the latter. Thus some 66.7 per cent of the admittedly restricted native flora (146 species) on the very isolated Juan Fernandez islands in the eastern Pacific are endemic, and these islands also have several endemic plant genera, and one endemic plant family (Good, 1974). Some 580 of the 1,600 to 1,700 native plant species of the Canary Islands are endemic either to the group, or to the Macaronesian Islands (the Azores, Madeira, the Canary Islands and the Cape Verde Islands) as a whole, and these are especially interesting in that they include a large number of old, relict endemics, which have survived only in this locality from a former, much wider African distribution in Tertiary times (Bramwell, 1976). For instance, the laurel forest of Tenerife, the largest island in the group, contains four relict tree endemics, of which the best known is loro (Laurus azorica), as well as 94 endemic (and mainly relict) species, and seven endemic genera of carabid beetles (Machado, 1976). But by and large, the Canary Island group is unusual in having so many relict forms. On many of the larger islands of the world (Madagascar, Jamaica, Hispaniola, Cuba, New Caledonia, Kauai, Principe and Sao Thome), over 20 per cent of the land bird species are narrow endemics, and these are predominantly newly-evolved forms (Mayr, 1965). In the Galapagos, c. 26 per cent of plant species are endemic to the group as a whole, and some of the smaller islands therein may have 100 per cent endemicity (Johnson and Raven, 1973): all the endemic species here also appear to have emerged recently, under conditions of a changing climatic environment over the last 10,000 years (Colinvaux, 1972; Colinvaux and Schofield, 1976). On mountain tops, over the world as a whole, the degree of narrow endemism will vary according to the degree of competition received from more widely-ranging species: for native plant species in Europe, it is as low as 14 per cent in the high Pyrenees, but may reach 18 per cent in the Alps, and over 38 per cent in the central ranges of Corsica, in which latter case the figure is undoubtedly boosted by Corsica's physical separation from the European mainland (Faverger, 1972).

On occasions, narrow endemics may also assume considerable importance in floras and faunas elsewhere in continents than on mountain tops. Two cases provide suitable illustrations. California is part of the North American continent which in large measure is environmentally isolated by high mountain ranges to the north and east, and by the Pacific Ocean to the west; and it is noteworthy for

Figure 2.5: Major areas of endemism in the central California coast ranges (after Raven and Axelrod, 1978).

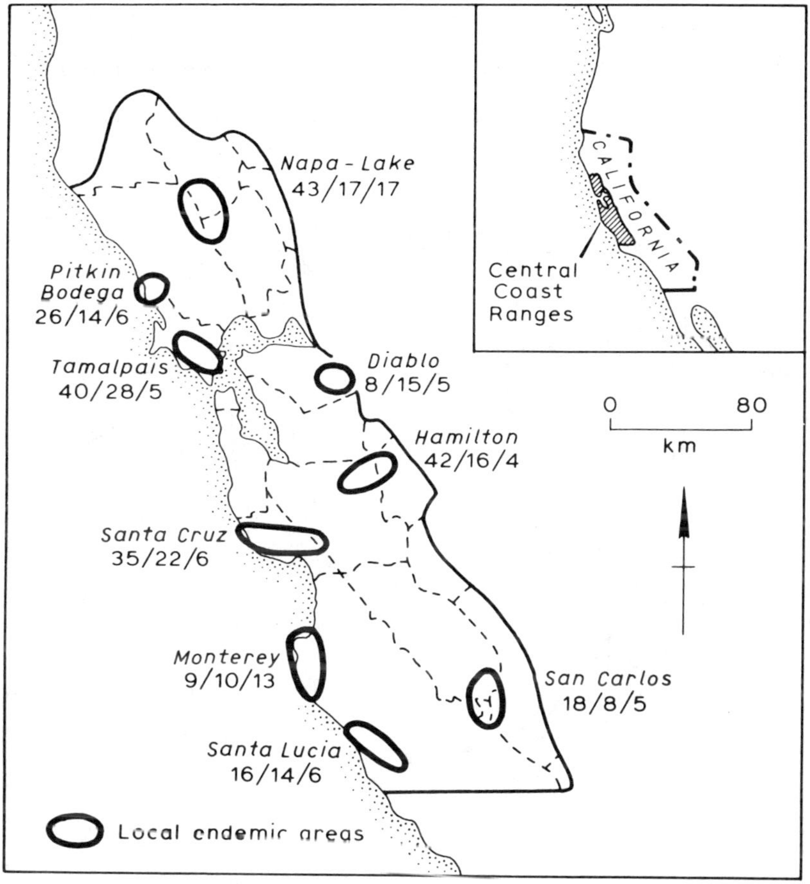

having a very rich flora of over 7,000 species, of which c. 38 per cent (Lewis, 1972) are endemic. A few Californian endemics, such as the Big Tree (Sequoiadendron giganteum (Lindl.), Buchh.) are relict, dating back to mid-Tertiary times, and surviving in the state because of the environmental peculiarities of moist, equable climates having been maintained almost by chance in topographic pockets of the lower western flanks of the Sierra Nevada; but most are newly-evolved, such as those in the Clarkia genus, the greatest proportion of which are confined to the state. Some mountain areas of central California are particularly rich in endemics, and these are indicated in Fig. 2.5. The numbers listed for each site represent, in sequence, species found in the local area which are endemic to the whole region; relict species; and endemic species which are restricted to the local area. Note that relict species are largely limited to the moister areas. A second example is located within the continental

Figure 2.6: Presumed forest refuges in the Amazon basin during the last phase of the 'dry' Pleistocene (after Turner, 1976).

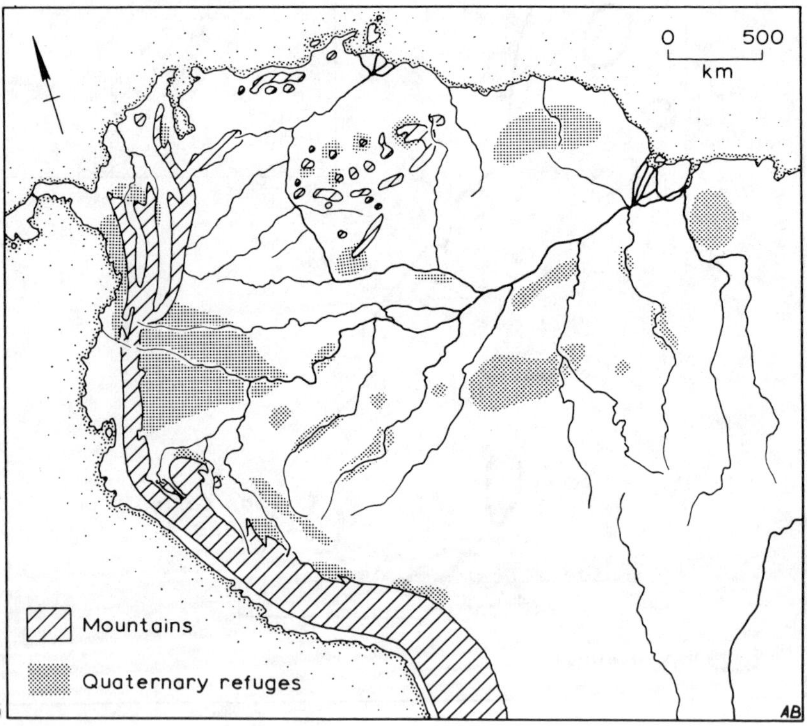

tropics. Here, in Pleistocene times, tropical lowlands became cooler by some 3° to 9° C during the glacial maxima of high latitudes, and much more arid. The consequence to vegetation was that the areal extent of rain forest was considerably diminished within these phases, so much so that it seems to have been maintained only in small refuges, established where there was abundant sub-surface, or standing water; and it was from these refuges that the rain forest subsequently expanded again, most recently during the last 7,000 years (van der Hammen, 1974; Flenley, 1979), when temperatures once again were ameliorated, and precipitation increased. The

Figure 2.7: Mean ranges of 979 species of African passerine birds (after Rapoport, 1975). Abscissa: range size. Ordinate: number of observed cases.

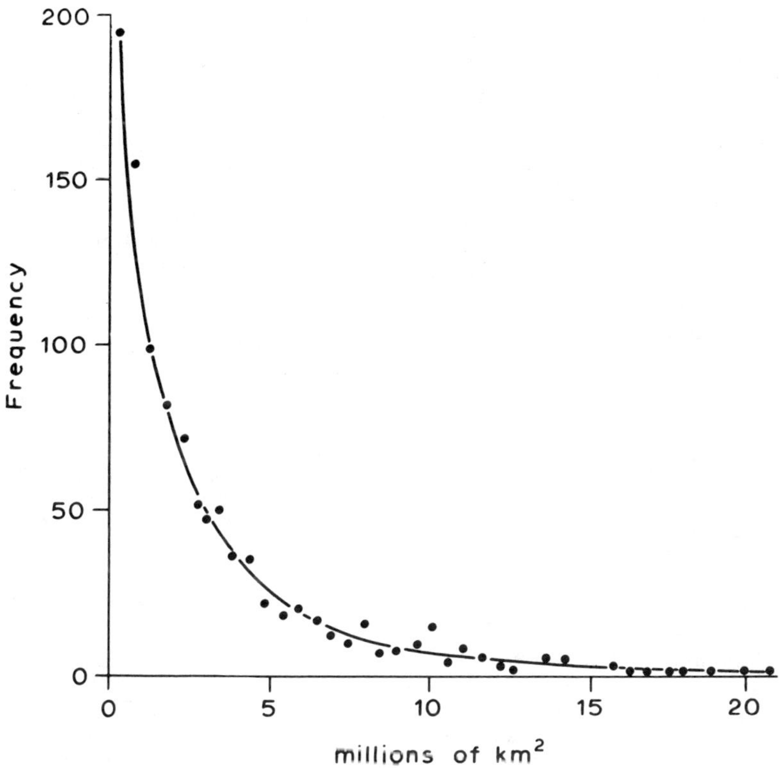

location of presumed forest refuges in the Amazon basin, as determined from a wide variety of evidence, is given in Fig. 2.6; and there is little doubt that similar refuges also existed within the 'dry' Pleistocene in tropical Africa and south-east Asia (Prance, 1982). Such refuge areas are of considerable biogeographic importance in that they possess a high degree of narrow endemism among certain types of organisms, including not only plants but also forest birds (Haffer, 1969), terrestrial vertebrates (Muller, 1973) and butterflies (Turner, 1976; Brown, 1982: see also pp. 50-52).

It was Willis (1922) who first pointed out that, in respect of their ranges, most species could be classified either as 'wides' (with a large range) or 'endemics', and that there were many more of the latter than the former. In reconfirming this by means of modern data analysis techniques, Valentine (1972) has further concluded that a dynamic balance exists between the two. Often a smooth-curve relationship between them may be displayed, as for African passerine birds, in Fig. 2.7. Why this should be so, and whether it is likely to be maintained in the face of the continued and massive disturbance of global habitats by man, remain two of the key questions of modern biogeography.

Discontinuous Ranges

Species ranges of both plants and animals are often not internally continuous; and where breaks in range extent are of a sufficient size for them seemingly to be incapable of being overcome by the normal, present-day means of dispersal of the species, or taxa concerned, they are referred to as range disjunctions.

Some disjunctions are intracontinental in scale, as in the case of European Arctic-Alpine plants, such as *Salix herbacea*, purple saxifrage (*Saxifraga oppositifolia*) and in the European distribution of *Poa alpina* (Fig. 2.8a). All these species are found both on Arctic and sub-Arctic lowlands or coasts, and high in the Alps, but nowhere in between, a distribution pattern which reflects the increasing amelioration of temperatures in the intervening areas since the final decline of the Wurm ice-sheets, some 10,000 years ago. Other disjunctions are intercontinental, and five major instances of these may be delimited (Solbrig, 1972: Figs. 2.8b to 2.8f). One is an amphi-Atlantic interruption of range, already noted previously (pp. 31-32) and exemplified in Fig. 2.8b by the case of the cowberry (*Vaccinium vitis-idaea*). Note, however, that the *sensu stricto* subspecies is restricted to Eurasia alone. A full examination of species with amphi-Atlantic discontinuities in their range is given by Love and Love (1963). A second major intercontinental disjunction involves species and genera which are present in eastern Asia and eastern North America alone: noted as long ago as the mid-nineteenth century (Gray, 1859), this is particularly the case at the generic level (e.g. for *Liriodendron*, the tulip tree, Fig. 2.8c; *Liquidambar*, sweet gum; *Hamamelis*, witch hazel; *Catalpa*; *Wisteria*; and *Magnolia*: see also Table 2.3 for species examples). These predominantly moisture-living genera and species are likely to have been driven out of western North America by increasing aridification since mid-Tertiary times, and failed to survive in Europe and western Asia because of cold conditions in the Pleistocene. Third and fourth intercontinental disjunction types are those present between tropical Africa and tropical South and Central America (e.g. the red mangrove, Rhizophora mangle, Fig. 2.8d: see also p. 32), and between South America and Australia (Migadopini beetles, Fig. 2.8e: see also Flenley, this volume, in respect of the Nothofagus range pattern). Both of these disjunctions owe a good deal to the influence of continental drift on plant and animal distributions (Darlington, 1965; Good, 1974; Flenley, this volume), although in the case of the red mangrove its presence on both sides of the Atlantic reflects more its present-day ability to disperse seeds over long distances in salt water. The fifth intercontinental disjunction is an amphitropical one. There are several well-documented examples of this, especially in the Americas (Raven, 1963) in respect of the woody genera Ficus and Acacia, and the woodland ground genus Ozmorrhiza). But the most complex amphitropical range conundrum relates to the distribution of the creosote bush (Larrea divaricata: Fig. 2.8f), which is frequently the only shrub found in the Monte desert of Argentina, and the Sonora desert of Mexico. No adequate resolution of the reasons for its absence from intervening areas has yet been forthcoming.

Ranges: Some Further General Considerations

It should perhaps be re-emphasised that there is nothing sacrosanct about the existing ranges of any species, which may be modified with changes in environment, rates of competition or predation, human activity or, on a longer time-scale, the influences of continental drift. Within a species range, the territory of individuals or pairs can also be altered quite quickly: for tree sparrows, Weeden (1965) has noted that the mean size of territory is reduced as population density increases, even over the course of one or two years, and this seems to be the case for many other bird and animal species as well (Andrewartha, 1961; Soja, 1971).

Moreover, under certain circumstances, species' ranges may disappear entirely from particular areas, either temporarily or permanently, as they are taken over by other life forms, through factors of competition. Little is known about the patterns of species turnover which are implicit in these events, though both for plants and animals the rates of turnover for some species may be quite high. Species turnover phenomena have been recorded over periods of months (Simberloff and Wilson, 1969) and years (Diamond, 1969). On Krakatoa, five bird species were lost, and another five gained during the ten years following 1920 (MacArthur and Wilson, 1967); and off the coast of Scotland, under very different environmental conditions, Lack (1942) has concluded that local extinctions and replacements of bird species can occur within a few decades. Fewer comparative data are available for plants, though Holland (1978) has found that between 1969 and 1976, 20 per cent and 14 per cent, respectively, of

Figure 2.8: Major world range disjunction types. For sources and explanations, see text.

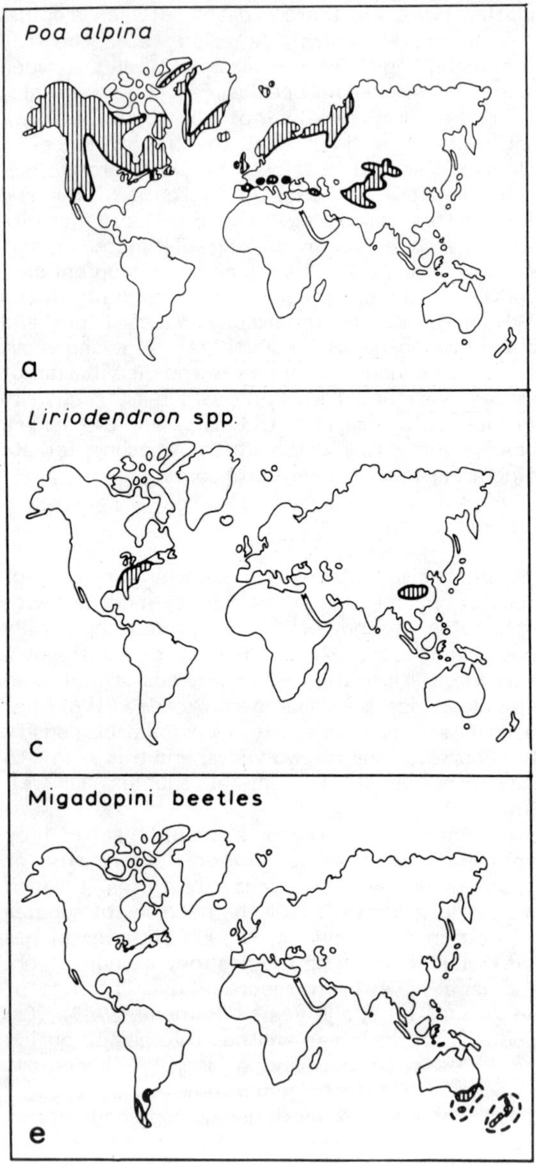

Figure 2.8 cont'd: Major world range disjunction types. For sources and explanations, see text.

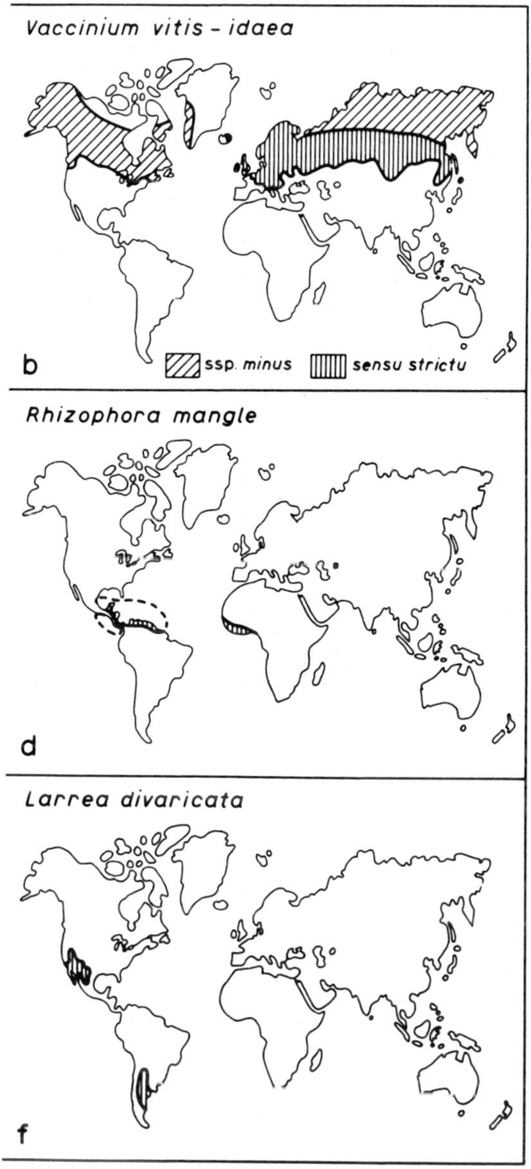

the plant species present in two quadrats of mature deciduous forest in southern Quebec were replaced. In general, there appears to be a tendency for smaller populations to have the highest turnover rates (Terborgh, 1974); and these may be further accelerated in very isolated situations, such as the remotest oceanic islands (Gilbert, 1980).

Figure 2.9: Simplified model of the increment of the species area with latitude (after Rapoport, 1975).

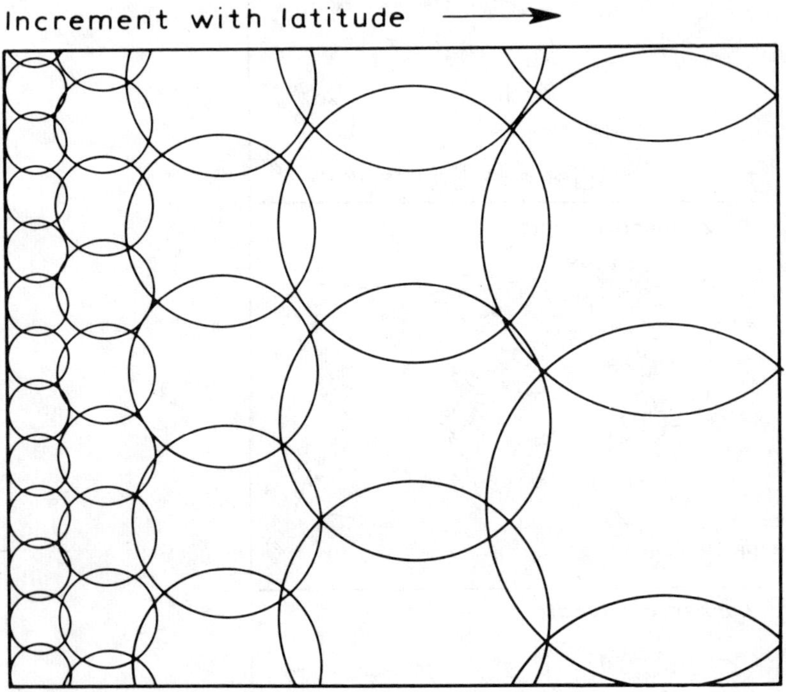

A final general consideration is that mean range sizes for all species tend to be greater in higher latitudes than elsewhere, and this is neatly expressed in the model displayed in Fig. 2.9, the further biogeographical implications of which are discussed later in this chapter. For animals, the most common, general distortion to this model relates to their position within the grazing food chain. Thus, for North American birds (Table 2.4), Robbins et al (1966) have indicated that the mean ranges of predators are greater than for non-predators, though they do not give any explanation for this: the cause may however simply be that non-predators are often restricted to the environments of particular range-limited foods, while predators normally are much less species-specific in their food requirements. Other, smaller distortions would be expected to arise from local conditions of environment, competition, predation or human interference.

Table 2.4: Mean ranges of groups of birds in North America (after Robbins *et al*, 1966, and Rapoport, 1975)

Groups of birds	Mean range size (10^4 km^2)
Strigiformes (owls, predators)	600 (*n* = 19)
Falconiformes (falcons, eagles: predators)	505 (*n* = 32)
Passeriformes, Fringillidae (sparrows, chaffinches)	343 (*n* = 37)
Galliformes (turkeys, partridges, etc.)	309 (*n* = 16)

Local Ranges

At the more local level, range patterning may exhibit rather different qualities from those so far exemplified at regional and continental scales for, although the essential determinants of range distribution are the same, their emphases in relation to each other can assume quite different proportions. Additionally, it is very likely that stochastic influences will play a more important role at the local scale in the determination of whether a species is, or is not present in a particular place. In other words, there is no perfect array of controls which leads one to suppose that a given local habitat will tend towards the support of a specific suite of plants or animals. Rather, at this scale, vegetation and animal communities may be more predisposed to become, in Gleason's words, 'fortuitous juxtapositions of individuals'. In particular, the patterns of arrival of plant propagules, or of animals, is likely to be random to a considerable degree, and evidence for this is beginning to accumulate from many parts of the world.

In his study of the effects of Quaternary climatic changes in Africa, Livingstone (1975) has suggested that there is a good deal of individualistic (i.e. random) behaviour in the determination of local plant associations in rain forest; and Flenley (1969) has demonstrated that, on mountain peaks in New Guinea, each summit has its own vegetational peculiarities, some of which at least might be due to the chance arrival of particular propagules. Local plant groupings on environmental gradients in the Great Smoky mountains in eastern North America display a large measure of individualism (Whittaker, 1956); and on the very small, red mangrove islands of the Florida Keys, chance factors have played an important role in the colonisation patterns of the eleven arboricolous ant species which occupy them (MacArthur and Wilson, 1967). Elsewhere, Connor and Simberloff (1978) have proposed that stochastic models can account for a large part of the between-island variation of both fauna and flora of the Galápagos group (see also Simberloff, 1978); and on many small islands in the West Indies, Linhart (1980) has concluded that chance factors relating to species arrival are significant to any explanation of plant distributions there. Possibly, where species turnover is rapid, stochasticism contributes even more substantially to the array of species distributions, though the evidence for this so far has been slight. But certainly, the old view that the species composition of local communities of plants and animals is determined by environmental controls such as climate (the climatic climax community model: Clements, 1936) is increasingly hard to swallow, as the antithetic evidence piles up (see also Flenley, this volume). Indeed, even the supposed stability of well-defined climax communities is often more apparent than real, for while such communities may well reject invaders for a while, even in severely disturbed environments such as the smaller West Indies (Watts, 1970), there is no good reason for supposing that this situation should continue indefinitely, or indeed that chance factors might not have had a role in the establishment of the community species composition in the first instance. On a broader perspective, Kellman (1980) has argued that a flexible conceptualisation of community species composition is required to fit the evidence, for, although repetitiveness in such composition may be clearly discerned in regions of low species diversity (for example, in high latitudes), this is not the case where high species densities prevail. The concept of the climax community perhaps is more an artefact of the mid-latitudinal origin and training of most ecologists and biogeographers thus far, rather than one which has any good basis in reality.

Species and Range Densities: Patterns and Variations

For both individuals and species, and their ranges, the numbers present per unit area may be expected to vary widely. At the level of individuals, the highest densities for many species groups (e.g. carabid beetles, birds and plants in north-west Europe (Hengeveld and Haeck, 1982) are customarily found close to range centres, and the lowest at range peripheries. Overall, more species per unit area are located in the humid tropics than anywhere else: here, Myers (1979) has estimated that between 40 to 50 per cent of the world's plant and animal species reside. Many of these are rare, and have very limited ranges (see Fig. 2.9 for the general diminution in range size as one approaches the tropics). The point has been made by Flenley (1979) that rare plant species may have an advantage in the hot, wet tropics in that they are perhaps less vulnerable to attacks by insect or other pests than those which have a wider distribution: and they may also encourage, through their very isolation, genetic drift among sister populations, and thus the increased genetic variability which can enhance the survival of the species (Federov, 1966). Moreover, rarity may be further stimulated by the root exudates produced by certain tropical rain forest trees, which inhibit seedling development: one good example is _Grevillea robusta_ in Queensland (Webb _et al_, 1967). Under these circumstances, the rarer a species is, the greater the chances that a seedling becomes established in the area beyond the immediate zone of inhibition beneath the parent tree. For animals, species densities in the humid tropics can be particularly accentuated by the phenomenon of _range superimposition_, in which organisms with similar niche requirements develop ranges which to a large extent are superimposed one on top of the other in the same small district. Simberloff (1970) has suggested that this is a relatively common occurrence among related, congeneric species; and recent research among bird populations in Colombia and Ecuador (Terborgh and Winter, 1982) makes it clear that over 25 per cent of the endemic bird species present in those countries, which have very small ranges of less than 50,000 km^2, cluster in the same, small local district, so increasing _range densities_ substantially.

The general case for decreased species ranges and increased species densities from high to low latitudes has been made for many land organisms. An increase in the number of bird species per unit area of 300 miles^2 along the same longitudinal boundaries, of from c. 17 to the north of Hudson Bay in Canada, to 61 at the southern end of the same Bay, to 142 in the vicinity of the Great Lakes, to 300 to 500 in southern Mexico, has been demonstrated by Blake (1953) and Robbins _et al_ (1966). In this instance, the higher densities in low latitudes appear to be in large measure due to the greater vertical extent and complexity of tropical vegetation communities, within which many more food specialists (insect-eaters, fruit and nut-eaters, nectar-feeders, etc.) can survive, especially at tree-top level, than elsewhere. A similar increase in the number of passerine bird species in Argentina, which is correlated strongly with increases in mean annual temperature, is especially well marked as one crosses the border of the sub-tropical rain forest from south to north (Rabinovitch and Rapoport, 1975). From these examples, it might be anticipated that, where the vertical stratification of vegetation _is_ similar, species densities of birds should be very closely related, and this does seem to be the case, certainly in the Americas (Karr, 1971, 1976). But it is not necessarily true for other species groups which are also dependent on vegetation to some extent: thus, for New World desert grasshoppers, regional Pleistocene history is more important than present vegetation patterning in explanations of species density variations (Otte, 1976). Evidence from other land animal groups for a high-to-low latitude species-density cline remains rather scanty; but the number of species per unit area of frogs, snakes and lizards within particular altitudinal divisions on mountains has been demonstrated to increase between California (Lat. 40°N) and Costa Rica (Lat. 10°N), as also did their turnover rates (Huey, 1978).

For continental plants, Cailleux (1953) has shown that mean species densities reach their peak in Central America (380-420 species per 10,000 km^2), with eastern Brazil (370) and Cuba (350) not far behind. Why the Americas should dominate in this respect is not immediately clear. The only other regions with mean species densities of over 300 are Borneo (320) and the Cape of Good Hope (320), the latter, of course, lying some distance away from the tropics, and possessing many local and endemic species of particular genera, including well over 600 of the 700 known species of heather (Erica). Regions which have mean species densities of over 250 include south-east Asia as a whole, the Malay peninsula, Java, Madagascar, the Phillipines and Burma, in descending order. Outside the tropics, the greatest plant species densities normally are found in those regions which undergo some periodic or annual environmental stress, most commonly a shortage of available water: additional to the Cape of Good Hope (above), good examples would be many parts of lowland California, and the endemic-rich grasslands and mountains of central California (Stebbins, 1972: Fig. 2.5). Other major centres of plant species endemism outside the tropics, as, for example, the Canary Islands (p. 36), also are often characterised by high species densities.

Within the major water bodies of the world, particular species density trends may also be observed. Some relate conspicuously to the high-latitude - low-latitude cline, which has already been described. Thus, Taylor and Taylor (1977) have recorded a massive increase in the number of predatory gastropod species present on the eastern North Atlantic shelf, as one moves from north to south, in the vicinity of Lat. 40°N. Other animal forms (crustaceans, worms, clams) have their highest species densities not on continental shelf areas, but in middle ocean depths, where environmental stability is accentuated, and where there are still sufficient energy reserves for a considerable variety of organisms to be supported (Sanders, 1968). Pleistocene history also has caused some regional variation in oceanic species densities. For example, there are many more oceanic faunal and floristic species in the mid-latitudes of the North Pacific than of the North Atlantic, consequential to the greater number of Pleistocene extinctions in the latter (Briggs, 1974); and for the same reason, more seaweed species (Rhodophyta, Phaeophyta and Chlorophyta), with more overlapping ranges, exist on the Pacific, than on the Atlantic coasts of North America (Pielou, 1977).

In respect of the smaller, motile organisms, many land phytopathogens (viruses, bacteria, fungi) have maximum species densities outside the tropics, within zones of intensive cultivation, such as the south-eastern USA, central California, north-central Europe, southern Japan, New South Wales and New Zealand (Rapoport et al, 1976). In contrast, the major centres of diversity of insect pests are in the tropics, and especially in the islands of south-east Asia, and in the Caribbean: species densities in this group of organisms are also high in southern Europe (Ezcurra et al, 1978).

Perhaps the most interesting biogeographical question which arises from a study of species density patterns is how one may account for them. Do they have their origins in past, or present-day conditions, or in some combination of both? It has been intimated thus far that some at least appear to have resulted from past events. Much, though not all, of narrow endemism may be accounted for by the adaptive radiation of some species groups over time, under

Figure 2.10: Allopatric and parapatric ranges. Subgroups of the same species, or closely-related species, exist under conditions of total physical separation in allopatry (a); in parapatry (b), they are either enclosed within one major population (Sg_{1-2}), or exist as contiguous but connected subpopulations (Sg_{3-4}).

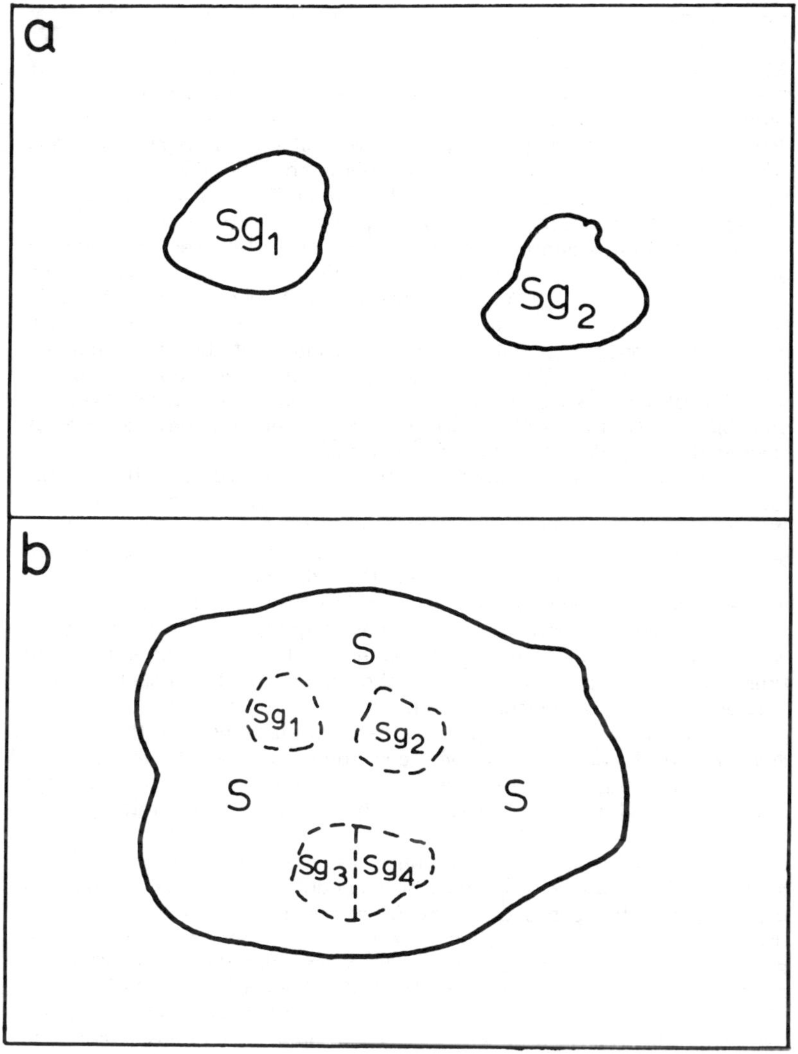

conditions of relative isolation and in changing environmental circumstances. The past subdivision of former, wide-ranging ancestor populations, under conditions of allopatry or parapatry (Fig. 2.10), can give rise, through genetic differentiation, to many new subgroups of the same species, or many new closely-related species, so adding to species densities. Reference to the splitting of old populations in this way has been termed the vicariance paradigm (Latin vicarius = substitute) by Croizat, Nelson and Rosen (1974). Although, technically, vicariance may refer to as simple a matter as the transfer of one small animal population from one host to another, usually in biogeography it is directed towards larger-scale events, and especially to the consequences, for plant and animal distributions and densities, of continental drift and Pleistocene climatic change. As far as the former is concerned, Rosen (1975) has indicated how modern plant and animal distributions, and to some extent densities, in the West Indies can be correlated with past plate tectonic events; and there are other, broader-scale examples (see Croizat, Nelson and Rosen, 1974; Cox, 1974). As regards climatic change, both in temperate latitudes and in the tropics, Pleistocene events clearly have had a considerable impact on the range distribution of certain plants and animals (see, for example, Fig. 2.6), and no doubt stimulated some speciation rates. But how far either type of vicariance activity has given rise generally to present-day patterns of species diversity and density is not quite so certain.

Important new evidence which may throw some light on this latter point has emerged from three studies of plant and animal taxa. Within the Amazonian rain forest, and for two groups of forest butterflies (Helicoiini and Ithomiinae), involving a data set of 866 subspecies, Brown (1982) has noted that distinctions may be made between zones of genetic purity, in which there is little variation in subspecies type per unit area, and zones of great genetic diversity, in which there are very many subspecies; and that in large measure these zones reflect differences in the environmental mosaic of the hot, wet tropics in this region. The zones of genetic purity, in which particular endemic species prevail, tend to correspond with areas of environmental homogeneity, whereas zones of diversity are located along environmental gradients in which there is a considerable degree of habitat variation. More especially, subspecies diversity, either through active hybridisation, or other means of differentiation, will often reach maximum levels along environmental gradients in areas of rich soils. Centres of endemism, and centres of existing maximum diversity at the subspecies level, accordingly may be spatially differentiated: the essential areal relationships between the two are given in Fig. 2.11. The zones of maximum diversity lie between the centres of endemism, and centres of hybridisation. Brown concludes that, for these animal groups, the oldest genetic forms are located in the zones of genetic purity, which in turn coincide to some extent with the supposed rain forest refuges of the Pleistocene (Fig. 2.6): in contrast, the present-day zones of maximum diversity are not correlated with the areal distribution of such refuges.

Figure 2.11: The relationship between zones of endemism and zones of maximum species (or subspecies) diversity in the Amazon rainforest (after Brown, 1982).

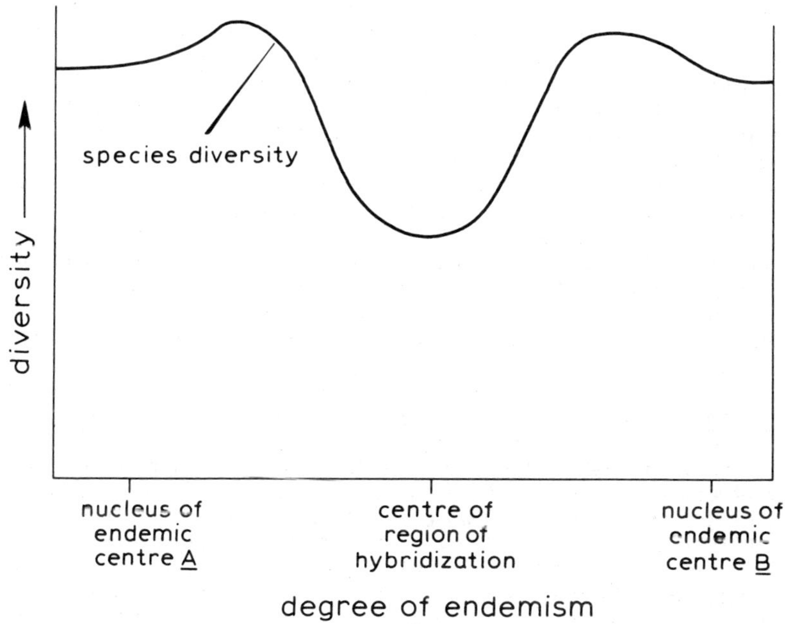

Using a different array of evidence, and noting that species diversity patterns for many plant and animal phyla in the Amazon basin look broadly similar, Endler (1982) has agreed with many of Brown's conclusions. He concurs with the view that the major centres of endemism in this region are to be found in areas of environmental uniformity. Subspeciation in the contact zones between the centres of endemism, along environmental gradients, is largely attributed to active (i.e. present-day) parapatric speciation among small local populations of closely-related forms, and not to past allopatry; and the extent to which this is taking place is likely to depend on the rates of migration of individual organisms and groups away from the major current centres of endemism, and on the width of particular contact zones, and their consequent ability to support a range of small sub-populations of closely-related forms in parapatric separation.

Both authors, accordingly, strongly support the idea that species diversity and density patterns in the Amazon rain forest relate more to the conditions of present-day ecology than to

Pleistocene events, the influence of which in this region, they say, has long since been expunged, as far as these organic life forms are concerned. It remains to be seen, however, whether these ideas prove to be capable of extension to situations in other parts of the world.

On a broader scale, the potential importance of current geographical conditions to organic distribution patterns has also been underlined by Flessa (1981), in a consideration of mammalian fauna. Using cluster and regression analysis, he has identified three geographically-distinctive groups among the 969 terrestrial genera: a western-hemisphere group; one based in Australia and New Guinea; and another in Eurasia and Africa. The degree of faunal similarity between these groups is, he suggests, highly correlated with the overland distance between them: indeed, three significant contemporary controls (distance, longitudinal separation - a measure of the extent of oceanic barriers - and areal extent) account for over 70 per cent of the observed variation in group data. Historical events are of minor significance, except in explanations of the distribution of particular mammals, in specific areas.

The Effects of Man

It is well known that plant and animal ranges, and range-density patterns, have been modified considerably by man in recent years. Perhaps two consequences of man's activities are of particular importance to spatial themes in biogeography at the present time. The first concerns the ever-accelerating rates of reduction in the areal extent of natural ecological systems, and of the ranges of species associated with them, as more and more parts of the world are cleared of forest, and/or put under agriculture. Often, species diversity patterns are also affected by these changes. The desirability of maintaining a responsible balance between land reserved for the preservation of natural plant and animal communities, and land allocated for agricultural and other developments, is a theme which has been treated extensively by many ecologists and biogeographers (e.g. Terborgh, 1974; Diamond and May, 1976; Higgs, 1981; Lovejoy, 1982), and it will not be considered further here. But arising from this is a need also to conserve adequate corridors of natural, or semi-natural vegetation between reserves, in which movements of native species, and gene flow, within a broader region than that of the boundaries of the immediate reserves, may be encouraged; and this problem, having received less attention thus far, will need to be looked at increasingly in the future. Network analysis, a tool familiar to most biogeographers, is likely to be important in the determination of how spatially extensive such corridors in particular areas should be.

The second consequence, linked to the first, is a general observable trend towards an inadvertent but general range extension among those organisms which are especially closely associated with man's activities, at the expense of others. At a continental scale, the result may be two very different types of range, and species density patterning, for the two broad groups of organisms. This may be seen clearly in the case of Africa south of the Sahara, where the mean range size of 979 species of native passerine birds is 2,978,500 km^2 (S.D. 1,665,000 km^2), while that for phytopathogens is larger, at 4,228,500 km^2 (S.D. 3,979,500 km^2) (Rapoport and Excurra, 1979: Fig. 2.12). At first sight, both of these sets of values are high, but they do correspond with the known range patterns of highly-motile organisms elsewhere. Species densities among the native passerine birds are relatively uniform, with only one area of major clustering, in the vicinity of Lake Victoria, where several biomes interconnect; and here over 30 per cent of the total number of bird species may be found. In contrast, the degree of clumping of phytopathogen species is much greater, reaching a peak on favoured agricultural land in the Cape of Good Hope region and in East Africa; and, secondarily, along the Guinea coast, where patches of intensive farming also exist.

Figure 2.12: Species richness patterns for passerine birds and phytopathogens in Africa south of the Sahara (after Excurra et al, 1978). In each case, figures on the lines of isodensity represent percentages of the total number of species enumerated.

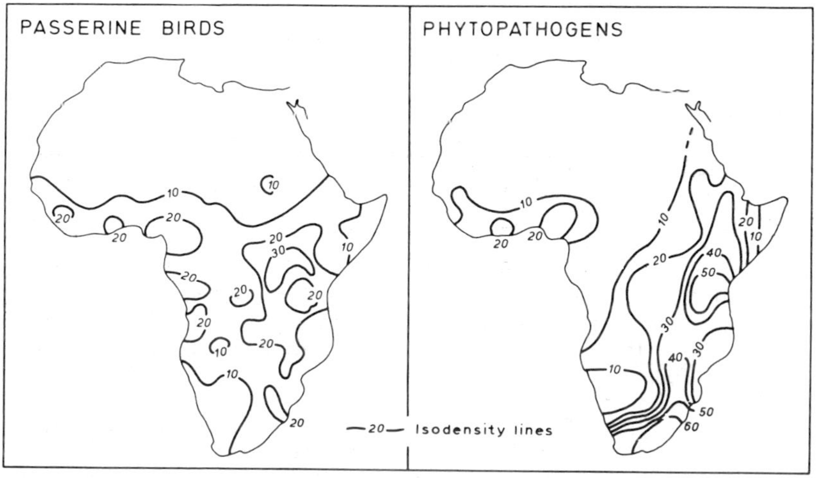

Ultimately, as world biotas become more and more mixed through human interference, one may perhaps expect new and simplistic biogeographical regions to emerge, dominated by cosmopolitan and semi-cosmopolitan species, and in which the overall gene-pool resources of the globe are considerably diminished. For insect pests, Excurra et al (1978) have forecast that these will consist of cold, temperate and tropical regions, the first two of which will be further subdivided into northern and southern hemisphere subregions. Fortunately, it will be some considerable time before such a biogeographically uninteresting and undesirable state of affairs could develop in its entirety, even if we were foolhardy enough to allow it to do so.

REFERENCES

Amadon, D. (1950), 'The Hawaiian honeycreepers', *Bulletin of the American Museum of Natural History*, **95**, 1-262.

Amerson, A.B. (1975), 'Species richness in the non-disturbed northwestern Hawaiian islands', *Ecology*, **56**, 435-444.

Andrewartha, H.G. (1961), *Introduction to the Study of Animal Populations*, Methuen, London.

Baker, H.G. (1972), 'Migrations of weeds' in Valentine, D.H. (ed.), *Taxonomy, Phytogeography and Evolution*, Academic Press, London, pp. 327-47.

Blake, E.R. (1953), *Birds of Mexico*, University of Chicago Press, Chicago, USA.

Bramwell, D. (1976), 'The endemic flora of the Canary Islands' in Kundel, G. (ed.), *Biogeography and Ecology in the Canary Islands*, W. Junk, The Hague, pp. 207-40.

Briggs, J.C. (1974), *Marine Zoogeography*, McGraw-Hill, New York, USA.

Brown, K.S. (1982), 'Palaeoecology and regional patterns of evolution in neotropical forest butterflies', in Prance, G.T. (ed.), *Biological Diversity in the Tropics*, Columbia University Press, New York, USA, pp. 255-308.

Bush, G.L. (1974), 'The mechanism of sympatric host race formation in the true fruit flies (Tephritidae)', in White, M.J.D. (ed.), *Genetic Mechanisms of Speciation in Insects*, Australia and New Zealand Book Co., Sydney, pp. 3-23.

Cailleux, A. (1953), *Biogéographie Mondiale*, Presses Universitaires de France, Paris.

Cain, S.A. (1944), *Foundations of Plant Geography*, Harper and Bros., New York, USA.

Clements, F.E. (1936), 'The nature and structure of the climax', *Journal of Ecology*, **24**, 252-84.

Colinvaux, P.A. (1972), 'Climate and the Galapagos Islands', *Nature*, London, **240**, 17-20.

Colinvaux, P.A. and Schofield, E.K. (1976), 'Historical ecology in the Galapagos Islands', Journal of Ecology, **64**, 989-1028.

Connor, E.F. and Simberloff, D.S. (1978), 'Species number and compositional similarity of the Galapagos flora and avifauna', Ecological Monographs, **48**, 219-48.

Cox, C.B. (1974), 'Vertebrate palaeo-distributional patterns and continental drift', Journal of Biogeography, **1**, 75-94.

Croizat, L., Nelson, G.J. and Rosen, D.E. (1974), 'Centers of origin and related concepts', Systematica Zoologica, **23**, 265-87.

Darlington, R.J. (1965), Biogeography of the Southern End of the World, Harvard University Press, Cambridge, Massachusetts, USA.

Diamond, J.M. (1969), 'Avifauna equilibria and species turnover rates on the Channel Islands of California', Proceedings of the National Academy of Science, **64**, 57-63.

Diamond, J.M. and May, R.M. (1976), 'Island biogeography and the design of natural reserves', in May, R.M. (ed.), Theoretical Ecology: Principles and Applications, Blackwell Scientific Publications, Oxford, pp. 163-86.

Endler, J.A. (1982), 'Pleistocene forest refuges: fact or fancy?' in Prance, G.T. (ed.), Biological Diversity in the Tropics, Columbia University Press, New York, USA, pp. 641-57.

Exell, A.W. and Stace, C.A. (1972), 'Patterns of distribution in the Combretaceae', in Valentine, D.H. (ed.), Taxonomy, Phytogeography and Evolution, Academic Press, London, pp. 307-23.

Ezcurra, E., Rapoport, E.H. and Marino, C.R. (1978), 'The geographical distribution of insect pests', Journal of Biogeography, **5**, 149-58.

Faverger, C. (1972), 'Endemism in the montane floras of Europe' in Valentine, D.H. (ed.), Taxonomy, Phytogeography and Evolution, Academic Press, London, 191-204.

Federov, An. A. (1966), 'The structure of tropical rain forest, and speciation in the humid tropics', Journal of Ecology, **54**, 1-11.

Flenley, J.R. (1969), 'The vegetation of the Wabag region, New Guinea highlands: a numerical study', Journal of Ecology, **57**, 465-90.

Flenley, J.R. (1979), The Equatorial Rain Forest: a Geological History, Butterworths, London.

Flenley, J.R. (1983), 'Time scales in biogeography' in Taylor, J.A. (ed.), Themes in Biogeography, Croom Helm, London, pp. 63-105.

Flessa, K. (1981), 'The regulation of mammalian faunal similarity among the continents', Journal of Biogeography, **8**, 427-38.

George, W. (1962), Animal Geography, Heinemann, London.

Gilbert, F.S. (1980), 'The equilibrium theory of island biogeography: fact or fiction?', Journal of Biogeography, **7**, 209-36.

Gleason, H.A. (1926), 'The individualistic concept of the plant association', Bulletin of the Torrey Botanical Club, **53**, 7-26.

Good, R. (1974), The Geography of the Flowering Plants, 4th edition (1st edition, 1947), Longmans, London.

Gray, A. (1859), 'Memoir on the botany of Japan, in its relation to that of North America, and of other parts of the northern temperate zone', Memoirs of the American Academy, **II:** (6), 1-66.

Grime, J.P. (1979), Plant Strategies and Vegetation Processes, John Wiley & Sons, New York, USA.

Haffer, J. (1969), 'Speciation in Amazonian forest birds', Science, **165**, 131-7.

Haffer, J. (1974), Avian Speciation in Tropical South America, Nuttall Ornithological Club, Cambridge, Mass., USA.

Hara, H. (1972), 'Japan and the Himalayas', in Valentine, D.H. (ed.), Taxonomy, Phytogeography and Evolution, Academic Press, London, pp. 61-72.

Hengeveld, R. and Haeck, J. (1982), 'The distribution of abundance. I. Measurements', Journal of Biogeography, **9**, 303-16.

Heyer, W.R. and Maxson, L.R. (1982), 'Distributions, relationships and zoogeography of Lowland frogs: the Leptodactylus complex in South America, with special reference to Amazonia' in Prance, G.T. (ed.), Biological Diversity in the Tropics, Columbia University Press, New York, USA, pp. 375-88.

Higgs, A.J. (1981), 'Island biogeographic theory and nature reserve design', Journal of Biogeography, **8**, 117-24.

Hill, T.A. (1977), The Biology of Weeds, Edward Arnold, London.

Holland, P.G. (1978), 'Species turnover in deciduous forest vegetation', Vegetatio, **38**, 113-18.

Hosie, R.C. (1969), Native Trees of Canada, Canadian Forest Service, Ottawa.

Huey, R.B. (1978), 'Latitudinal pattern of between-altitude faunal similarity: mountains might be 'higher' in the tropics', American Naturalist, **112**, 225-9.

Johnson, N.K. and Raven, P. (1973), 'Species numbers and endemism: the Galapagos revisited', Science, **179**, 893-5.

Juvik, J.O. and Austring, A.P. (1979), 'The Hawaiian avifauna: biogeographic theory in evolutionary time', Journal of Biogeography, **6**, 205-24.

Karr, J.R. (1971), 'Structure of avian communities in selected Panama and Illinois habitats', Ecological Monographs, **41**, 207-33.

Karr, J.R. (1976), 'Seasonality, resource availability and community diversity in tropical bird communities', American Naturalist, **110**, 973-94.

Kellman, M.C. (1974), 'The viable weed seed content of some tropical agricultural soils', Journal of Applied Ecology, **11**, 669-78.

Kellman, M.C. (1978), 'Microdistribution of viable weed seed in two tropical soils', Journal of Biogeography, **5**, 291-300.

Kellman, M.C. (1980), Plant Geography, 2nd edition, Methuen, London and New York.

Kornas, J. (1972), 'Corresponding taxa and their ecological background in the forests of temperate Eurasia and North America' in Valentine, D.H. (ed.), Taxonomy, Phytogeography and Evolution, Academic Press, London, 37-59.

Lack, D. (1942), 'Ecological features of the bird faunas of British small islands', Journal of Animal Ecology, **11**, 9-36.

Lewis, H. (1972), 'The origin of endemics in the California flora', in Valentine, D.H. (ed.), Taxonomy, Phytogeography and Evolution, Academic Press, London, 179-189.

Linhart, Y.B. (1980), 'Local biogeography of plants on a Caribbean atoll', Journal of Biogeography, **7**, 159-72.

Livingstone, D.A. (1975), 'Late Quaternary climatic change in Africa', Annual Review of Ecology and Systems, **6**, 249-80.

Löve, A. and Löve, D. (1963), North American Biota and Their History, Pergamon Press, Oxford, UK.

Lovejoy, T.E. (1982), 'Designing refugia for tomorrow', in Prance, G.T. (ed.), Biological Diversity in the Tropics, Columbia University Press, New York, USA, 673-80.

MacArthur, R.H. and Wilson, E.O. (1967), The Theory of Island Biogeography, Princeton University Press, Princeton, New Jersey, USA.

Machado, A. (1976), 'Introduction to a faunal study of the Canary Islands' laurisylva, with special reference to the ground beetles' in Kunkel, G. (ed.), Biogeography and Ecology in the Canary Islands, W. Junk, The Hague, 347-412.

Major, J. and Pyott, W.T. (1966), 'Buried, viable seeds in two California bunchgrass sites, and their bearing on the definition of a flora', Vegetatio, **13**, 253-82.

May, R.M. (1981), 'The world's worst weeds', Nature, London, **290**, 85-6.

Mayr, E. (1965), 'Avifauna: turnover on islands', Science, **150**, 1587-8.

Melzack, R.N. and Watts, D. (1982), 'Cold hardiness in the Yew (Taxus baccata, L.) in Britain', Journal of Biogeography, **9**, 231-41.

Mirov, N.T. (1967), The Genus Pinus, Ronald Press, New York, USA.

Muller, P. (1973), 'The dispersal centres of terrestrial vertebrates in the neotropical realm', Biogeographica, **2**, 1-244.

Myers, N. (1979), The Sinking Ark, Pergamon Press, Oxford.

Otte, D. (1976), 'Species richness patterns of new world desert grasshoppers in relation to plant diversity', Journal of Biogeography, **3**, 197-210.

Parsons, J.J. (1972), 'The spread of African grasses to the American tropics', Journal of Range Management, **25**, 12-17.

Pianka, E.R. (1970), 'On r- and k- selection', American Naturalist, **104**, 592-7.

Pielou, E.C. (1977), 'The latitudinal spans of seaweed species and their patterns of overlap', Journal of Biogeography, **4**, 299-311.

Pielou, E.C. (1979), Biogeography, John Wiley & Sons, New York, USA.

Polunin, W. (1950), Introduction to Plant Geography, Longmans, London.

Prance, G.T. (ed.) (1982), Biological Diversity in the Tropics, Columbia University Press, New York, USA.

Rabinovitch, J.E. and Rapoport, E.H. (1975), 'Geographical variation of diversity in Argentine passerine birds', Journal of Biogeography, **2**, 141-57.

Raikow, R.J. (1976), 'The origin and evolution of the Hawaiian honeycreepers (Drepanididae), The living bird, **15**, 95-119.

Rapoport, E.H. (1975), Areografia: Estrategias Geograficas de las Especies, Fondo de cultura economica, Mexico.

Rapoport, E.H. and Ezcurra, E. (1979), 'Natural and man-made biogeography in Africa: a comparison between birds and phytopathogens', Journal of Biogeography, **6**, 341-8.

Rapoport, E.H., Ezcurra, E. and Drausal, B. (1976), 'The distribution of plant diseases: a look into the biogeography of the future', Journal of Biogeography, **3**, 365-72.

Raven, P.H. (1963), 'Amphitropical relationships in the floras of North and South America', Quarterly Review of Biology, **38**, 151-77.

Raven, P.H. and Axelrod, D.I. (1978), 'Origin and relationships of the Californian flora', University of California Publications in Botany, **72**, 1-134.

Robbins, C.S., Braun, B., Zim, H.S. and Singer, A. (1966), Birds of North America, Golden Press, New York, USA.

Rosen, D.E. (1975), 'The vicariance model of Caribbean biogeography', Systematica Zoologica, **24**, 431-64.

Sagar, G.R. and Harper, J.L. (1964), 'Biological flora of the British Isles, Plantago major, L., P. media, L. and P. lanceolata, L.', Journal of Ecology, **52**, 189-221.

Salisbury, E.J. (1962), Weeds and Aliens, Collins, London.

Sanders, H.L. (1968), 'Marine benthic diversity: a comparative study', American Naturalist, **102**, 243-82.

Sarukhan, J. (1974), 'Studies in plant demography: Ranunculus repens, L., R. bulbosus, L. and R. acris, L.: II. reproductive strategies and seed population dynamics', Journal of Ecology, **62**, 151-77.

Schmidt, K.P. (1954), 'Faunal realms, regions and provinces', Quarterly Review of Biology, **29**, 322-31.

Simberloff, D.S. (1970), 'Taxonomic diversity of island biotas', Evolution, **24**, 23-47.

Simberloff, D.S. (1978), 'Using island biogeographic distributions to determine if colonisation is stochastic', American Naturalist, **112**, 713-26.

Simberloff, D. and Wilson, E.O. (1969), 'Experimental zoogeography of islands: the colonisation of empty islands', Ecology, **50**, 278-96.

Solbrig, O.T. (1972), 'New approaches to the study of disjunctions, with special emphasis on the American amphi-tropical desert disjunctions', in Valentine, D.H. (ed.), Taxonomy, Phytogeography and Evolution, Academic Press, London, 85-100.

Soja, E.W. (1971), 'The political organisation of space', Commission on College Geography, Resource Paper, **8**, 1-64. Association of American Geographers, Washington DC, USA.

Stebbins, G.L. (1972), 'Ecological distribution of centers of major adaptive radiation in angiosperms', in Valentine, D.H. (ed.), Taxonomy, Phytogeography and Evolution, Academic Press, London, pp. 7-34.

Stone, B.S. (1967), 'A review of the endemic genera of Hawaiian plants', Botanical Review, **33**, 216-59.

Szafer, W. (1966), The Vegetation of Poland, PWN, Warsaw.

Taylor, J.D. and Taylor, C.N. (1977), 'Latitudinal distribution of predatory gastropods on the eastern Atlantic shelf', Journal of Biogeography, **4**, 73-81.

Terborgh, J. (1974), 'Preservation of natural diversity: the problem of 155 extinction-prone species', Bioscience, **24**, 715-22.

Terborgh, J. and Winter, B. (1982), 'Evolutionary circumstances of species with small ranges', in Prance, G.T. (ed.), Biological Diversity in the Tropics, Columbia University Press, New York, USA, 577-600.

Tivy, J. (1982), Biogeography: a Study of Plants in the Ecosphere, 2nd edition, Longman, London.

Turner, J.R. (1976), 'Forest refuges as ecological islands: disorderly extinction and the adaptive radiation of muellerian mimics' in Deseimon, H. (ed.), Biogéographie et Evolution en Amérique Tropicale, École Normal, Paris, pp. 98-117.

Udvardy, M.D.F. (1969), Dynamic Zoogeography, Van Rostrand-Reinhold, New York, USA.

Valentine, D.H. (1972), 'Introduction' in Valentine, D.H. (ed.), Taxonomy, Phytogeography and Evolution, Academic Press, London, 1-6.

Van Der Hammen, T. (1974), 'The Pleistocene changes of vegetation and climate in tropical South America', *Journal of Biogeography*, **1**, 3-26.

Voous, K.H. (1962), *Die Vögelwelt Europas und ihre Verbreitung*, Verlag P. Parey, Hamburg.

Wallace, A.R. (1876), *Island life*, Macmillan, London.

Watts, D. (1970), 'Persistence and change in the vegetation of oceanic islands: an example from Barbados, West Indies', *Canadian Geographer*, **14**, 91-109.

Webb, L.J., Tracey, J.G. and Haydock, K.P. (1967), 'A factor toxic to seedlings of the same species associated with living roots of the non-gregarious subtropical rain forest tree *Grevillea robusta*', *Journal of Applied Ecology*, **4**, 13-25.

Weedon, J.S. (1965), 'Territorial behaviour of the tree sparrow', *Condor*, **67**, 193-209.

Whittaker, R.H. (1956), 'Vegetation of the Great Smoky Mountains', *Ecological Monographs*, **26**, 1-80.

Willis, J.C. (1922), *Age and Area*, Cambridge University Press, Cambridge, UK.

Zimmerman, E.C. (1960), 'Possible evidence for rapid evolution in Hawaiian moths', *Evolution*, **14**, 137-8.

Zimmerman, E.C. (1970), 'Adaptive radiation in Hawaii, with special reference to insects', *Biotropica*, **2**, 32-8.

FURTHER READING

Cain, S.A. (1944) *Foundations of Plant Geography*, Harper & Bros., New York City, USA.

Cox, C.B. and Moore, P.D. (1980) *Biogeography: an Ecological and Evolutionary Approach*, 3rd edition, Blackwell Scientific, Oxford, UK.

Darlington, R.J. (1965) *Biogeography of the Southern End of the World*, Harvard University Press, Cambridge, Mass., USA.

Furley, P.A. and Newey, W.W. (1983) *Geography of the Biosphere*, Butterworths, London, UK.

Gleason, H.A. and Cronquist, A. (1964) *The Natural Geography of Plants*, Columbia University, New York, USA.

Good, R. (1974) The Geography of the Flowering Plants, 4th edition, Longmans, London, UK.

MacArthur, R.H. and Wilson, E.O. (1967) The Theory of Island Biogeography, Princeton University Press, Princeton, NJ, USA.

Pielou, E.C. (1979) Biogeography, John Wiley & Sons, New York, USA.

Polunin, N. (1950) Introduction to Plant Geography, Longmans, London, UK.

Rapoport, E.H. (1982) Areography: Geographical Strategies of Species, Pergamon, Oxford, UK.

Tivy, J. (1982) Biogeography: a Study of Plants in the Ecosphere, 2nd edition, Longmans, London, UK.

Watts, D. (1971) Principles of Biogeography, McGraw-Hill, Maidenhead, UK and New York, USA.

Chapter 3

TIME SCALES IN BIOGEOGRAPHY

J. R. Flenley

Introduction

Time is different from the three other dimensions of the universe in that it moves in one direction only. This does not, however, inhibit the progress of investigations relating to it. Just as in spatial analysis, where the aim is to make generalisations which lead to hypotheses about points in space other than those analysed, so in temporal analysis, the aim is to reach generalisations which make possible the formation of hypotheses about the future (predictions) or about the past (retrodictions). The difference is that whereas spatial hypotheses can often be tested, predictions far into the future are untestable, and retrodictions are testable only in favourable circumstances, i.e. those in which some record of the past has been preserved. Those analysing temporal phenomena, therefore, would do well to bear in mind Occam's Razor: *Essentia non sunt multiplicanda praeter necessitatem* (Lindsay, 1911) which, very loosely translated, means that it is not appropriate to multiply hypotheses beyond what is needed. Also important is Popper's addition to this dictum: that science progresses more by attempts to disprove hypotheses than by the accumulation of evidence in their favour (Popper, 1968).

Time is important to geographers because the spatial patterns which they analyse are frequently explicable only in historical terms. In biogeography, time is important for at least four major reasons. First, time allows evolution to occur. Second, time allows migration to happen. Third, time permits environmental change to take place. Fourth, time allows autogenic changes to come about. Let us examine each of these in turn.

Evolution

Biogeographers distinguish sharply between evolution and development; these terms are used synonymously by most geographers. Development, in biogeography, covers the changes occurring in a single individual organism during its lifetime. Evolution means the appearance of new species of organisms from those already existing. The concept of evolution was known from the time of the Greeks, but the idea gained little credence, partly because of the lack of a plausible mechanism, even after the 'age of enlightenment' in the eighteenth century had made it possible to question religious beliefs.

Darwinism. When the theory of natural selection was put forward, perhaps first by Wells (1818), there was little interest. The breakthrough came with Darwin's comprehensive amassing of the evidence in The Origin of Species (1859). The argument was, in essence, surprisingly simple: individuals of a species vary and some of these variations are inherited; organisms tend to over-produce their offspring, yet numbers remain generally stable; therefore most offspring must die; those individuals best adapted to their environment will have a greater chance of survival; therefore natural selection will occur.

Neo-Darwinism. Darwin wrote without knowledge of the pioneer genetic work of Mendel (1866) and Darwinism was modified in this century after the re-discovery of Mendel's work in 1901. Neo-Darwinism recognises that what is evolving is the gene-pool of a population, and that the variations available for natural selection to act on include not only the small inherited variations emphasised by Darwin, but also larger, if usually rarer, mutations. In many cases, the rate of occurrence of mutations may be the factor limiting the rate of evolution. It is, then, important to realise that environmental change could have produced variations in mutation rates in the past. When the magnetic polarity of the earth reverses - which there is now good evidence that it has done on many occasions - there is a period of perhaps 2,000 years when the strength of the magnetic field is greatly reduced. This could lead to heavy bombardment of the earth by cosmic rays, which are normally diverted away by the magnetic field. Cosmic rays are powerful mutagenic agents. Investigation of the fossil record shows some tendency for evolutionary change to occur preferentially at these times, for instance in marine radiolaria (Fig. 3.1). Whether this is a general phenomenon is still under investigation.

Neo-Darwinism has been questioned recently by those who point out that, despite examples like the demonstration of industrial melanism in moths (Ford, 1940) and the experimental partial reconstruction of bread wheat by hybridisation (Hutchinson, 1965), there are few actual demonstrations of the evolution of new species, and almost none of new genera. Furthermore, it is pointed out that satisfactory fossil sequences demonstrating evolution are somewhat rare. They do occur, however: for example, the evolution of *Sonneratia* pollen in the Neogene of Malesia (Muller, 1972). But in many cases new genera or families just seem to appear *de novo* in the fossil record.

Figure 3.1: The relationship between magnetic reversals and the extinction and appearance of radiolaria, as observed in a deep-sea sediment core from the Antarctic (after McElhinny, 1973)

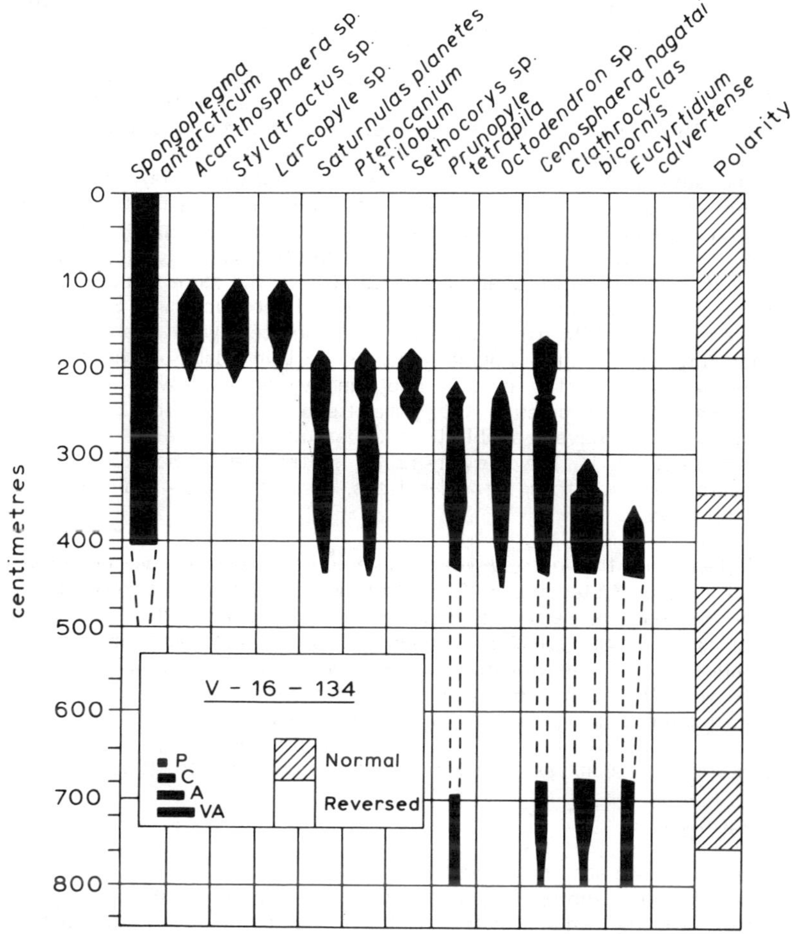

Saltation. This gave support to the idea of evolution by saltation (literally 'jumping'), which means the appearance of major new groups at a stroke, as it were, by the occurrence of one or more massive mutations. Botanical evidence from extant taxa for such a mechanism was quoted by Good (1956). He studied in particular the

Asclepiadaceae, the flowers of which have a complex structure, including an elaborate pollination mechanism. It is difficult to derive such structures by the progressive selection of small heritable variations, because nothing but the finished mechanism could function. Any half-way-house would have no selective advantage, and would tend to be eliminated. Similar arguments may be advanced concerning animals. Take the cuckoo (Cuculus canorus), for example. The successful parasitism of this species depends on: large eggs, rapid hatching, presence of a hollow in the back of the chick, the chick's instinct to eject other eggs and young, the parent's instinct to place eggs in the nests of appropriate species. None of these alone has much selective advantage, and some may be strongly disadvantageous. Only the complete complement of characters has survival value. These arguments are not watertight, partly because we know little of what is or is not of selective value to an organism. Sometimes a harmful character is carried by the same gene as, or is strongly genetically linked to, a useful character. Sickle-cell anaemia, a debilitating disease, is associated with resistance to malaria, and is thus advantageous in certain parts of the tropics. The examples of the cuckoo and the Asclepiad flower also both involve other organisms: the parasitised bird and the pollinating insect. These have probably been undergoing change also. What has been evolving is the flower-insect system and the cuckoo-host system. Despite these considerations, there may be a grain of truth in saltation.

Cladism. In the last few years some controversy has been generated by the idea of cladism. This is a method of numerical taxonomy and was originally aimed at phylogenesis. Some cladists felt that their results supported ideas of sudden change in evolution against the 'gradualism' of Darwinism. It was even argued that cladism was in line with the revolutionary ideas of Marxism, while the progressive 'improvement' suggested by Darwinism was seen as a capitalist notion. This parallel with politics seems irrelevant, and the cladists' results are not at variance with neo-Darwinism (Berry, 1982).

Migration

The second major way in which time-scales influence biogeography is through migration of organisms. One is referring here not so much to the well-known seasonal migrations of some birds, fish, butterflies and other animals (although these are striking and important phenomena), as to the extensions of range which are, or presumably have once been, exhibited by all organisms. If we accept monophylesis, then almost every species which has survived must have extended its range by a little at least. It is tempting to suggest that the age of a taxon might therefore be correlated, within related groups with similar dispersal mechanisms, with the area of its range: this was the age and area hypothesis of Willis (1922). Although the general tendency towards expansion is probably reasonable, there are so many complications and exceptions that the hypothesis is not widely applicable. For example, restricted distributions are often not those of new taxa but are the relict distributions of ancient taxa, once more widely distributed. The classic example is the genus Metasequoia, the Dawn Redwoods, now restricted to a small area of western China, but known as fossils in many parts of the northern hemisphere in early Tertiary times.

Some restricted distributions could well be those of young taxa. For example, Senecio cambrensis is a herbaceous plant confined to a small area of Flintshire, North Wales, UK. It is an allotetraploid, with twice as many chromosomes as Senecio vulgaris x squalidus (Groundsel x Oxford Ragwort) which is closely similar in appearance. Probably S. cambrensis has arisen from S. vulgaris x squalidus by chromosome doubling fairly recently, and has not yet spread very far (Clapham, Tutin and Warburg, 1962).

Disjunct distribution. The most puzzling distributions are the disjunct ones. Where the disjunction lies across a continent, it is sometimes possible to postulate an explanation in terms of climatic change. For instance, Darwin (1859) and Wallace (1869) explained the isolated occurrences of temperate genera of plants on tropical mountains in terms of migration during periods of Quaternary cooling. Recent palynological results have given some support to this idea (Flenley, 1979). Oceanic disjunctions are more difficult, and three main explanations have been offered: continental drift, land bridges and long distance dispersal. Continental drift will be considered in the next section under environmental change. The land bridge theory (van Steenis, 1962) has received scant support from modern geology and has lost credibility. A modified form of it lives on, however, as the 'stepping stone theory'. When sea-level was at least 100 m lower in the Pleistocene, many more islands must have been exposed, especially in the Pacific (Menard and Hamilton, 1963). These could have provided 'stepping stones' for migrating organisms. The third idea, long distance dispersal, must be at least somewhat effective: isolated islands have all acquired a biota. Easter Island, over 3,000 km from South America and 2,250 km from Pitcairn Island, had acquired at least 46 native species of vascular plant during its 15 million years of existence. Palynological work shows that, before human disturbance, the flora was somewhat larger (Flenley, 1981). On the other hand, it is clear that long-distance dispersal is not universally effective, or the floras of continental areas with similar environments would long ago have become much more similar than they are.

Rates of migration. There have been few measurements of rates of successful migration, but rates seem to vary from almost nil to high values. This is true even if we define migration as requiring the dispersal of propagules, thus excluding mere vegetative spread. The hoop pine (Araucaria cunninghamii) in Australia is believed to be capable of migrating at only 60 m per year (Havel, 1971). The Canadian pondweed, Elodea canadensis, introduced into Britain about 1842, spread outwards at a rate of 70 km/yr (Fritsch and Salisbury,

1938). Aquatic plants (and small aquatic animals) are in general rapid migrators, perhaps being carried by water birds. Darwin (1859) found the seeds of many species of plants adhering to the feet of birds. Obviously, the rates depend not only on dispersal mechanisms (Ridley, 1930), but on many factors such as the ecology of establishment of the propagules. For instance, the mussels inhabiting the rocky coasts of the Middle East vary in their toleration of sedimentation. *Brachidontes variabilis* does not tolerate heavy sedimentation and requires hard substrata; the reverse is true of *Modiolus auriculatus* (Safriel, Gilboa and Felsenberg, 1980).

Another way of approaching the measurement of rates of migration is to study the rates of passage of species across a barrier. An example here is provided by the island group of Krakatau, sterilised by the eruption of 1883. The barrier in this case is the sea. Since neighbouring islands were probably also sterilised, the source of immigrating organisms was either Java or Sumatra, both over 40 km away. When Rakata, the main surviving fragment of Krakatau, was visited by Verbeek a few months after the eruption, the only living thing found was a single spider. The subsequent record of animal colonisation of Rakata is incomplete, but the plant record is fuller (Flenley and Richards, 1982). By 1886 there were 24 species of higher plants recorded; by 1897 this had increased to 50. In 1906, 99 species were recorded, and by 1936 this had increased to about 180. The inclusion of the other islands of the Krakatau group gave a total of 217 species at this time. By 1979, 96 years after the eruption, it was possible to find about 200 species on Rakata (Fig. 3.2). The cumulative total was much higher (at least 371), as many species found by previous expeditions were no longer recorded. This may be because turnover is occurring. The rate of arrival of new species, averaged over the last 50 years, seems to be about 2.3 to 2.6 species per year. Extinction seems to occur at about 1.2 to 1.7 species per year, giving a small net gain of species (cf. Chapter 1).

It is, of course, a far cry from the *c*. 40 km journey needed for reaching Krakatau to the thousands of km needed to reach oceanic islands. Clearly, the probability of success would be much reduced; but the frequency of successful migration needed to explain the biota is not high. If the native flora of Easter Island is only 46 species and the island is approximately 15 million years old, then only about one species every 300,000 years would need to immigrate in order to accumulate the total flora. This is a very low rate indeed, and seems likely to fall within the rate achievable by long-distance dispersal. If turnover occurs, a higher rate would be needed, but given the larger area of Easter Island at times of eustatically lowered sea-level, and given the probable existence of more stepping stones, then it seems possible that long-distance dispersal could still account for the flora.

There is no need to decide between plate tectonics, the stepping stone theory and long-distance dispersal as the means of explaining the biota of the world on a macro-scale. All three have their part to play and all three require an understanding of temporal phenomena to explain their role. There are, of course, many other methods of migration, particularly dispersal by man and animals.

Unfortunately, there is not space to consider them here but they are, none the less, important.

Figure 3.2: Growth of the flora of Rakata (Krakatau) 1883-1979, plotted on a semi-log scale (after Whittaker and Flenley, 1982)

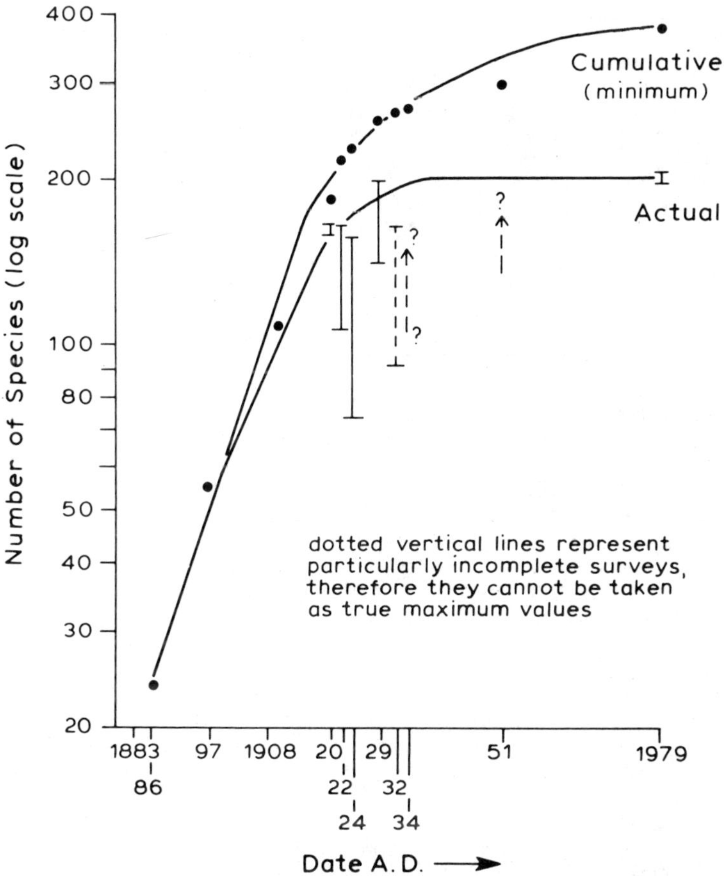

Environmental Change

Most significant of all is perhaps the way in which time works through environmental change to influence biogeography. Environmental change comes in many forms: continental drift and plate tectonics, climatic change, geomorphological change, edaphic

Figure 3.3: The distribution of six genera of Coniferales. Pinus, Picea and Sequoia are northern genera which have always been northern. Lebachia is a fossil genus, also exclusively northern. Podocarpus is a southern genus which has always been southern. Araucaria is exclusively southern today but was formerly northern also (after Florin, 1963)

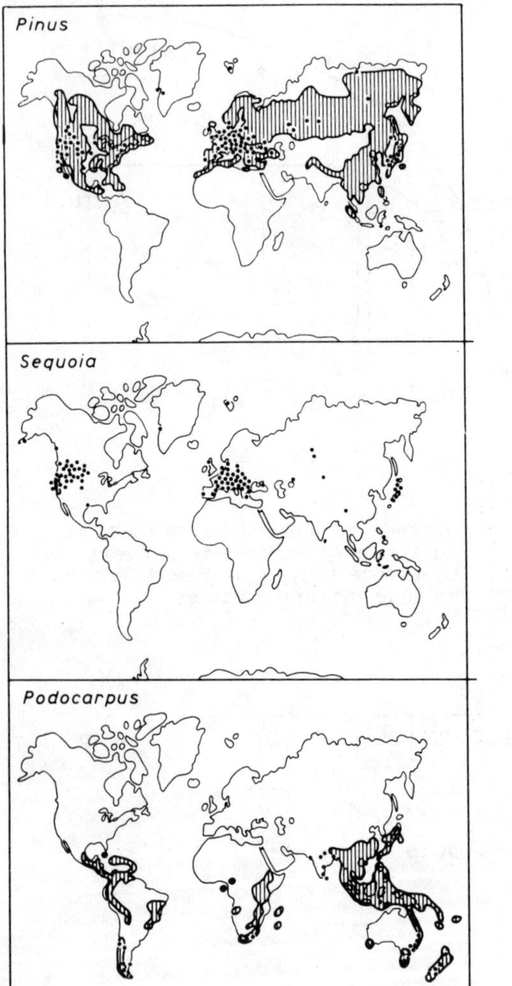

Figure 3.3 cont'd.

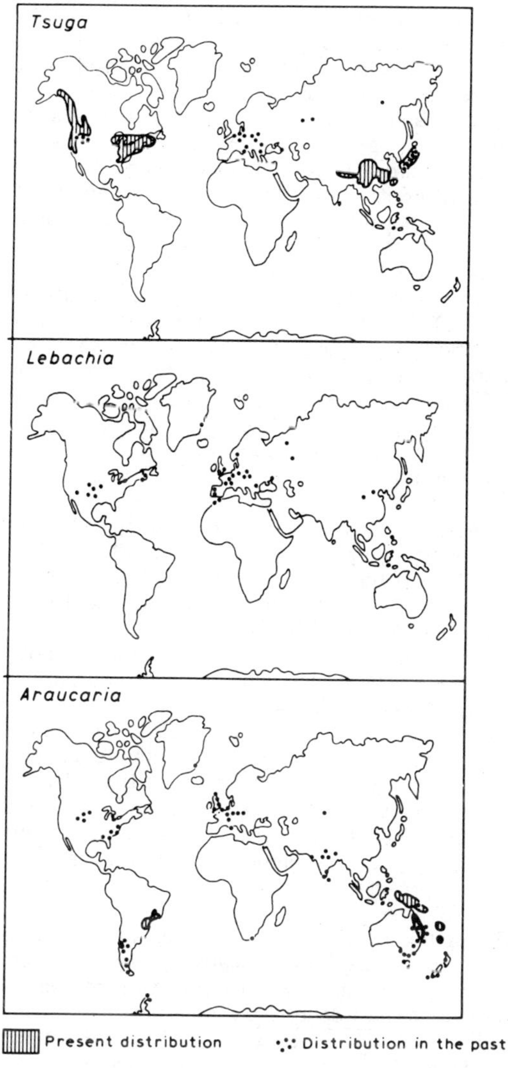

change, etc. Many are interdependent; for instance, a plate movement could lead to a geomorphological change which might itself bring about a climatic change and hence an edaphic change. It is often very difficult to be certain of the ultimate causes of the changes in biota.

Continental drift. Since continental movement is often a primary factor in explaining biogeographical distributions, it is appropriate to consider that first. Three decades ago we would have considered under this heading the biogeographical evidence *for* continental drift (e.g. Good, 1947). Now, however, the situation is reversed. The theory of plate tectonics, the successor to continental drift, is so well established that it no longer needs biogeographical evidence to bolster it. Instead, we may consider what biogeographical data are explicable by the movement of continents. The major movements of concern to us are those which took place in the last 200 million years, since the Jurassic. In Jurassic times the present southern continents, plus India, were assembled together near the south pole as Gondwanaland; the northern continents similarly formed Laurasia (Smith, Briden and Drewry, 1973). This may be compared with the present-day and fossil distributions of some conifers, which were the dominant plants in the Jurassic (Fig. 3.3). Conifers today fall into two groups: northern and southern, the latter including India. The northern genera include *Pinus*, *Abies*, *Cedrus*, *Tsuga*, *Larix*, *Cupressus*, *Sequoia*, *Sequoiadendron*, *Metasequoia*, etc. A few penetrate as far as the tropics (e.g. *Pinus* in Sumatra). The principal southern genera are *Podocarpus*, *Araucaria*, *Agathis*, *Acmopyle*, *Callitris*, etc. Actually, several of these extend slightly into the northern hemisphere: the conifer equator is at about 20°N. The fossils show that this distribution has always been so: the northern conifers were always northern, the southern ones always southern. Even totally fossil genera such as *Lebachia* preserve this pattern. The single exception is *Araucaria*, which formerly occurred in both hemispheres, but is now extinct in the north: its metamorphosed wood, however, is familiar to Yorkshiremen as Whitby jet.

The southern continents are now widely separated from each other. Are their present conifer floras sufficiently similar to suggest that they were formerly closer together? We can test this idea by principal components analysis, as has been done by Sneath (1967). He projected the loadings of different land areas on the three major components outwards on to an imaginary globe. The result was that the continents took up their pre-drift positions, with the southern continents grouped near the 'S' pole.

This separation of northern and southern floras is the more distinct if we consider even earlier, totally fossil groups. Carboniferous floras of Europe and North America are often rich in remains of the giant club-mosses (*Lepidodendron*) and giant horsetails (*Calamites*). The Coal Measures consist partly of their metamorphosed remains. These taxa are absent from most parts of the southern continents. In their place, in deposits of the same age, extending from Australia, South America, South Africa, Antarctica

Figure 3.4: Above and Centre: the present and fossil distributions of the southern beeches, Nothofagus (southern hemisphere), and of the northern beeches, Fagus (northern hemisphere). Below: the three pollen types in Nothofagus (after Soepadmo, 1972; and Cranwell, 1963)

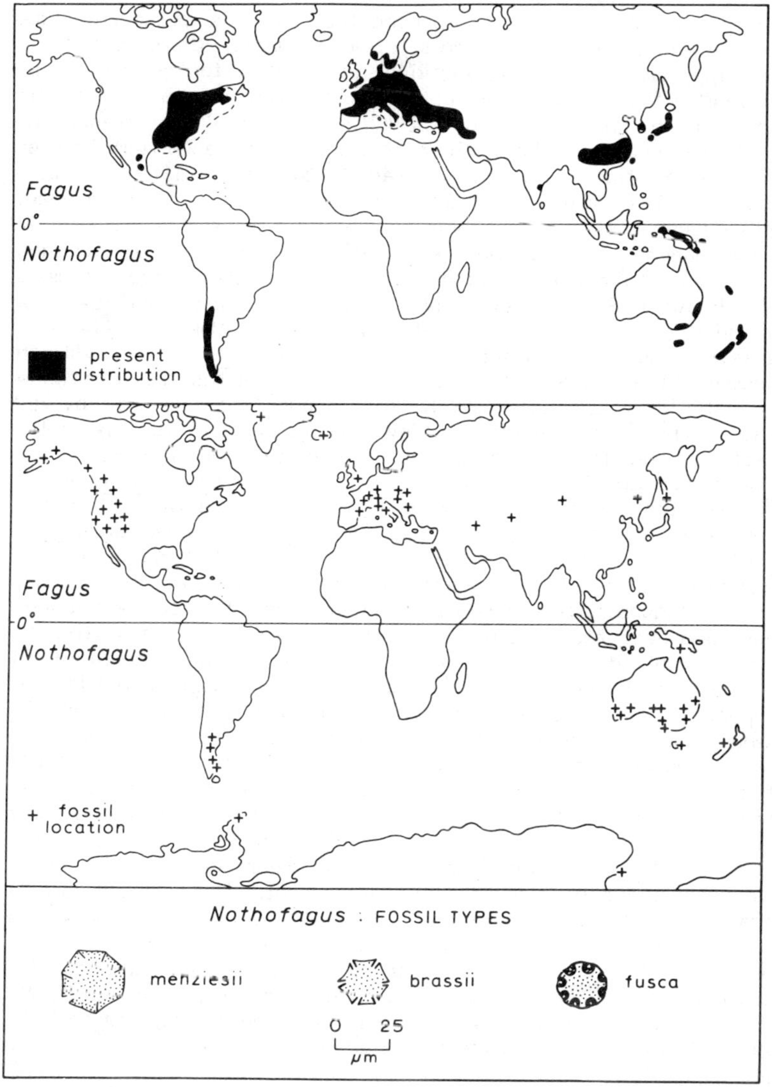

to India, are abundant remains of the probable pteridosperm (seed-fern) Glossopteris, which has never been found in northern continents. In fact, four distinct floras can be recognised at this stage (Plumstead, 1973). By the time the angiosperms appear clearly in the fossil record, which is only from the early Cretaceous (Wolfe et al, 1975; Hughes, 1976), the northward drift of Gondwanaland was well under way. So it is not surprising that there is less north-south differentiation in these plants. Nevertheless, there are some strikingly disjunct southern families. Proteaceae, for instance, occur in South Africa, South America and Australasia, but nowhere in the northern hemisphere (Good, 1947). Reported occurrence of Proteaceae pollen in the northern hemisphere (in the Tertiary lignites on Mull, Scotland) has been discredited by Martin (1968). Similar southern distributions in extant families are found in Epacridaceae, Restionaceae, etc. An example of particular interest is provided by the Fagaceae. Quercus (the oaks), Fagus (the beeches) and Castanea (the chestnuts) are northern; Nothofagus (the southern beech genus) is well-known in South America, New Zealand and Australia (Fig. 3.4). In the 1930s a new species of tree from New Caledonia was placed in a new genus of Euphorbiaceae, named Trisyngyne. When Erdtman examined the pollen, it became clear to him that this tree was really a Nothofagus. Subsequently, the genus was also discovered abundantly in the mountains of New Guinea. Malesia (roughly South-East Asia and New Guinea) is also the home of some other Fagaceae. Quercus occurs there, and also the related genus Lithocarpus, which differs from Quercus partly by having erect male catkins rather than pendulous ones, and by being insect-pollinated. Also present is Castanopsis, a relative of Castanea. In 1964, the Royal Society expedition to Mt. Kinabalu in North Borneo discovered yet another genus, Trigonobalanus, which is in some ways intermediate between oaks and beeches. This has since also turned up on the South-East Asian mainland. The occurrence of so many genera of the family in Malesia might suggest that region as the point of origin of the family, especially since Trigonobalanus is in some ways rather primitive. To check this we may refer to the fossil pollen record for Nothofagus, which has much more distinctive pollen grains than the rest of the family. Nothofagus pollen falls into three morphological groups. The brassii pollen group occurs at present in New Guinea and New Caledonia, and has now been recognised in South America (van Steenis, 1971). The fusca and menziesii groups occur in Australia, New Zealand and South America at present. The fossil record is shown in Fig. 3.5. Occasional, probably wind-blown, occurrences have been omitted from this record in view of the well-known property of Nothofagus pollen to be carried long distances, for example the 4,000 km from South America to Tristan da Cunha (Hafsten, 1960). Claims for northern hemisphere Nothofagus (e.g. Zaklinskaya, 1964) are possibly based on mis-identification (Petrov and Drazheva-Stamatova, 1974) and have also been disregarded. Perhaps the most striking features of the remaining record are the two occurrences in Antarctica (McMurdo Sound and Seymour Island).

Although the McMurdo Sound records are unspecified as to

Figure 3.5: The temporal record of the genus Nothofagus (after van Steenis, 1971 and 1972)

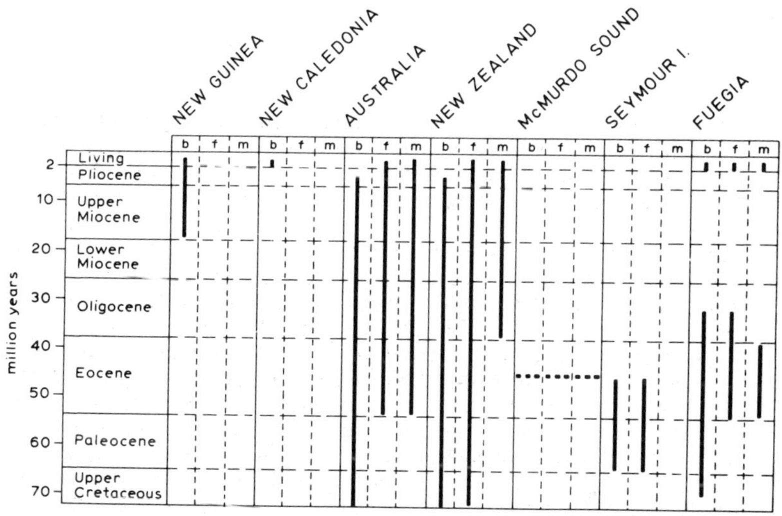

type, the Seymour Island records include both brassii and fusca types. Antarctica was never far from the pole in Tertiary time, and it is a little difficult to visualise trees surviving the long Antarctic winter, even in the absence of glaciation. Other points of interest are the lengthy record, especially of the brassii type (which is morphologically the most primitive), in New Zealand, Australia and South America. On the other hand, the record from New Guinea is rather short, going back only to the Upper Miocene (Khan, 1974). This is possibly due to the inadequacy of the New Guinea fossil record. Clearly, the fossil record does not yet support the idea that Malesia was the place of origin of Nothofagus. The record is well explained by the former closeness of the southern continents, given a southern origin for Nothofagus.

Numerous animal groups have distributions apparently explicable by continental movement. The lung-fishes (Dipnoi) are good examples: they occur in fresh water in South America, Africa and Australia (Darlington, 1957). The South American (Lepidosiren sp.) and African (Protopterus spp.) species belong to the same family, Lepidosirenidae. The Australian lung-fish, Epiceratodus forsteri, is placed in a separate family, Ceratodontidae. The fossil record reveals, however, that things are not as simple as they seem. The extinct genus Ceratodus (Ceratodontidae) dates from the Triassic and was more or less cosmopolitan at the time, presumably in fresh water. The present distribution of the family therefore probably results from extinction over most of its range. The African and South American lung-fish have fossil relatives in their home continents, but in view of the incompleteness of the fossil record it would be premature to cite them as examples attributable to continental drift.

The Dinosaurs, since they flourished while gymnosperms dominated the flora, might be expected to show evidence of similar distinction into northern and southern groups. To some extent this is true of some dinosaur groups at some stages. During the Jurassic many dinosaur groups were widespread, but during the Cretaceous the only reliable evidence for connection between northern and southern landmasses is the presence of a hadrosaurine hadrosaur in the Upper Cretaceous of Argentina (Cox, 1974).

Geomorphological change. Although it seems inevitable that any major geomorphological change will affect the biogeography, there are relatively few documented cases of this. We do, however, have some historical (pollen) evidence for changes during the origin of a major mountain range: the Andes of Colombia. It comes from the Sabana de Bogotá, an intermontane basin which contains an intermittent record back to the Pliocene. Today it stands at 2,600 m altitude, but the oldest deposits (Salto Tequendama) are dominated by pollen of lowland tropical elements (Fig. 3.6). These occur at present below 1,000 m. In the Facatativo 13 sequence there is a dominance of Andean and Subandean elements, which occur today principally between 1,000 and 3,500 m. The species present suggest that, given a climate like today's, the site lay at about 1,500 to 2,300 m. The Pliocene record includes the first occurrences of *Hedyosmum* and *Myrica* in South America. Presumably these (at least the *Myrica*) had recently migrated in from North America. The Late Pliocene Choconta 4 sequence already includes some pollen from the Paramo (i.e. from above the forest limit at 3,500 m). In the Early Pleistocene sequence (Choconta 1 and Bogotá Cuy) Páramo pollen expands to 50 per cent or more, although it is still poor in species. This implies a vegetation at present found well above the Sabana de Bogotá, and perhaps indicates an early cold phase of the Quaternary. The taxa of this primitive Páramo appear to have been Gramineae, *Polylepis/Acaena*, *Aragoa*, *Hypericum*, *Miconia*, *Umbelliferae*, *Borreria*, *Jussiaea*, *Polygonum*, *Valeriana*, *Plantago*, *Ranunculaceae*, *Myriophyllum* and *Jamesonia*. Some of these presumably evolved from the surrounding tropical lowland vegetation, while others, for example *Plantago* and *Polygonum*, since they are of temperate affinities, probably arrived along the Andean ranges from the northern or the southern temperate regions. The forest elements were also being added to at this time, and *Alnus* makes its first sporadic appearance, probably by migration across the mountains of the Isthmus of Panama during a phase of cool climate. Geological evidence that such a migration, by *Alnus*, *Myrica* and *Juglans*, did take place, has been assembled by Graham (1973).

Figure 3.6: Summarised pollen diagram from the Pliocene and Lower Pleistocene of the high plain of Bogota (Colombia), demonstrating the uplift of the area and the Early Pleistocene glacials and interglacials (after van der Hammen, 1974)

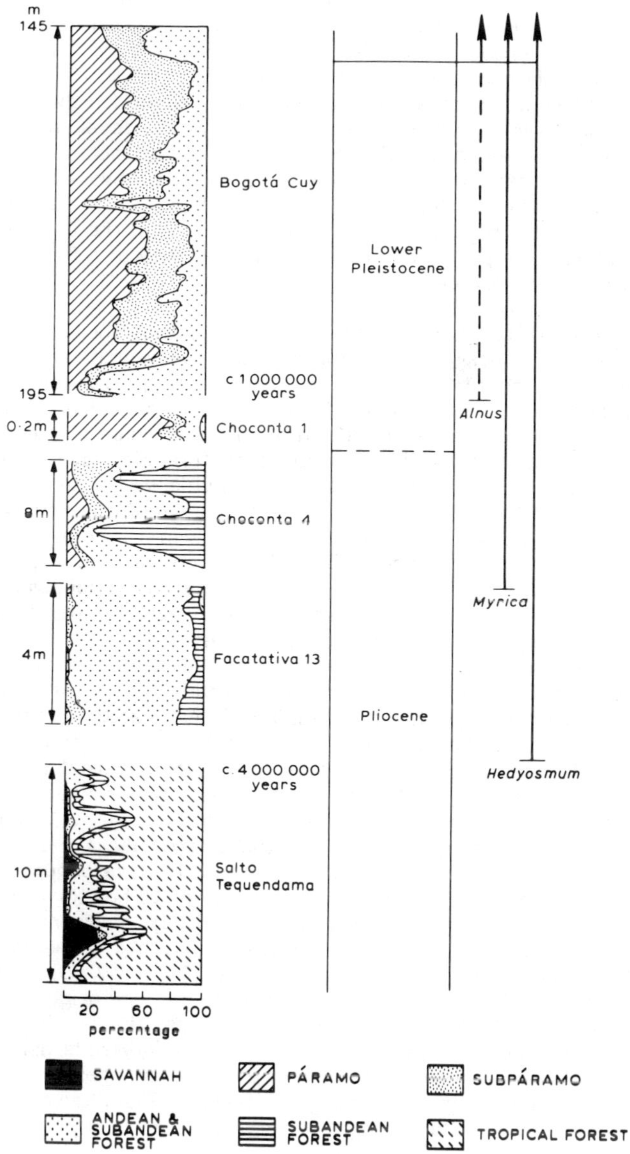

One of the most significant geomorphological changes for biogeography is an alteration in sea-level. A striking example is provided by the history of the Baltic Sea. When the ice first retreated to Scandinavia at the end of the last glaciation, sea-level was greatly depressed, for eustatic reasons. In the Baltic Basin a fresh-water lake formed in the early Late Pleistocene. Later, the sea flooded in, and abundant fossils of the marine mollusc *Yoldia arctica* are found. The land then rose (isostatically) faster than the sea and isolated the basin as a lake containing abundant freshwater mussels, the *Ancylus* lake. The final eustatic rise of sea-level in the Holocene led to the breaking in of the sea through the Kattegat, so that the *Ancylus* lake became brackish and was invaded by periwinkles (*Littorina*) to form the Littorina Sea. The latter gradually changed into the present-day Baltic, but land in the area is still rising at up to 1 cm per year, so there is a possibility of a further change to fresh water eventually.

Climatic change. It is generally agreed nowadays that there have been pronounced changes in the climate of the earth during geological time. Much of the evidence often quoted, however, is itself biogeographical. The use of this in the present connection (i.e. when we are trying to explain distribution by climatic change) would involve a circular argument. It is desirable, therefore, first of all to outline the climatic shifts indicated by other evidence. The best continuous record for the last two million years has been provided by oxygen isotope analysis of the calcium carbonate in foraminiferal tests from the bed of the deep ocean (Shackleton and Opdyke, 1973). Variations in the record are believed to result partly from changes in the temperature of the surface water of the ocean, where the foraminifera lived, and partly from the variations in the amount of ice in the earth's major ice sheets (Shackleton, 1967). The results, from a core obtained in the tropical Pacific, extending back through the complete Quaternary, are shown in Fig. 3.7. It is clear that there have been multiple fluctuations in ^{18}O, suggesting many cool or glacial phases in temperate regions. Fluctuations are apparent even in the Pliocene. During the Quaternary there appear to have been at least eleven cool or glacial phases (the even numbers in Fig. 3.7), and twelve temperate or inter-glacial phases (the odd numbers). It is noteworthy that times as warm as the present seem to have occupied only a small proportion of Quaternary time. It is not to be expected that the changes would be of the same magnitude everywhere. Taking all the evidence for the peak of the last temperate glaciation at 18,000 BP (before present, i.e. years before 1950 AD), Climap (1976) have assembled a world map showing likely deviations of surface temperatures from present values. Since this includes biogeographical evidence, however, it is not strictly admissible here. Its general indications - rather small or non-existent changes in the tropical oceans, with much greater changes poleward - may, however, be borne in mind. This polar exaggeration is not very surprising, since the large ice sheets of glacial times would be expected to increase the albedo of poleward areas and thus exaggerate cooling

Figure 3.7: Coarse-fraction record (centre), oxygen isotope record (left) and palaeomagnetic record (right) in core U28-239 from the equatorial Pacific Ocean. The accumulation rate is approximately 1 cm per 1,000 years, so that the core covers approximately the last 2.1 million years (after Shackleton and Opdyke, 1973)

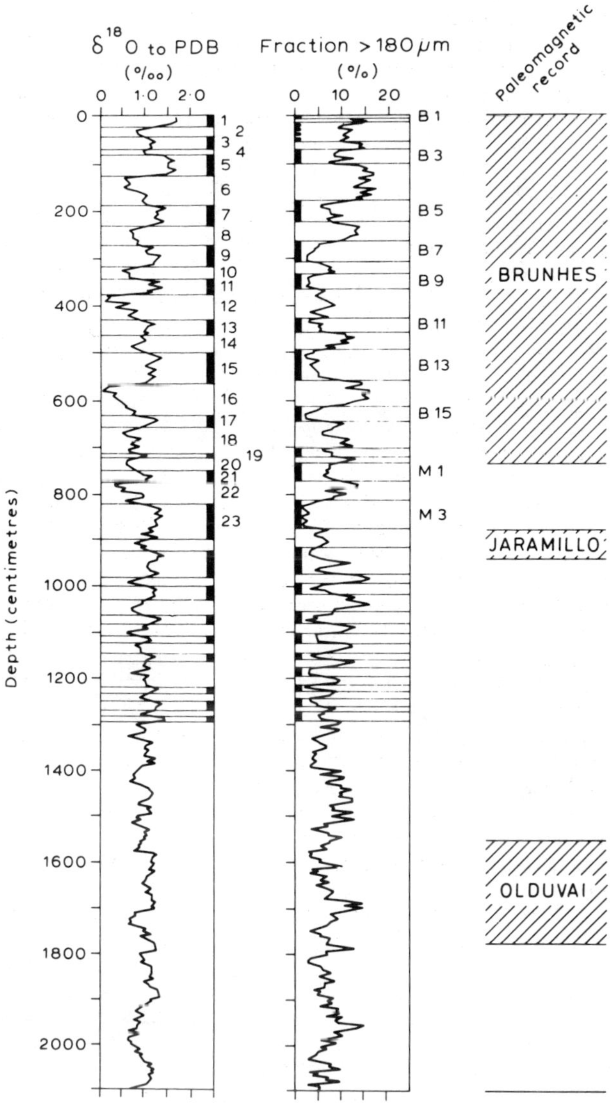

Figure 3.8: Radiocarbon dates of lake deposits in tropical Africa and south-west USA (after Goudie, 1977 and Grove et al, 1975)

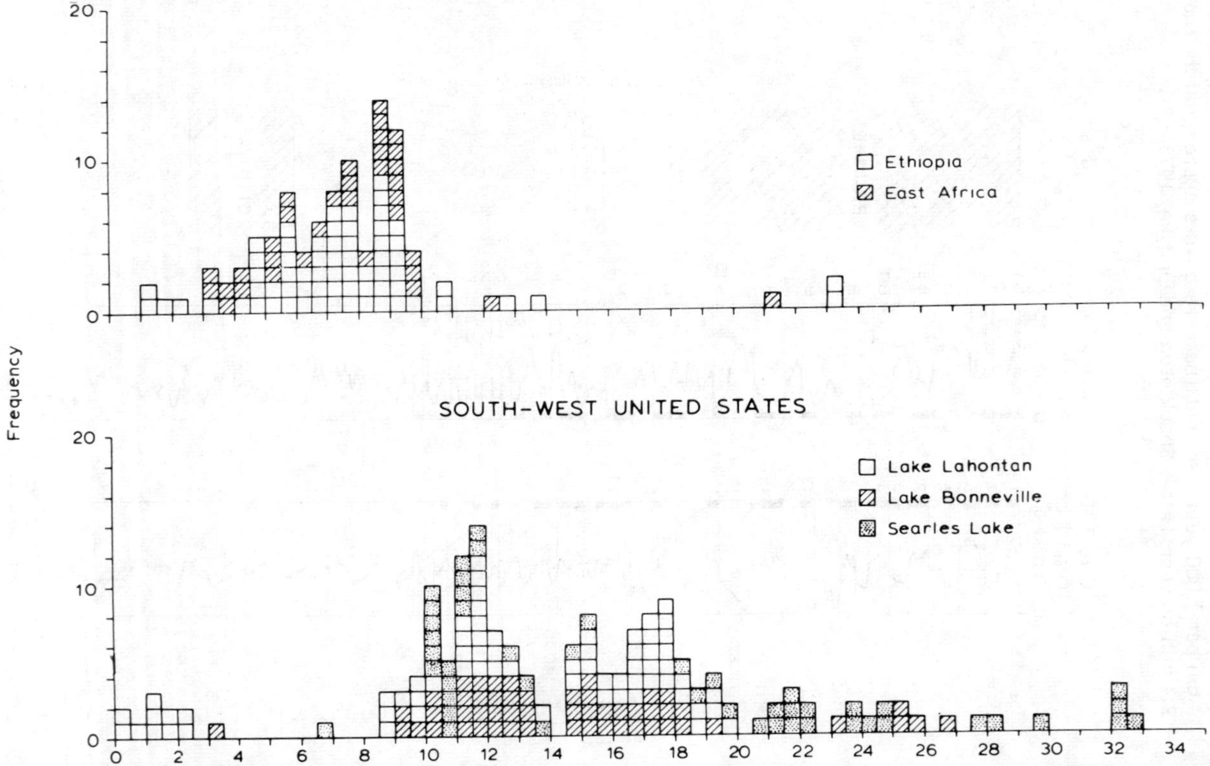

there due to the greater reflection of insolation.

It is known that temperature changes have occurred long before the Quaternary. There is good geological evidence of earlier glacial phases, particularly one in the Permo-Carboniferous of South Africa and Australia, when these continents would have been closer together near the South pole. The evidence consists of tillite (believed to be a compacted till) and striated rocks similar to those found in the Quaternary.

Temperature is not the only climatic parameter of importance to organisms. Hydrological variations are also very significant. The best records of hydrologic change come from centres of internal drainage. In semi-arid regions, lakes in such situations may dry up completely in periods of low precipitation, and are thus sensitive indicators of change. The occurrence of moist phases is not by any means synchronous everywhere: e.g. in tropical Africa the wettest phase in the last 30,000 years was around 9,000 BP; in the south-west USA it was near 12,000 BP (Fig. 3.8). Geological evidence of former arid phases is familiar to us from desert-laid sandstones, salt deposits, etc.

As an example of the response of vegetation to climatic (and other) change, we may consider the Late Devensian and early Flandrian vegetational changes in England (Godwin, 1975; Walker and West, 1970). There are many excellent published pollen records illustrative of this, but the one selected here comes from The Bog at Roos, North Humberside (Beckett, 1981), illustrated in Fig. 3.9. For convenience the diagram is divided into zones. Zone RB1 (c. 13,000-12,000 BP) is dominated by pollen of Cyperaceae, arboreal *Betula*, Gramineae, *Hippophae*, *Juniperus*, *Helianthemum*, *Thalictrum* and *Betula cf. nana*. Pollen influx values (not shown in the diagram) are generally low, so this zone probably represents an open vegetation with few trees but quite a range of shrubs. The climate may have been too cold for much tree growth and/or too dry. Dryness is suggested by the presence of *Hippophae*, which is drought tolerant. In Zone RB2 (c. 12,000-11,000 BP) tree pollen increases to 60 per cent or more, chiefly *Betula*, suggesting an open birch woodland. There is a peak of *Filipendula*, the northern limit of which coincides with the 14°C isotherm today (Seddon, 1962). A pronounced increase in temperature is thus suggested. This is part of the Windermere interstadial, as recorded in the Lake District (Pennington, 1977). In Zone RB3 (c. 11,000-10,000 BP) tree pollen declines and herb pollen dominates. The birch woodland must have thinned out again, although isolated trees could have survived. This was probably due to a much colder climate at the time, and probably corresponds with the Loch Lomond readvance of the Scottish icecap. In the Flandrian section of the record the local zones have been correlated with regional pollen assemblage zones (RPAZ). The first of these, the *Betula-Pinus* RPAZ is characterised by *Betula* values rising to 80 per cent, which implies a closed birch woodland. The only reasonable explanation of this is a great increase in temperature compared with the Late Devensian. In the succeeding *Corylus/Myrica-Ulmus* RPAZ, pollen of *Corylus/Myrica* (the two are not readily distinguishable)

Figure 3.9: Pollen diagram from the Late-glacial and Flandrian deposits at The Bog, Roos, North Humberside, UK, covering approximately the last 13,000 years. Only selected taxa are shown (after Beckett, 1981)

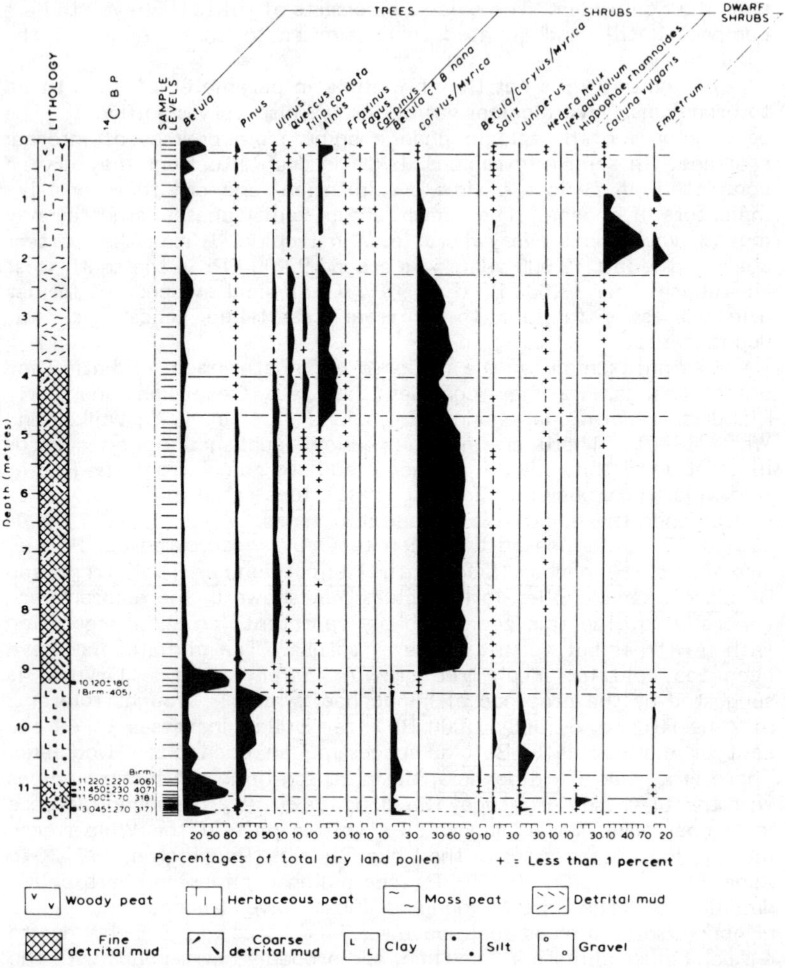

Figure 3.9 cont'd.

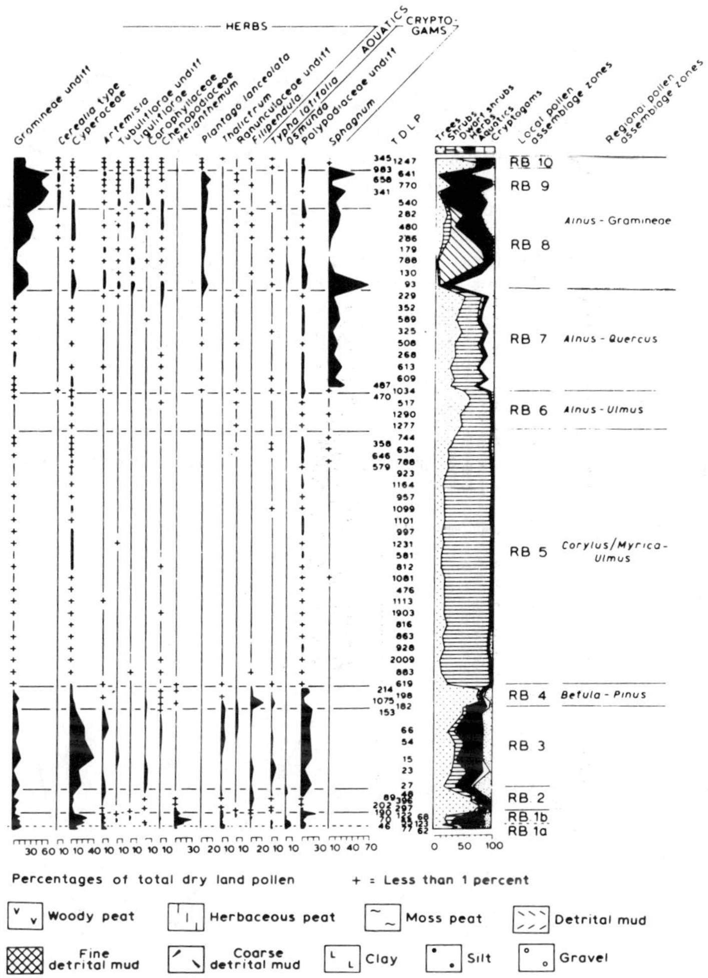

dominates throughout. Pollen of Ulmus, Quercus, Alnus, Tilia cordata and Fraxinus appears successively through the zone. This could be explained either by a further increase in temperature, or by successive immigrations during a stable climate. The next zone, the Alnus-Ulmus RPAZ, is marked by a rise of Alnus to high values; Tilia reaches its peak value late in the zone. The alder rise was once interpreted as the result of an increase in wetness of the climate, and this may be partially the explanation. In Scotland, however, the Alnus rise has been dated at several sites and occurs progressively later further north (Birks, 1977). It seems likely that the effect of a climatic change would either be synchronous everywhere or that, if it were an increase in wetness, it might be felt first in the north and west of Scotland which are already moister. The expansion of Alnus later in the north of Scotland rather than earlier suggests, therefore, that some other explanation must be found. Either a further increase in temperature, or delayed immigration of the tree seem possible and would also apply to Tilia cordata. Further research may be needed to distinguish between these explanations, and to see whether they apply to England as well as Scotland.

Figure 3.10: Variations in the mean July temperature for lowland Britain during the Late-glacial period, inferred from some fossil pollen records (pecked line) and from fossil Coleoptera (continuous line) (after Coope, 1970)

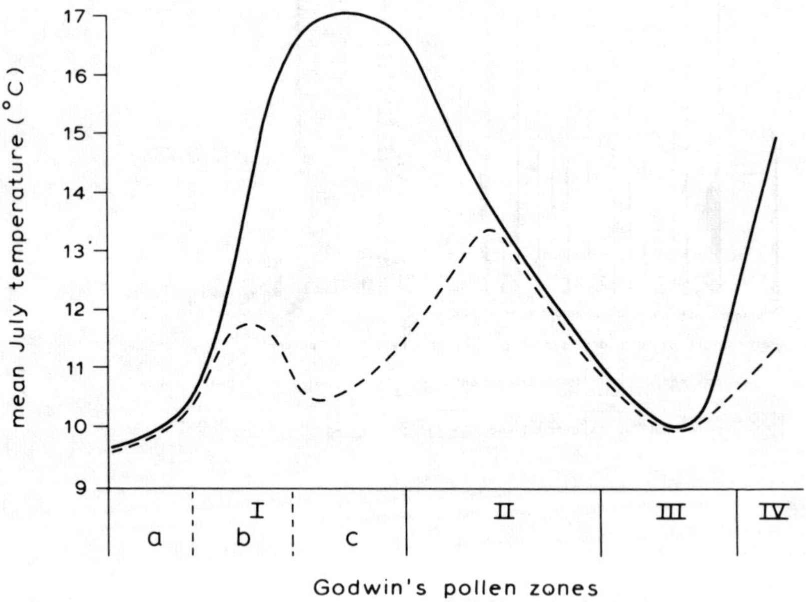

We have now described the vegetational history up to about 5,000 BP. Changes since this date are at least partly explicable by the activities of man, and will be described in the next section. Even within the time period already described, there is considerable discussion as to the speed of response of vegetation to climatic change. This problem was highlighted when Coope (1970 and 1977) found that, before the main Late Devensian expansion of trees (Zone RB2), there was a period in which occurred, at several sites, remains of beetles at present found in climates even warmer than the present British one. These were species not tied to a particular vegetation type, but general feeders able to take advantage rapidly of the areas quite recently vacated by ice. Trees could very probably have grown in this environment, but had simply not yet immigrated (see also Chapter 11).

Such problems must constantly be borne in mind when biogeographical evidence is used in palaeoenvironmental reconstruction. In the present example, the palaeotemperature curves suggested by pollen and Coleoptera are strikingly different (Fig. 3.10). Actually, some of the botanical evidence suggests warmer conditions in the Late Devensian. Typha latifolia, regularly present at that time, today grows no further north than central Norway (in north-west Europe) (Fitter, 1978). So far we have considered changes of vegetation and flora through time. It is desirable also to consider them spatially. For eastern North America, the spatial distribution of various taxa has been mapped for different phases of the last 11,000 years (Bernabo and Webb, 1977), employing pollen evidence. Using these maps (Fig. 3.11) it is possible to trace the migrations of taxa out from their glacial refugia. The results also provide evidence of some retreat of taxa during the latter part of the Post-glacial. Similar work in Europe (Huntley and Birks, 1983) has shown that different elements of present-day vegetation types did not necessarily share the same refugium. Such findings are not in accordance with the 'holistic' concept of vegetation accepted by Clements (1916) and Tansley (1920) but criticised by Gleason (1926).

The impact of man. We shall start this section by continuing our consideration of the pollen diagram from Roos, North Humberside (Fig. 3.9), as an example of vegetational change in the British Flandrian (Post-glacial or Holocene). The three RPAZs so far described took up approximately 5,000 years, the first half of the Flandrian. The second half begins with the Alnus-Quercus RPAZ. The opening of this zone is marked by the decline of Ulmus (and to some extent of Tilia also). This 'elm' decline has been noted widely over north-western Europe, usually dated to about 5,000 BP ± 250. Such synchroneity might suggest a climatic change, but several other explanations have been advanced. The recent attack of the elm disease in western Europe exemplifies one obvious possibility. Other explanations involve Neolithic man, who is believed to have introduced farming to north-west Europe about this time. Farmers might have felled elm and lime selectively because these taxa grew

Figure 3.11: Changes in the Holocene pollen record of north-eastern North America. Percentage values are of total terrestrial pollen (after Bernabo and Webb, 1977)

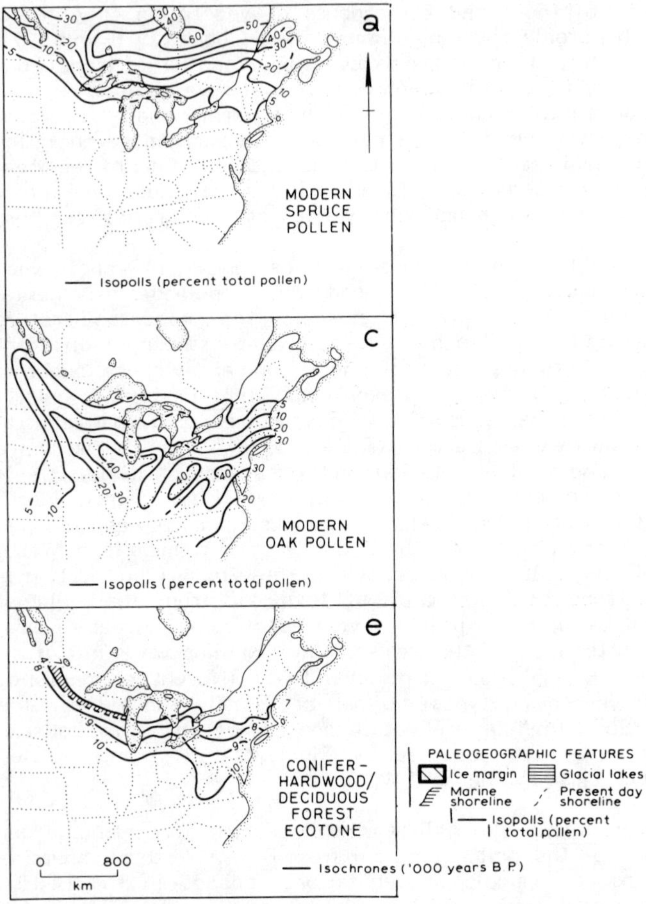

A. Percentage isopoll map showing modern distribution of pollen of Picea. This corresponds closely with the distribution of Picea trees.

C. Percentage isopoll map showing modern distribution of Quercus pollen. This corresponds closely with the distribution of Quercus trees.

E. Isochrones in millenia BP showing the movement of the conifer-hardwood/deciduous forest ecotone. The shaded area shows southerly retreat after maximum northward extension.

Figure 3.11 cont'd.

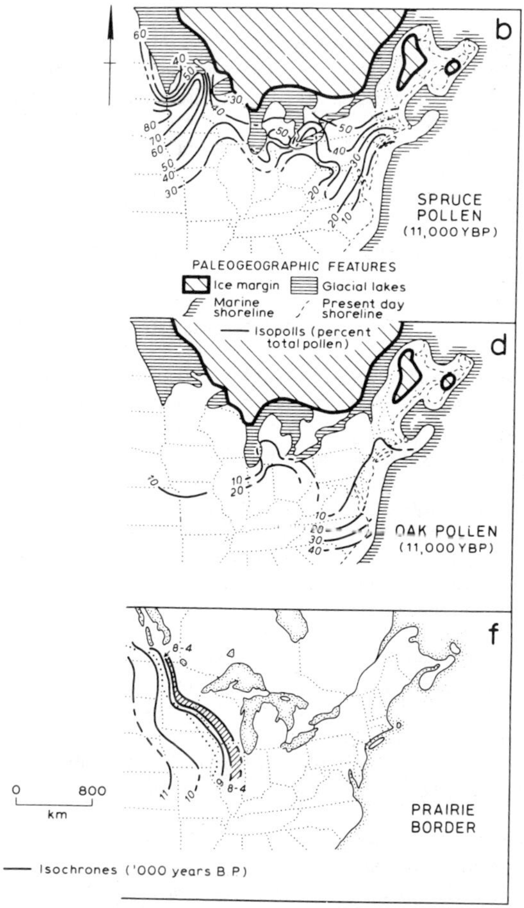

B. Percentage isopoll map showing distribution of pollen of Picea at 11,000 BP.

D. Percentage isopoll map showing distribution of pollen of Quercus at 11,000 BP.

F. Isochrones in millenia BP showing the movement of the prairie/forest ecotone. The shaded areas show westerly retreat after maximum eastward extension.

on the more nutrient-rich soils (Morrison, 1959). Another possibility is that regular lopping of branches was practised to provide twigs, the bark of which can be eaten by stock in winter (Troels-Smith, 1956). This technique is still used today in many areas, such as the Alps and the Himalayas.

The next zone, the Alnus-Gramineae RPAZ, shows a dramatic decline in tree pollen and undoubtedly represents the clearance of forest by man. There is a great expansion of Gramineae and Calluna pollen and of many herbaceous types, especially Plantago lanceolata which is a common weed of pastures. The zone is subdivided in the Roos record, and a later phase of the zone saw the elimination of Calluna and a further rise of Gramineae, with Cerealia pollen present. This could indicate a change from pastoral to arable farming. The most recent phase, indicated by the rise of Betula and the decline of Plantago lanceolata, is probably the time since the Agrarian Revolution. Improved agricultural practices meant cleaner land with fewer weeds; drainage of the bog led to planting of trees on it.

Changes of vegetation during the last 5,000 years in the British uplands are not so certainly the result of human influence alone, and they will be considered in the next section on edaphic changes. The impact of man on vegetation has now been recognised in pollen records from many parts of the world. In what is now the Rajasthan Desert of north-west India, Singh et al (1974) found Gramineae pollen of cereal type dating back to 9,260 ± 115 BP. They were accompanied by the first significant quantities of carbonised plant fragments, which were taken to indicate burning of scrub by man (Singh, 1971).

From New Guinea there is now clear evidence for forest disappearance back to 5,000 BP in two montane areas (Walker and Flenley, 1979; Powell et al, 1975), and there is archaeological evidence from Kuk Swamp at c. 1,600 m for several agricultural phases, extending back possibly to 9,000 BP. At 6,000 BP the swamp was apparently inundated by a mineral inwash layer, resulting possibly from soil erosion after major forest clearance (Golson and Hughes, 1976; Golson, 1977).

It is rare for man to have deforested completely all the land available to him. This appears, however, to have occurred within the last 1,000 years, on Easter Island. The resulting shortage of timber may even have contributed to the demise of the megalithic culture there (Flenley, 1982).

The effect of man on vegetation need not always have been through felling, burning or other direct means. In the Somerset Levels, for example, there are changes which seem to be associated with man but they are probably not a direct result of his actions. There is evidence that man was active from Neolithic times in this area of lowland mires, surrounded by the partly calcareous Mendip and Polden Hills. Wooden trackways constructed at various times across the mires are preserved in the peat (Dewar and Godwin, 1963). It originally appeared that the trackways fell into two distinct groups, dating from the Neolithic and Late Bronze ages. The

Neolithic trackways occur at a level where the peat changes from brushwood peat (suggesting fen carr) to Sphagnum peat (suggesting raised bog). The Late Bronze age trackways were associated with a level where the Sphagnum peat is interrupted by a layer of Cladium peat. Cladium is a large sedge usually found in alkaline fens. Both these changes were explained in terms of climatic deterioration (increasing wetness), the later change involving a flooding of the bog surfaces by calcium-rich runoff from the surrounding alkaline hills (Godwin, 1948). The construction of trackways was thus seen as a response to the increased difficulty of crossing the mires. More recent work suggests that the trackways are actually of a wider age range (e.g. Coles, Hibbert and Orme, 1973). Knowing what we now do of the effect of deforestation of hill land in causing the flooding of lowlands by siltation of water courses with eroded soil material (e.g. Ward, 1978; Goudie, 1977 and 1981), it seems possible that at least the upper change (to Cladium peat) was the result of deforestation of the slopes of the Mendip and Polden Hills. Eventually, the peat grew above the new water table and Sphagnum was able to re-establish.

Edaphic changes. The interaction between soils and vegetation is a two-way one: soils influence vegetation, and vegetation affects soils. It is, therefore, sometimes difficult to be certain which is the prime mover in causing a change in the soil-vegetation system. That such changes do occur, however, under otherwise stable conditions, is undoubtedly true, and it is, in many cases, likely that edaphic changes were the underlying cause. An interesting, if controversial, example is provided by the vegetational changes in upland Britain during the Flandrian. There is evidence that, in the early or middle part of this period, forest covered the uplands, to about 1,500 m or more in the Pennines and the Lake District, c. 1,000 m in the Southern Uplands and 750 m in the central Scottish Highlands (Birks, 1977). The present vegetation above 500 m of all these areas is a mixture of Callunetum, blanket bog and other non-forest vegetation types (Taylor, 1983). The present-day soils of the uplands are usually acid and either podzolised or peaty and waterlogged. In most cases they will not support tree growth today without considerable artificial drainage and/or mechanical disruption of the iron pan. We know, however, that the soils were not always like this. Dimbleby (1962) found that the soils preserved beneath Bronze age barrows on the North York Moors were of brown earth type, such as would support trees, and indeed they contained pollen of birch, hazel and other woodland taxa. The strong possibility therefore emerged that the soils, having been progressively leached during the Flandrian, eventually became so podzolised as to cause a decline in the vegetation from forest to Callunetum. That this should have occurred particularly in upland regions made the explanation more plausible for two reasons. First, rainfall is higher in the uplands, and leaching therefore presumably occurs faster. Second, many lowland glacial deposits were highly alkaline because of the presence of chalk or limestone in the deposit, but upland soils were often derived from the country rocks, which were only locally alkaline.

This edaphic theory of the origin of upland vegetation has often been strongly linked with ideas of climatic change. Godwin (1956) argued that many blanket bogs were initiated at the time of the *Alnus* rise (c. 7,500 BP in England) and were associated with the increasing wetness of the climate which he believed this rise indicated. A second group of bogs were believed to have started growth at about 2,500 BP when another decline in climate was suggested. The dates of forest decline and of the onset of peat growth have now been obtained in a number of places by radiocarbon assay. A wide range of ages is recorded, but synchroneity is suggested in some areas. For instance, in northern and western Scotland, Birks (1977) found that forest decline and peat initiation tended to occur about 4,000 BP which might argue in favour of a regional change to moister climate at that time.

The chief alternative theory has been to attribute the vegetational change to man's destruction of the forest by felling and burning, and by the introduction of grazing animals. It is argued that the opening up of the forest allowed the expansion of *Calluna*, which is believed to have the property of accelerating podzolisation of soils (Dimbleby, 1962; Taylor and Smith, 1981: also see Chapter 7). Increased runoff following deforestation would have led to raised water tables in basins and the initiation of peat growth there, later spreading outside them. Evidence in favour of the burning of upland forest has been put forward by Moore (1973). Evidence that the destruction of forest was at any rate catastrophic rather than gradual has accrued from the North York Moors. Rapid forest removal often results in soil erosion, and the eroded matter can be carried into bogs. Layers of such inwashed material were found in bogs on the North York Moors (Simmons *et al*, 1975), and were associated with pollen suggesting forest removal.

The conflict between the theories may be partly resolved by reference to the very detailed studies carried out in the English Lake District (Pennington, 1965). The deposits in lakes can be regarded as a series of inwashed soils, and chemical analysis of such deposits showed that the soils of the Lake District mountains were indeed becoming leached quite early in the Flandrian, before there is any evidence of substantial influence upon the vegetation by man. Thus, when man did burst upon the scene, he attacked a vegetation already in disequilibrium with its environment. The forest was able to do little in the way of regeneration, and became permanently degraded. This is, in contrast with the lowland situation, where regeneration was frequently very effective (e.g. Blackham *et al*, 1981). In the uplands, therefore, it appears that edaphic change (possibly encouraged by climatic decline) and man may both have been involved in the degradation of vegetation.

It seems logical to ask next whether the lowlands, given leaching for long enough, would have experienced changes similar to those of the uplands without the intervention of man. We can answer this by reference to pollen diagrams from interglacials, when the influence of man was probably slight. A good example is that from Nechells in Birmingham, which is of Hoxnian (penultimate interglacial) age. We know from other evidence (Shackleton and Turner, 1967) that this was probably a long interglacial, perhaps 30,000 years or more. The Nechells diagram (Fig. 3.12) shows that the early part of the interglacial was dominated at this site by mixed oak forest (Quercus, Ulmus, Tilia, Alnus) like the Flandrian (although with differences). This was followed, however, by a phase in which Taxus reached a considerable peak. Later, there were successive peaks of Picea, Abies and Pinus. There is little evidence that these indicate changes in climate. The proportion of tree pollen remains fairly high throughout these phases and is higher than when Quercus was dominant. The final peak, of Pinus, is accompanied by a peak of Ericaceae. This suggests the development of heathlands on podzolised soils. The possibility cannot be excluded that the changes in the upper part of the diagram are the continued result of progressive immigration and/or soil maturation, rather than of climatic change. It seems likely that, in the absence of man, similar changes would eventually occur in the Flandrian, were it to be sufficiently prolonged.

Autogenic Changes and Related Topics

When communities of living organisms change through time, they usually do so gradually, via a sequence of changes: a succession. Some successions are brought about by external changes in the environment. These allogenic successions include many of those brought about by climatic, geomorphological and edaphic changes and by man, which have already been discussed. Other successions are autogenic, i.e. brought about by the action of the organisms themselves. The classical examples are: the hydrosere, in which the accumulation of the dead remains of plants in water reduces water depth so that a new community of plants can immigrate; and the xerosere, in which plants progressively colonise a bare surface, resulting eventually in a complex vegetation.

The climax theory. According to the classical ideas of Clements (1916) and Tansley (1920), autogenic vegetational successions (seres) in any one region had a tendency to end up with a particular vegetational type, known as the climax. The climax for a region was dependent principally - perhaps exclusively - on the climate: in fact, it is usually called the climatic climax. Later the concept was modified to include also edaphic climaxes, where the soil prevented the development of the climatic climax. Plagio-climaxes were also recognised, in which human impact is a quasi-permanent, arresting factor (e.g. heather moorlands maintained by regular firing). In addition, preclimax, postclimax and disclimax communities were defined, to accommodate other apparent anomalies. It was an essential part of the climax theory that each community in a sere was of fairly fixed composition, and that these necessarily succeeded each other in a fixed sequence, leading up to the climax. This idea fitted in closely with the concept that all communities were highly integrated. Each species was so dependent on other species that the

Figure 3.12: Pollen diagram from the Hoxnian interglacial at Nechells, Birmingham, UK. Only selected taxa are shown (after Kelly, 1964)

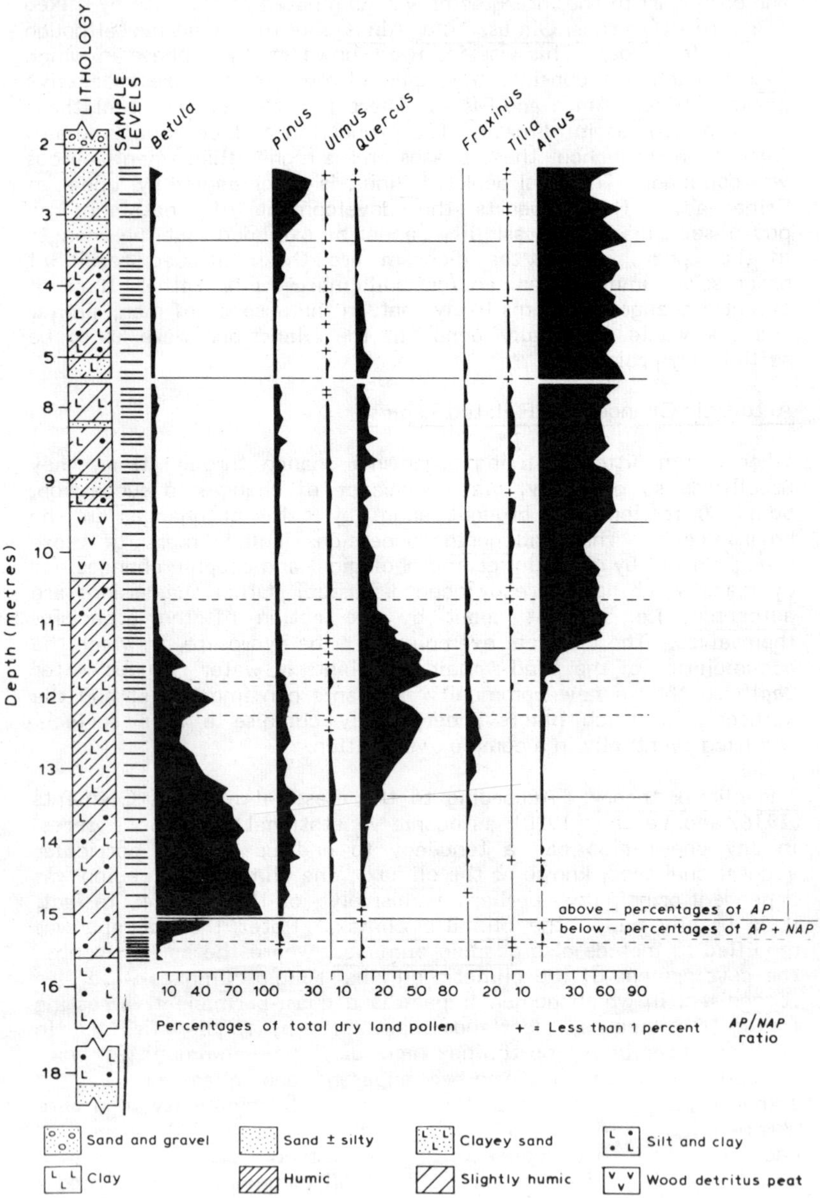

Figure 3.12 cont'd.

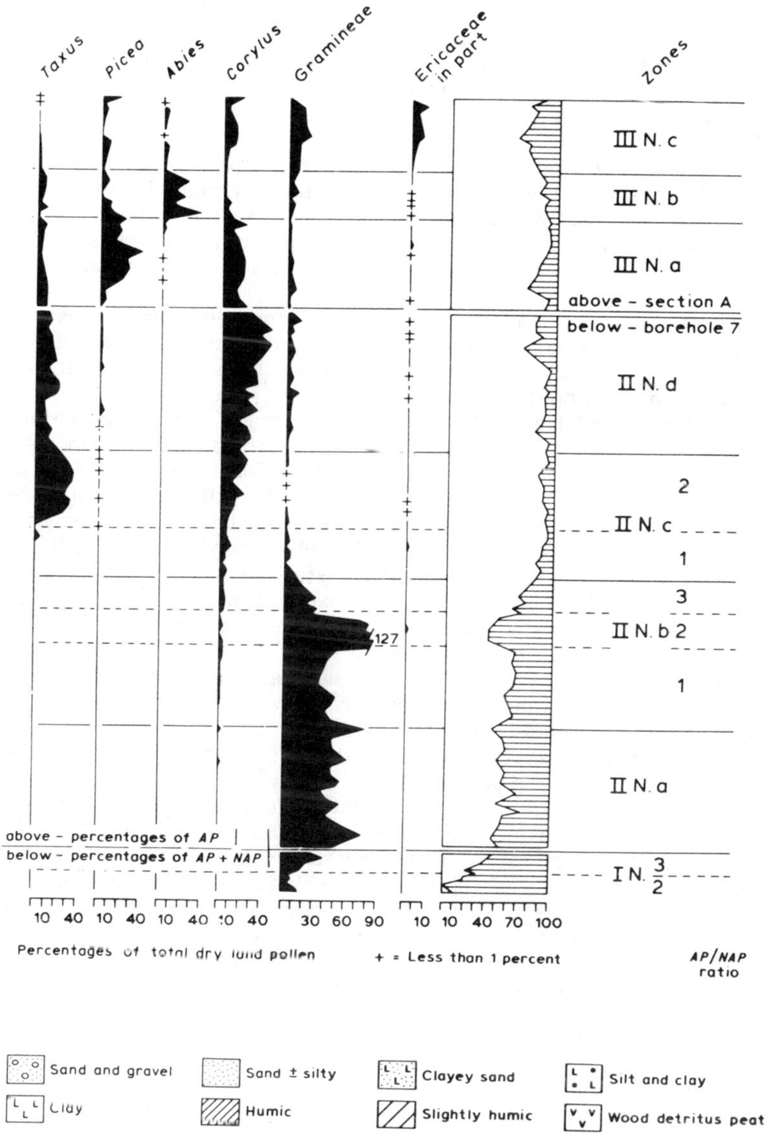

community became a 'quasi-organism' (Tansley, 1920). Removing an individual tree from a forest was thus like wounding an organism. Scar tissue (young trees) would grow up to heal the wound (close the canopy). These ideas also fitted in very well with the classifying of vegetation for mapping and other purposes, and with the concept of the plant association (Braun-Blanquet, 1932).

Criticisms of the climax theory. Although it is clear that there is some degree of integration in communities (the mistletoe must have its host tree), the theory seemed to Gleason (1926) to go too far. He argued instead that:

> The vegetation of an area is merely the resultant of two factors, the fluctuating and fortuitous immigration of plants and an equally fluctuating and variable environment. As a result, there is no inherent reason why any two areas of the earth's surface should bear precisely the same vegetation.

Much later, Curtis and McIntosh (1951) proposed that, instead of vegetation being considered as a set of distinct communities, it should be regarded as a continuum; rather than classifying it, we should be ordinating it.

These revolutionary ideas seemed to argue against the climax theory. If there were no communities of fixed composition, why should there be a climatic climax of uniform composition? If vegetation was a continuum, how could it fall neatly into the successive stages leading to a climax? Perhaps these criticisms were not all that serious: climate itself was continuously variable after all. But stronger criticisms were made as historical evidence was brought to bear. Raup (1964) found that the Harvard forest in Connecticut, USA was not the stable climax it had once been assumed to be. A study over several decades showed that forest composition was always changing in response to natural catastrophes such as storms or insect attack, and would therefore probably never reach the stable composition embodied in the ideal of the climax.

Over the longer time-scale of the Post-glacial period, Walker (1970) investigated the progress of British hydroseres by studying the stratigraphic evidence derived from both macrofossils and pollen. He found that there was no stage when a hydrosere was committed to a particular sequence of changes; rather, each successive change depended on a variety of factors from which random chance could by no means be excluded. Nor was the succession at all certain to end with the 'climatic climax' for Britain which was supposed (at least in the lowlands) to be oak forest. Some hydroseres (e.g. some of the Fenland) may possibly end up with this. Others become raised bogs, many of which show no sign of turning into anything else. So the idea of a fixed 'climatic climax' is not supported.

Further historical evidence may be used to test the idea of the climax over a longer time scale. If oak forest is the British lowland climatic climax, then it ought to persist as the dominant, at least throughout an interglacial. Reference to Fig. 3.12, showing the

Nechells pollen diagram through the Hoxnian (penultimate) interglacial, will demonstrate that this is not the case. Oak does dominate near the start, perhaps for several thousand years. Then there is general dominance by Alnus, accompanied successively by peaks of Taxus, Corylus, Picea, Abies and Pinus. As discussed earlier, these changes may well result not just from climatic change but also from progressive leaching of the soil, or may be the product of successive immigrations. If this is so, considerable doubt is cast on the idea of the 'climatic climax'.

It has long been assumed that the ultimate example of a climatic climax is the tropical rain forest. This was believed to have survived unchanged from the Tertiary, under a stable climate to which it had become fully adapted. In recent years, however, it has become clear that this is not the case. The tropical climate has changed, and the vegetation has altered dramatically during the Quaternary (Flenley, 1979). Many rain forests are only a few thousand years old or less, which is only a relatively small number of tree generations. It is unlikely that a stable equilibrium has yet been reached in such cases. The climax theory is therefore, to biogeography, what the Davisian cycle of erosion was to geomorphology: a great classical ideal, but one which turns out not to be true in practice. For the climax theory, the problem is one of time-scales. Judged over a period of a hundred or so years, a stable climax may well appear to exist. But when the generation time of the individual species is often greater than this, surely a much longer time scale is appropriate. As soon, however, as one increases the time scale to the thousands of years needed to take account of tree life-cycles, one is in the realm of major climatic changes, of pedogenesis, and of major migrations and colonisations by plants, animals and man. It therefore seems likely that, at least during the Quaternary, very few vegetation types even approximating to a stable climax have ever been reached. Succession, of course, undoubtedly occurs; but it is like Tantalus, always striving but never arriving. The climax theory is not necessarily wrong: it is simply irrelevant to many actual situations and particular time-scales and biogeographers would do well to abandon it completely.

Stability of communities. We have touched already upon the matter of stability. Stability implies lack of change through time, and is therefore a topic of relevance to the time factor. A popular ecological theory has been that those ecosystems with the most species (i.e. the highest diversity) would be the most stable (MacArthur, 1955). There is some sense in this: for instance, the populations of a predator and its prey in a simple community often fluctuate wildly, but in a more complex one the predator can use various prey and fluctuations are thus damped down. Studies of the most complex and diverse community known, the tropical rain forest, now show that it is not particularly stable (Flenley, 1979), so that the theory is not supported. Perhaps stability is more a property of the individual species than of the community as a whole.

Diversity of communities. The species diversity of communities is extremely variable, being generally (but not uniformly) greater at low latitudes. The more one thinks about this, the less certain appear the reasons for it, and many explanations have now been advanced. Most of them are summarised in Table 3.1.

Table 3.1: An outline of some of the main hypotheses to explain greater species diversity in the tropics than in temperate regions (after Ricklefs, 1973, and Flenley, 1979)

---

1 Tropical environments are older and more stable, hence:

(a) tropical communities have had time to accumulate more species (non-equilibrium hypothesis)

(b) the constant physical environment allows smaller populations to persist, so extinction rates are lower

(c) biological communities are more completely integrated, therefore smaller populations can persist, so extinction rates are lower

(d) tropical populations are more sedentary, facilitating geographical isolation

2 It is an advantage to be rare in the tropics (and many rare species means more can be packed in) because:

(a) 'pest pressure' is greater in the tropics because of more rapid breeding of pests; rarity prevents epidemics

(b) allelopathy is relatively a more important means of controlling seedling survival in the tropics; in temperate areas non-selective control occurs by climate

(c) genetic drift increases the rate of speciation once rarity has been established

3 Speciation proceeds faster in the tropics because:

(a) life cycles are shorter

(b) greater productivity leads to greater turnover of populations and hence more selection

(c) great importance of biological factors in the tropics enhances selection

4 Extinction rates are lower in the tropics because of less competition resulting from:

(a) more resources in the tropics

(b) more spatial heterogeneity in the tropics

(c) competing populations are checked by predators in the tropics (cf 2a above).

Some of these explanations involve the time factor and are therefore of relevance here. The number of species in an area is the resultant of two factors: the rate of acquisition of species by immigration or speciation, and the rate of loss of species by emigration or extinction. A balanced but dynamic equilibrium is achieved when these two rates are equal (MacArthur and Wilson, 1967). The first group of hypotheses in Table 3.1 depends on the idea that tropical environments are older and more stable. On the whole, the historical evidence mentioned previously is against this idea: older, perhaps; more stable, probably not. The advantage is therefore given to the other groups of hypotheses.

Conclusion

It will be clear from the foregoing that the time factor is important in biogeography but is effective in different ways depending on the time scale which is under consideration. At very short time scales, seasonal or even diurnal changes are important. These include striking variations in rates of plant growth and nutrient cycling and the seasonal shedding of leaves of deciduous trees. In many short-lived organisms the entire life cycle falls within this time span. Some migrations take place on this scale: even, in the case of certain birds, world-wide ones.

Extending the time scale a little, to the order of a decade or a century, we find ourselves in the realm of the life-cycles of many of the larger organisms and also of some of the more rapid successions, such as those on abandoned agricultural land. Many significant migrations have been observed on this time scale, and undoubtedly climatic and pedogenic change can occur in less than a century. Even natural selection on a small scale (micro-evolution) can occur on this time scale for organisms with annual or shorter life-cycles.

Changes at the above time scales are perhaps those best known to man because they fall within the compass of his own life time. For spans of the order of a millenium, man must rely on historical evidence to show what happens. The evidence demonstrates clearly that, at this time scale, successions and pedogenesis are in full swing. Climatic fluctuations can be increasingly significant. Evolution is still largely at the micro-level, except perhaps at times of very rapid environmental change.

The 10,000 to 100,000 years time scales are particularly associated with Quaternary climatic change, with all its important biogeographical effects. Only above this level, perhaps nearer to the million year time scale, do we approach that necessary for the evolution of most new species by natural selection. This is fortunate for, if it were not so, we could not make much use of biological evidence in Quaternary reconstructions. The evolution of taxonomic groups much higher than species may need even longer, the ten million year scale. We are now in the realm of major geological changes such as orogeny, continental movement, polar wandering and the maximal climatic fluctuations, such as the 200 million-year cycle of ice ages. One further power of ten brings us to a time scale within which is included the origin of life and eventually of the earth itself.

ACKNOWLEDGEMENTS

I am most grateful to my wife for typing the first draft, to Miss J.P. Bell for typing the final draft, to Mr K. Scurr for drawing the diagrams, and to the Editor for many helpful suggestions, and for inviting me to write this chapter.

REFERENCES

Beckett, S.C. (1981), 'Pollen diagrams from Holderness, N. Humberside', J. Biogeogr., **8**, 177-98.

Bernabo, J.C. and Webb, T. (1977), 'Changing patterns in the Holocene pollen record of Northeastern North America: a mapped summary', Quaternary Research, **8**, 64-96.

Berry, R.J. (1982), Neo-Darwinism, Studies in Biology No. 144, Arnold, London.

Birks, H.J.B. (1977), 'The Flandrian forest history of Scotland: a preliminary synthesis', in Shotton, F.W. (ed.) (1977) British Quaternary Studies: Recent Advances, Clarendon Press, Oxford, UK, 119-36.

Blackham, A., Davies, C. and Flenley, J.R. (1981), 'Evidence for Late Devensian landslipping and Late Flandrian forest regeneration at Gormire Lake, North Yorkshire', in Neale, J. and Flenley, J.R. (eds.) (1981) The Quaternary in Britain, Pergamon Press, Oxford, UK, 184-93.

Braun-Blanquet, J. (1932), Plant Sociology: the Study of Plant Communities, McGraw-Hill, New York and London.

Clapham, A.R., Tutin, T.G. and Warburg, E.F. (1962), Flora of the British Isles, 2nd edition, CUP, Cambridge, UK.

Clements, F.E. (1916), 'Plant Succession: an Analysis of the Development of Vegetation', Publs. Carneg. Instn., **242**, Washington DC, USA.

Climap Project Members (1976), 'The surface of the Ice-Age Earth', Science, NY, **191**, 1131-37.

Coles, J.M., Hibbert, F.A. and Orme, B.J. (1973), 'Prehistoric roads and tracks in Somerset, England: 3, The Sweet Track', Proc. Prehist. Soc., **39**, 256-93.

Coope, G.R. (1970), 'Climatic interpretation of late Weichselian Coleoptera from the British Isles', Rev. Geogr. phys. Geol. dyn. (2), **12**, (2), 149-55.

Coope, G.R. (1977), 'Fossil Coleopteran assemblages as sensitive indicators of climatic changes during the Devensian (Last) cold stage', Phil. Trans. R. Soc. Lond., B, **280**, 313-40.

Cox, C.B. (1974), 'Vertebrate palaeodistributional patterns and continental drift', J. Biogeogr., **1**, 75-94.

Cranwell, L.M. (1963), 'Nothofagus - Living and Fossil', in Gressitt, J.L. (ed.) (1961), Pacific Basin Biogeography. Symposium of the 10th Pacific Science Congress, Hawaii, Bishop Museum Press, Hawaii.

Curtis, J.T. and McIntosh, R.P. (1951), 'An upland forest continuum in the prairie-forest border region of Wisconsin', Ecology, **32**, 476-96.

Darlington, P.J. (1957), Zoogeography: the Geographical Distribution of Animals, Wiley, New York, USA.

Darwin, C. (1859), The Origin of Species by Means of Natural Selection, John Murray, London.

Dewar, H.S.L. and Godwin, H. (1963), 'Archaeological discoveries in the raised bogs of the Somerset Levels, England', Proc. Prehist. Soc., **39**, 17-49.

Dimbleby, G.W. (1962), The Development of British Heathlands and Their Soils, Oxf. For. Mem., No. **23**, Oxford, UK.

Fitter, A.H. (1978), An Atlas of the Wild Flowers of Britain and Northern Europe, Collins, London.

Flenley, J.R. (1979), The Equatorial Rain Forest: a Geological History, Butterworths, London.

Flenley, J.R. (1981), 'The Late Quaternary palyno-flora of Easter Island', XIII International Botanical Congress, Sydney, Australia, 21-28 August 1981. Abstracts, p. 104.

Flenley, J.R. (1982), 'Histoire de la vegetation de l'Ile de Pâques au Quaternaire récent: quelques indications palynologiques préliminaires', in Valenta A. (ed.) (1982) Nouveau Regard sur l'Ile de Paques, Moana, Paris, 109-16.

Flenley, J.R. and Richards, K. (eds.) (1982), The Krakatoa Centenary Expedition: Final Report, University of Hull, Department of Geography, Miscellaneous Series No. 25, Hull, UK.

Florin, R. (1963), 'The Distribution of conifer and taxad genera in time and space', Acta Horti Bergiani, **20**, No. 4, 1-178.

Ford, F.B. (1940), 'Genetic research in the Lepidoptera', Ann. Eugenics, **10**, 227-52.

Fritsch, F.E. and Salisbury, E.J. (1938), Plant Form and Function, Bell, London.

Gleason, H.A. (1926), 'The individualistic concept of the plant association', Bull. Torrey Bot. Club, **53**, 7-26.

Godwin, H. (1948), 'Studies of the Post-glacial history of British vegetation. X. Correlations between climate, forest composition, prehistoric agriculture and peat stratigraphy in Sub-boreal and Sub-atlantic peats of the Somerset Levels', Phil. Trans. Roy. Soc. B., **233**, 275-86.

Godwin, H. (1956), The History of the British Flora: a Factual Basis for Phytogeography, CUP, Cambridge, UK.

Godwin, H. (1975), The History of the British Flora, 2nd edition, CUP, Cambridge, UK.

Golson, J. (1977), 'No room at the top: agricultural intensification in the New Guinea Highlands', in Allen, J., Golson, J. and Jones, R. (eds.) (1977), Sunda and Sahul: Prehistoric Study in Southeast Asia, Melanesia and Australia, Academic Press, London.

Golson, J. and Hughes, P.J. (1976), 'The appearance of plant and animal domestication in New Guinea', Paper prepared for the IXth Congress Union Internationale des Sciences Prehistoriques et Protohistoriques, Nice, France, Sept. 1976.

Good, R. (1947), The Geography of the Flowering Plants, Longmans, London.

Good, R. (1956), Features of Evolution in the Flowering Plants, Longmans, London.

Goudie, A.S. (1977), Environmental Change, OUP, Oxford, UK.

Goudie, A.S. (1981), The Human Impact: Man's Role in Environmental Change, Blackwell, Oxford, UK.

Graham, A. (ed.) (1973), Vegetation and Vegetational History of Northern Latin America, Symposium held at Bloomington, Ind., USA, 1970, Elsevier, New York.

Grove, A.T., Street, F.A. and Goudie, A.S. (1975), 'Former lake levels and climatic change in the rift valley of southern Ethiopia', Geogr. J., **134**, 194-208.

Hafsten, U. (1960), 'Pleistocene development of vegetation and climate in Tristan da Cunha and Gough Island', Arbok Univ. Bergen, Mat-Naturv. Serie 20, 1-48.

Hammen, T. van der (1974), 'The Pleistocene changes of vegetation and climate in tropical South America', J. Biogeogr., **1**, 3-26.

Havel, J.J. (1971), 'The Araucaria forests of New Guinea and their regenerative capacity', J. Ecol., **59**, 203-14.

Hughes, N.F. (1976), Palaeobiology of Angiosperm Origins, Cambridge Earth Sciences Series, CUP, Cambridge, UK.

Huntley, B. and Birks, H.J.B. (1983), An Atlas of Past and Present Pollen Maps for Europe: 0-13,000 Years Ago, CUP, Cambridge, UK.

Hutchinson, J. (1965) (ed.), Essays on Crop Plant Evolution, CUP, Cambridge, UK.

Kelly, M.R. (1964), 'The Middle Pleistocene of North Birmingham', Phil. Trans. Roy. Soc., B., **247**, 533-92.

Khan, Asrar M. (1974), 'Palynology of Neogene sediments from Papua (New Guinea). Stratigraphic boundaries', Pollen Spores, **16**, 265-84.

Lindsay, T.M. (1911), 'Occam, William of', Encyclopaedia Britannica, 11th edition, **19**, 965-6, Encyclop. Britann. Company, New York, USA.

MacArthur, R.H. (1955), 'Fluctuations of animal populations and a measure of community stability', Ecology, 36, 533-6.

MacArthur, R.H. and Wilson, E.O. (1967), The Theory of Island Biogeography, Princeton University Press, Princeton, New Jersey, USA.

Martin, A.R.H. (1968), 'Aquilapollenites in the British Isles', Palaeontology, **11**, 549-53.

McElhinny, M.W. (1973), Palaeomagnetism and Plate Tectonics, CUP, Cambridge, UK.

Menard, H.W. and Hamilton, E.L. (1963), 'Paleogeography of the tropical Pacific' in Gressitt, J.L. (ed.) (1961), Pacific Basin Biogeography. A Symposium of the Tenth Pacific Science Congress, Honolulu, Bishop Museum Press, Honolulu, Hawaii.

Mendel, G.J. (1866), 'Versuche uber Pflanzen-Hybriden', Verh. Naturforschenden Verein Brunn, **4**, 3-47.

Moore, P.D. (1973), 'The influence of prehistoric cultures upon the initiation and spread of blanket bog in Wales', Nature, London, **241**, 350-3.

Morrison, M.E.S. (1959), 'Evidence and interpretation of 'landnam' in the north-east of Ireland', Bot. Notiser, **112**, 185-204.

Muller, J. (1972), 'Palynological evidence for change in geomorphology, climate and vegetation in the Mio-Pliocene of Malesia' in Ashton, P. and Ashton, M. (eds.) (1972), Transactions of the Second Aberdeen-Hull Symposium on Malesian Ecology, University of Hull, Department of Geography, Miscellaneous Series No. 13, Hull, UK, 6-16.

Opdyke, N.D., Glass, B., Hays, J.D. and Foster, J.H. (1966), 'Paleomagnetic study of Antarctic deep-sea cores', Science, **154**, 349-57.

Pennington, W. (1965), 'The interpretation of some post-glacial vegetation diversities at different Lake District sites', Proc. Roy. Soc. Lond., B, **161**, 310-23.

Pennington, W. (1977), 'The Late Devensian flora and vegetation of Britain', Phil. Trans. Roy. Soc. Lond., B, **280**, 247-71.

Petrov, S.I. and Drazeva-Stamatova, Ts. (1972), 'Reevesia Lindl. fossil pollen in the Tertiary sediments of Europe and Asia', Pollen Spores, **14**, 79-95.

Plumstead, E.P. (1973), 'The Late Palaeozoic Glossopteris flora', in Hallam, A. (ed.) (1973), Atlas of Palaeobiogeography, Elsevier, Amsterdam, 187-205.

Popper, K.R. (1968), The Logic of Scientific Discovery, Hutchinson, London.

Powell, J.M., Kulunga, A., Moge, R., Pono, C., Zimike, F., and Golson, J. (1975), Agricultural Traditions of the Mount Hagen Area, University of Papua, New Guinea, Department of Geography, Occasional Paper No. 12, Papua, New Guinea.

Raup, H.M. (1964), 'Some problems in ecological theory and their relation to conservation', J. Ecol., **52** (Suppl.), 19-28.

Ricklefs, R.E. (1973), Ecology, Nelson, London.

Ridley, H.N. (1930), The Dispersal of Plants Throughout the World, Reeve, London.

Safriel, U.N., Gilboa, A. and Felsenburg, T. (1980), 'Distribution of rocky intertidal mussels in the Red Sea coast of Sinai, the Suez Canal and the Mediterranean coast of Israel, with special reference to recent colonizers', J. Biogeogr., **7**, 39-62.

Seddon, B. (1962), 'Late-glacial deposits at Llyn Dwythwch and Nant Ffrancon, Caernarvonshire', Phil. Trans. Roy. Soc., B., **244**, 459-81.

Shackleton, N.J. (1967), 'Oxygen isotope analyses and Pleistocene temperatures reassessed', Nature, London, **215**, 15-17.

Shackleton, N.J. and Opdyke, N.D. (1976), 'Oxygen-isotope and paleomagnetic stratigraphy of Pacific core V28-239, Late Pliocene to Latest Pleistocene', Mem. Geol. Soc. Am., **145**, 449-64.

Shackleton, N.J. and Turner, C. (1967), 'Correlation between marine and terrestrial Pleistocene successions', Nature, London, **216**, 1079-82.

Simmons, I.G., Atherden, M.A., Cundill, P.R. and Jones, R.L. (1975), 'Inorganic layers in soligenous mires of the North Yorkshire Moors', J. Biogeogr., **2**, 49-56.

Singh, G. (1971), 'The Indus Valley Culture seen in the context of post-glacial climatic and ecological studies in North-West India', Archaeol. and Phys. Anthropol. Oceania, **6**, 177-89.

Singh, G., Joshi, R.D., Chopra, S.K. and Singh, A.B. (1974), 'Late Quaternary history of vegetation and climate of the Rajasthan Desert, India', Phil. Trans. Roy. Soc. Lond., B, **267**, 467-501.

Smith, A.G., Briden, J.C. and Drewry, G.E. (1973), 'Phanerozoic World Maps' in Hughes, N.F. (ed.) (1973), Organisms and Continents through Time, Special Papers in Paleontology No. 12, 1-42.

Sneath, P.H.A. (1967), 'Conifer distributions and continental drift', Nature, London, **215**, 467-70.

Soepadmo, E. (1972), 'Fagaceae', Flora Malesiana, Series I. Spermatophyta, **7** (2), 265-403.

Steenis, C.G.G.J. van (1962), 'The Landbridge Theory in Botany', Blumea, **11**, 235-372.

Steenis, C.G.G.J. van (1971), 'Nothofagus, key genus of plant geography, in time and space, living and fossil, ecology and phylogeny', Blumea, **19**, 65-98.

Steenis, C.G.G.J. van (1972), 'Nothofagus, key genus to plant geography' in Valentine, D.H. (ed.) (1972), Taxonomy, Phytogeography and Evolution, Academic Press, London, 275-88.

Tansley, A.G. (1920), 'The classification of vegetation and the concept of development', J. Ecol., **8**, 118-49.

Taylor, J.A. (1983), 'The peatlands of Great Britain and Ireland' in Gore, A.J.P. (ed.) (1983), Mires: Swamp, Bog, Fen and Moor, B. Regional Studies, Elsevier, Amsterdam, 1-46.

Taylor, J.A. and Smith, R.T. (1981), 'The role of pedogenic factors in the initiation of peat formation, and in the classification of mires', in Proc. 6th Inst. Peat Congress, Duluth, Mich., USA, 109-18.

Troels-Smith, J. (1956), 'Neolithic period in Switzerland and Denmark', Science, NY, **124**, 876.

Walker, D. (1970), 'Direction and rate in some British post-glacial hydroseres' in Walker, D. and West, R.G. (1970), Studies in the Vegetational History of the British Isles. Essays in honour of Harry Godwin, CUP, Cambridge, UK, 117-39.

Walker, D. and Flenley, J.R. (1979), 'Late Quaternary vegetational history of the Enga District of Papua New Guinea', Phil. Trans. R. Soc. Lond. B, **286**, 265-344.

Walker, D. and West, R.G. (eds.) (1970), Studies in the Vegetational History of the British Isles, CUP, Cambridge, UK.

Wallace, A.R. (1869), The Malay Archipelago, Macmillan, London.

Ward, R.C. (1978), Floods: a Geographical Perspective, Macmillan, London.

Wells, W.C. (1818), 'An Account of a Female of the White Race of Mankind, Part of Whose Skin Resembles That of a Negro; With Some Observations on the Causes of the Differences in Colour and Form Between the White and Negro Races of Men', appended to W.C. Wells, Two Essays: One upon Single Vision With Two Eyes; the Other on Dew ... (London, 1818), pp. 431-32. This volume also contains Wells' 'Memoir' of his own life. See also Richard H. Shryock, 'The Strange Case of Wells' Theory of Natural Selection (1813): Some Comments on the Dissemination of Scientific Ideas', in Studies and Essays on the History of Science and Learning Offered in Homage to George Sarton ... (New York, 1946), 197-207; Conway Zirkle, 'Natural Selection Before the 'Origin of Species'', Proc. Amer. Philos. Soc., LXXXIV (1941), 71-123.

Whittaker, R.J. and Flenley, J.R. (1982), Chapter 2, 'The flora of Krakatau' in Flenley, J.R. and Richards, K. (eds.) (1982), The

Krakatau Centenary Expedition, Final Report, University of Hull, Department of Geography, Miscellaneous Series No. 25, Hull, UK, 9-53.

Willis, J.C. (1922), Age and Area: a Study in Geographical Distribution and Origin of Species ... with chapters by H. de Vries and others, CUP, Cambridge, UK.

Wolfe, J.A., Doyle, J.A. and Page, V.M. (1975), 'The bases of angiosperm phylogeny: palaeobotany', Ann. Miss. Bot. Gdn., **62**, 801-24.

Zaklinskaya, E.D. (1964), 'On the relationships between Upper Cretaceous and Palaeogene Floras of Australia, New Zealand and Eurasia, according to data from Spore and Pollen Analyses', in Cranwell, L.M. (1961), Ancient Pacific Floras - The Pollen Story, 10th Pacific Science Congress Series, Honolulu, University of Hawaii Press, Honolulu, pp. 85-6.

FURTHER READING

Berry, R.J. (1982), Neo-Darwinism, Studies in Biology, No. 144, Arnold, London.

Birks, H.J.B. and Birks, H.H. (1980), Quaternary Palaeoecology, Arnold, London.

Colinvaux, P.A. (1973), Introduction to Ecology, Wiley, New York, USA.

Cox, B.C. and Moore, P.D. (1980), Biogeography. An ecological and evolutionary approach, 3rd edition, Blackwells, Oxford, UK.

Flenley, J.R. (1979), The Equatorial Rain Forest: a Geological History, Butterworths, London.

Godwin, H. (1975), The History of the British Flora, 2nd edition, CUP, Cambridge, UK.

Goudie, A.S. (1977), Environmental Change, OUP, Oxford, UK.

Goudie, A.S. (1981), The human impact: man's role in environmental change, Blackwells, Oxford, UK.

Stott, P.A. (1981), Historical Plant Geography. An Introduction, Allen and Unwin, London.

Chapter 4

BIOGEOGRAPHY AND ECOSYSTEMS

R. P. Moss

Introduction

In a recent review, a not undistinguished biologist has complained of 'the new wave of incubine biogeographers who rape ecology of her principles and claim them as their own' (Shimwell, 1983). The Oxford English Dictionary admits no adjective _incubine_ from the noun _incubus_, but whether biogeographers are seen by ecologists simply as hymenopterous parasites, or, more imaginatively, as evil spirits who descend upon unsuspecting female ecologists, seeking carnal intercourse with them in their sleep, the oppressive, nightmarish quality of their relationship to the humble, self-respecting ecologist cannot be mistaken! More seriously, however, the review continues to point out that when the reader is told that:

> '... the two main objectives are to illustrate the importance of the ecosystem as the core concept in biogeography, and to demonstrate the potential integrative role of biogeography within geography, the reader's wariness is more compounded. By the time he gets into Chapter 1, with its phalanx of thermostat-heater systems, feedbacks and homeostatic thresholds, and thence to the good old ecosystem, he is probably almost as disturbed and disrupted as the disturbing and disrupting beings which, the authors tell him, form the theme of the book.'

The rest of the book reviewed receives a better write-up. But the point is clear; biogeographers have frequently made pretentious and categorical claims for the efficacy of the ecosystem concept which move into the realm of the metaphysical, by implication, if not in fact. Too often such claims have masked an absence of clear consistent thought concerning the concept and its potential in solving problems of real substance, and it has been so emptied of content as to be scarcely more than an imprecise formalisation of inadequately analysed relationships.

In this essay an attempt will be made to reflect constructively on the notion of the ecosystem, and to outline some of the ways in which it has been, and may be, used in the general field of ecology in its application to geographical problems. In the discussion 'geographical problems' will be interpreted without philosophical overtones in the naive sense of 'problems which geographers have, or might tackle', and 'geographer' is also interpreted trivially in the sense of an individual with a basic training from, or a study location in, a geography department in an institution of higher education. Such initial caveats are unfortunately necessary in view of the strange predisposition of many geographers towards seeking profound philosophical justification for geography as a discipline before they feel easy about using the word without apology. Perhaps geographers are more in need of self-acceptance than an unassailable philosophical bastion.

The Ecosystem as a Concept

It is almost an ecological duty to initiate any consideration of the concept of the ecosystem with a deferential nod in the direction of Tansley (1935). But the relevant paragraph is his major application (Tansley, 1939) of the concept to practical biogeographical questions, even if the original article (1935) is not consulted. The 'key concept' in the scientific enterprise:

> '... to formulate the phenomena of vegetation in a rational system is the idea of progress towards equilibrium, which is never, perhaps, completely attained, but to which approximation is made whenever the factors at work are constant and stable for a long enough period of time...'.

In the empirical dimension this leads to an inclusion of:

> '... the units of vegetation in the general conception of physical systems of which the universe is composed - systems which mark positions of relative, if only temporary, stability in the general flux...'.

The systems may indeed, for various reasons, be wholly or partly destroyed, but:

> '... so long as conditions remain the same and the original components are present, these will always tend to re-establish a system of the same type as before...'.

Then follows the adumbration of the concept:

> A unit of vegetation considered as such a system includes not only the plants ..., but the animals habitually associated with them, and also all the physical and chemical components of the immediate environment or habitat which together form a recognisable self-contained entity. Such a system may be

> called an ecosystem ..., because it is determined by the particular portion ... of the physical world that forms a home ... for the organisms which inhabit it.

That particular portion of the physical world is the ecotope, from the Greek words τόποσ (place), and οἶκοσ (home). Tansley's concept is thus:

1. a theoretical concept, in the sense that it is not a direct denotation of an empirical entity, but is defined and used in the context of a particular theoretical view of the development of vegetation;

2. a concept involving heterogeneous elements, in that it tries to bring together objects which, in traditional studies of nature, are conventionally investigated separately, notably vegetation, fauna, soil and climate;

3. an eclectic concept, which seeks to unite these heterogeneous elements into a single object of study, a 'recognisable self-contained entity';

4. a focussed concept, in which the system is defined visibly as a 'unit of vegetation', which implies that the spatial extension of the system is defined in terms of floristic composition and structural arrangement or physiognomy of the plant community.

In his masterly description of the vegetation of these islands that is precisely the way in which he uses the concept. The units are seen in dynamic successional terms, as embracing diverse components of both plants and habitats, bringing them together in categories which are defined and delimited as floristic and physiognomic units of vegetation.

Table 4.1: Arrangement of 'The British Islands and Their Vegetation' (Tansley, 1939)

PART I	The British Islands as Environment of Vegetation
	Physical Features and Geological History
	Climate
	Regional Climates
	Soil
	Distribution of Rocks and the Soils they Produce
	The Biotic Factor
PART II	History and Existing Distribution of Vegetation
	Pre-History
	The Historical Period
	Distribution of the Forms of Vegetation

PART III **The Nature and Classification of Vegetation**

The Nature of Vegetation
Different Methods of Classifying Vegetation
  Raunkiaer, Moss, Crampton

PART IV **The Woodlands**

Nature and Status of British Woodlands
  Dominant and Other Trees
  The More Important Shrubs
Oakwood
  Introductory
  Pedunculate Oakwood (Quercetum Roboris)
  Sessile or Durmast Oakwood
    (Quercetum Petraeae or Sessiliflorae)
    English woods
    Welsh, Irish and Scottish woods
    Mixed oakwood on sandy soils
      (Quercetum Roboris et Sessiliflorae
      or Ericetosum)
Beechwood
  Introductory
  Beechwood on Calcareous Soil (Fagetum Calcicolum)
  Chalk Scrub and Yew Wood
  Seral Ashwood
  Beechwood on Loam (Fagetum Rubosum)
  Beechwood on Sands and Podsols
    (Fagetum Arenicolum or Ericetosum)
  Summary
Ashwood on Limestone (Fraxinetum Calcicolum)
Pine and Birch Woods
Alderwood (Alnetum Glutinosae)
Scrub Vegetation or Bushland (Fruticetum)

PART V **The Grasslands**

Nature and Status of British Grasslands
Acidic Grasslands
Basic Grasslands
Neutral Grassland

PART VI **The Hydroseres. Freshwater, Marsh, Fen and Bog Vegetation**

The hydroseres. Vegetation of ponds and lakes
The Cumbrian Lakes
The vegetation of rivers
Marsh and fen vegetation
  The East Anglian Fens
  North-East Irish Fens

It can scarcely be denied that Tansley's work was a major contribution to biogeography, in the sense in which many modern geographers would conceive it. But Tansley was, of course, a distinguished ecologist of biological pedigree. At the time, geographers were preoccupied with concerns other than the plant cover of the earth, which, despite its rather obvious presence in any view of landscape, received only the most cursory attention, if it was mentioned at all, in the geographical 'landscape studies' of the day; it fitted neither into the physical or the cultural categories. The only book (Newbigin, 1936) on biogeography produced in English by a geographer (although Marion Newbigin had initial training in biology) was hardly an example of the application of the ecosystem concept; Stamp's monograph on 'The Vegetation of Burma' was a pioneering study bearing a tenuous relationship to Tansley's notion. This the attraction of the original ecosystem concept for geographers can scarcely be discerned. Its popularity had to await further ecological development which fitted it to appeal, perhaps speciously, to the synthetic, integrative or holistic aspirations of professional geographers.

The transformation of the concept. In 1942 Lindemann provided the seed idea which changed the concept of the ecosystem from a tool for description into an analytical instrument; and, in so doing, paved the way for its abduction by the marauding geographers. He saw that the essence of the links between the heterogeneous elements essential to the Tansley concept was the transfer of energy, and that therefore the ecosystem could be studied thermodynamically as a system (Lindemann, 1942). The phenomenon of nutrient cycling was then seen as a further link between the elements, and the role of the plant-soil system in the hydrological cycle was similarly emphasised. The ecosystem was thus transformed from a theoretical concept in a developmental scheme focussed on vegetation, to a physical and biochemical notion which provided a whole new theoretical framework for study and research (Hutchinson, 1948). This in turn opened up the possibility of a new scale of study which eventually expanded to global concerns. The publication of Wiener's 'Cybernetics' in 1948 opened the way to a new dimension of investigation, and the jargon (which is neither meaningless nor useless in the right context) which induced the trauma in the reviewer with whose medieval libidinous allusion this chapter began, was incorporated into ecosystem study.

Meanwhile, as the theoretical dimensions of ecosystem study expanded in these ways, the early mathematical work of Lotka (1925), Volterra (1925) and Gause (1934) on populations of organisms, was being used as the foundation of a parallel theoretical edifice in biological science, that of population ecology. In ecology both have built imposing theoretical structures, and occasioned considerable controversy as to their relative merits in relation to the field of ecology as a whole (McIntosh, 1983). It is of some real philosophical interest that this division is significantly analogous to that in geography between those who have concentrated on the development of theory in the spatial modelling mode in human geography, and those who have tried to preserve a more traditional interest in the relations between human communities and the environments in which they live, amongst whom the proponents of the ecosystem in geography must be numbered (Tivy and O'Hare, 1981). The debates in ecology deploy strikingly similar arguments to those advanced by the protagonists in the geographical arena (McIntosh, 1983). The following might apply to geography as much as to its primary reference:

> Much of the 'revolution' in ecology in the 1950's and 1960's took the form of increasingly quantitative methodology, introduction of diverse external bodies of theory into ecology ... Some of the 'revolution' was more apparent than real as it became fashionable not to waste valuable time finding what was in the relevant ecological literature and to avoid the ordinary scientific reviewing process (McIntosh, 1983).

The rejection of much classical Clementsian climax theory and its replacement by the analysis of complex systems which effectively perpetuated the pseudo-organismic and developmental core of that theory, is closely paralleled by the widespread rejection of the regional concept and the traditional developmental man/land framework and its substitution by the systems paradigm and advocacy of the ecosystem model in geographical study. This too attempted to preserve, though perhaps unconsciously, the classical conceptual core, albeit arguably in a more rigorous form. There are indications of a new interest in traditional regional geography. The re-writing of some of the older regional geographies, such as Spate's 'India', represents a vital part of possible research work in the developing world. The growth of theoretical, predictive ecology in relation to population studies from the same quantitative roots closely parallels the development of 'positivist' geography since the 'quantitative revolution'. What was written of theoretical ecology in 1975 (and vehemently criticised!) would, without alteration, have been claimed by many in respect of geography:

> Within two decades new paradigms had transformed large areas of ecology into a structured predictive science that combined powerful quantitative theories with the recognition of widespread patterns in nature (Cody and Diamond (eds.), 1975).

These parallels between ecological and geographical development, which could be further developed, perhaps indicate an underlying affinity of approach which explains the very easy transference of concepts from one field to the other, which is not only in one direction.

The transfer of the concept. It is interesting that Haggett's classic work, Locational Analysis in Human Geography, which incorporated techniques which had been developed in ecology, appeared in the same year as the article by Stoddart which first argued the case for the use of the ecosystem concept as a geographical model (Haggett, 1965; Stoddart, 1965). Other articles at the same time, and quite independently conceived and written, pursued related ecological-geographical links, such as the community concept (Morgan and Moss, 1965; Moss and Morgan, 1967). Stoddart developed his theme from the energetics of Lindemann (1942), while others drew on the analogy between sociological and ecological concepts (Moss and Morgan, 1965). It is perhaps a reflection on the entrenched conservatism of geographers in their didactic mode that the ecosystem notion has to be forced into the straight-jacket of 'physical' geography, whereas the community notion experiences a no less Procrustean inclusion in cultural or 'human' geography. It is, quite clearly, the fact that these ecological concepts do not clearly fit into the inadequate categories of current geographical prejudice which makes them so attractive, philosophically and psychologically, to some geographers.

It is obvious, however, that the notion of the ecosystem is neither simple nor straightforward in its interpretation; it includes and implies a number of dimensions or aspects, and it can be used in a number of different ways in scientific praxis. Nor should it be expected that its use by geographers should in any way be different in essence from its employment by biologists. Even the common strategy of trying to base a distinction on spatial or locational emphases is trivial and effete, for the traditional use of the basic Tansleyian concept by ecologists inevitably involves a locational and spatial specificity by virtue of the incorporation of biotope into its essential content. This becomes even more obvious when the notion is expanded in scale to describe major biomes; indeed then the spatial extent and locational factors become paramount.

From this two necessary observations follow. First, that in speaking of the ecosystematic approach to biogeographical studies, we can refer only to a broad and diffuse set of approaches, and not to a closely-defined structured method of study. Second, we can refer only to a broad spectrum of interests and problems to which both biological and geographical ecologists have made a contribution. Such attention as is given here to the geographical contribution in no way implies a devaluation of the biological contribution, which, in many fields, has been significantly more distinguished than that of those working in the ambience of geography, particularly in North America. It is perhaps most helpful to view ecosystem studies as a developing field in broadly defined ecological science to which both trained biologists and competent geographers have a major contribution to make. There are encouraging signs of co-operation in some institutions even now.

The Ecosystem in Geography

The complexity of the concept contributes the versatility in its use. On the broadest scale it serves as an organising principle for systematising diverse theoretical material. At another level it can be treated as an object of study in itself in order to develop specific theory of ecosystems. Then it can be used as a specific concept defining a problem-solving approach. It can also be used as an empirical category denoting the visible expression of a specifiable set of ecological relationships. Then the term has been used in a much wider sense to denote almost any set of discernible relationships at the earth's surface. It might be argued that this is stretching meaning too far. But, as Chorley (1973) has pointed out, unless we are prepared to stretch the meaning in this way, the ecosystem approach has in fact little to offer contemporary geography as a synthesising principle, except in a rapidly decreasing set of particular situations. So long as the ecosystem concept is forced into providing 'spatial' or 'landscape' or even 'locational' explanations, its role is minor, a fact which lends extra force to the conclusions of this essay.

Before each of these is reviewed in a little more detail, some comment is necessary on the 'holistic' aspirations of the concept. Geographers have, and arguably quite mistakenly, supposed that the 'holism' claimed for the ecosystem concept is synonymous with the twin chimeras of integration and synthesis, so strenuously and ineffectually pursued by geographers for so long. Holistic claims for the concept of the ecosystem have three elements:

1. the assertion that the behaviour of the whole system is more than the sum of the behaviour of the individual parts, and is therefore worthy of study in its own right;

2. the complementary claim that in fact the behaviour of the

individual parts cannot be fully understood without reference to the behaviour of the whole;

3. the resultant implication that relationships between categories of objects conventionally studied as sets in themselves become of central importance for investigation.

There is no sense in which ecosystem study attempts to include everything, even the everything that geographers would wish to include. The elements are defined by the system selected for study; thus plants become autotrophs, or most do, and grazing animals and predators become different elements in the heterotrophs, and so on. The ecosystem concept provides a different focus for scientific investigation; it does not provide the rigorous answer to the regional or environmentalist geographer's dream. And few regional geographers would claim so, or even perhaps wish that it should. It would appear that the extravagant claims made can hardly be justified either rationally or in terms of empirical results.

The distinction between individualistic and reductionist studies and holistic approaches is not that one group is analytic and the other synthetic, the one divisive and the other integrative; rather, both are analytic and divisive the more rigorous they become. The difference is that the analysis and the reduction proceed from different presumptions and proceed to different goals. It is a question of the definition of the systems for study, and also, as a corollary, of the scale of investigation. And, of similar importance, the elements in the systems which are studied must be defined in terms of those systems, not in terms of other systems to which the central focus of interest may be more or less closely related. Thus, if energetics is the central focus, the proper categories are autotrophs, heterotrophs, and the like, rather than plants and animals, even though it may be convenient to use the species or some other taxon in the hierarchy; if cybernetic relations are the object of interest, then different, though not necessarily unrelated, categories are appropriate. Similarly for nutrient cycling, or any other particular aspects of ecosystematic relations.

It may be argued, therefore, that the claims made for the primacy and centrality of the ecosystem concept in biogeography (and still less in geography) are ill-founded; they are either trivial or extravagant. This is borne out by an examination of the ways in which geographers have in fact used the concept.

<u>The use of the ecosystem concept in geography</u>. Nevertheless, biogeographers have used the ecosystem concept extensively, both explicitly and implicitly. In the first place they have used it as <u>an organising principle</u>. There is a great deal of knowledge which is of interest to geographers, though not necessarily exclusive to them, which can be illuminatingly organised using the categories implied by the concept in its developed form. Watts' <u>Principles of Biogeography</u> is a well-known example of a substantial study of this kind. It develops a basic ecosystematic framework which serves as a basis for

the application of ecological understanding to distributional and spatial problems. Given this approach, it is not surprising that its Table of Contents is not dissimilar to that of a number of standard texts on ecology without geographical pretensions, such as Odum's well-known Fundamentals of Ecology, though it is inevitably totally different from other texts on ecology, based on population theory, which have spatial implications even in their titles, such as Krebs' Ecology: The Experimental Analysis of Distribution and Abundance. Here the ecosystem is a subsidiary and auxiliary notion, used rather as a descriptive than an analytical concept. The central notion is the dualism of population and community, which, in the view of the present writer, is an approach which has much to offer the biogeographer (Odum, 1971; Watts, 1971; Krebs, 1978) (Table 4.2).

The organising principle has also been used in a related but distinct way by directing it to problem areas with which geographers are concerned, such as land use and natural resources. Simmons' contributions (1966 and 1970) to this field have been a notable element in biogeographical study. Others have employed the notion of plant-soil systems as an approach to the organisation of available knowledge in a particular area of biogeography (Moss and Morgan, 1977). The idea of the ecosystem may even be detected behind work on environmental systems (Bennett and Chorley, 1978). Not least the volume to whose review reference was made earlier uses the ecosystem concept as an organising principle and in a didactic mode (Tivy and O'Hare, 1981). It is perhaps of significance that few books, addressing themselves specifically to biogeography as a distributional study, emphasise the ecosystem as a pre-eminent concept, though it is used in context as one tool for description and analysis; it may perhaps be important to realise that books of this type are generally written by biologists, like Dansereau (1957), and Cox, Healey and Moore (1976). This perhaps confirms the assertion made earlier that the ecosystem is a spatial concept only in a naive and trivial sense, and is in essence one tool which may be employed by both geographical and biological ecologists in addressing a particular set of questions about nature which are not the exclusive province of either field. Exclusivist claims of any kind are misplaced.

It is sometimes implied that the mere organisation of current understanding of knowledge in any particular field is in some sense an inferior activity to problem-solving research projects. Certainly there can be no knowledge to organise apart from continuing research, and the former is ineluctably dependent upon the latter. But the dependence is not wholly one way. A competently accomplished organisation of what we think we know in any field is in itself, or ought to be, not simply a stimulus and springboard for research, but a source and specification for particular ideas for exploration, and a pointer towards the precise posing of the questions which need to be asked. New perspectives ought always to illuminate new directions of exploration with perceptive clarity. The fact is, however, that such successful presentations require consummate skill of a high order, which is possessed and developed only rarely. Too often they degenerate into another variation on an old theme, a

Table 4.2: A comparison of the contents of textbooks by Watts (1971), Odum (1971) and Krebs (1972)

WATTS: Principles of Biogeography	ODUM: Fundamentals of Ecology	KREBS: Ecology
Introduction: Scope and origins of biogeography Definitions: nature of the biosphere, ecosystems, other assemblages and environment	Introduction: The Scope of Ecology: Ecology: its relations to other sciences and its relevance to human civilisation The sub-divisions of ecology About models	What is Ecology? Introduction to the science of ecology Definition History of ecology Basic problems and approach
	Principles and Concepts Pertaining to the Ecosystem: Concept of the ecosystem Biological control of the chemical environment Production and decomposition in nature Homeostasis of the ecosystem	The Problem of Distribution: Populations: Factors limiting distribution: dispersal Factors limiting distribution: behaviour Factors limiting distribution: inter-relations with other organisms Factors limiting distribution: temperature Factors limiting distribution: moisture Factors limiting distribution: other physical and chemical factors Complications: adaptation
Energy Controls of Ecosystems: Energy sources and transfer sources and availability laws of energy exchange measurement of energy exchange Food chains and webs techniques for tracing food chains examples of simple food chains ecological pyramids theory of energy exchange ecological efficiency of energy transfer Energy transfer within selected ecosystems an example of energy flow grazing and detritus food chains the importance of decomposers	Energy in Ecological Systems: Review of fundamental concepts The energy environment Concept of productivity Food chains, food webs and trophic levels Metabolism and size of individuals Trophic structure and ecological pyramids Ecoystem energetics	The Problem of Abundance: Populations: Population parameters Demographic techniques Population growth Species interactions: competition Species interaction: predation Natural regulation of population size Examples of population studies Applied problems: optimum yield Applied problems: biological control Distribution and Abundance at the Community Level: Community parameters The nature of the community Community change Community metabolism: primary production

Table 4.2 cont'd.

Biological productivity in ecosystems Primary and secondary productivity		Community metabolism: secondary production Species diversity
Biogeochemical Cycles within Ecosystems: Chemical elements in living organisms: the relative abundance; measurement of relative abundance; elements, cell growth, nutrition Cycles of element exchange phosphorus and sulphur carbon nitrogen the importance of soil organisms Man-induced modifications of element exchange patterns: environmental deterioration pesticides radioactive substances air pollution Patterns of element exchange in major world systems: causes of variability in element exchange patterns	Principles and Concepts pertaining to Biogeochemical Cycles: Patterns and basic types Quantitative study The sedimentary cycle Cycling of non-essential elements Cycling of organic nutrients Nutrient cycling in the tropics Recycle pathways	Community organisation Evolutionary ecology Human Ecology: Human population Food production Appendices, etc.
Environmental Limitations on Ecosystem Development: The concept of tolerance: techniques for measuring environmental conditions Limits of tolerance in terrestrial ecosystems: the light factor heat and temperature humidity and moisture: the hydrological cycle	Principles pertaining to Limiting Factors: Liebig's 'Law' of the Minimum Shelford's 'Law' of Tolerance Combined concept of limiting factors Conditions of existence as regulatory factors Physical factors of importance as limiting factors Ecological indicators	

Table 4.2 cont'd.

energy/temperature-water relationships wind topography edaphic considerations the biotic factor Population Limitations within Ecosystems: The demography of organisms: Methods of estimating populations Birth, death and growth rates Patterns of population growth The importance of age Genetic factors The spatial arrangement of organisms Migration Density-dependent and density-independent controls Competition between organisms: Intra-specific competition: territoriality Inter-specific competition: the ecological niche Positive and negative competition Competition and succession	Principles pertaining to Organisation at the Community Level: The biotic community concept Intra-community classification, and concept of ecological dominance Community analysis Species diversity in communities Pattern in communities Ecotones and the concept of edge effect Paleoecology: community structure in past ages Principles pertaining to Organisation at the Population Level: Population group properties Population density and indices of relative abundance Basic concepts regarding rates Natality Mortality Population age distribution Intrinsic rate of natural increase Population growth form and concept of carrying capacity Population fluctuations and 'cycle' oscillations	Contents completed

Table 4.2 cont'd.

	Population regulation; concepts of density-dependent and density-independent action in control Population dispersal Population energy flow (bioenergetics) Population structure: internal distribution patterns (dispersion) Population structure: aggregation and Allee's Principle Types of inter-action between two species Negative interactions: inter-specific competition Negative interactions: predation, parasitism, and anti-biosis Positive interactions: commensalism, co-operation, and mutualism The Species and the Individual in the Ecosystem: Concepts of habitat and ecological niche Ecological equivalents Character displacement: sympatry and allopatry Natural selection: allopatric and sympatric speciation Artificial selection: domestication Biological clocks Basic behavioural patterns Regulatory and compensatory behaviour Social behaviour
The Time Factor: Dynamic Aspects of Ecosystems: Ecological aspects of change: World patterns of distribution of organisms Modes of dispersal of organisms Climax and polyclimax succession Climatic change and equilibrium within ecosystems	Development and Evolution of Ecosystems: Strategy of ecosystem development Concept of the climax Relevance of ecosystem development theory to human ecology Evolution of the ecosystem Co-evolution Group selection

Table 4.2 cont'd.

Evolutionary aspects of change:
- Evolution as a reaction to changing environments: general
- Phyletic evolution and speciation
- The problem of extinction

Man in Ecosystems:

Some environmental restraints:
- Human origins and diversity
- Environmental limitations
- The web of disease

Man as an agent of ecosystem change:
- The use of fire
- The domestication of plants and animals, and the ecological status of agricultural systems

Systems Ecology: the Systems Approach and Mathematical Models in Ecology:
- Nature of mathematical models
- Goals of model-building
- Anatomy of mathematical models
- Basic mathematical tools in model-building
- Analysis of model properties
- Approaches to the development of models

(The Habitat Approach)
- Freshwater Ecology
- Marine Ecology
- Estuarine Ecology

Terrestrial Ecology:
- The terrestrial environment
- Terrestrial biota and biogeographic regions
- General structure of terrestrial communities
- The soil sub-system
- The vegetation sub-system
- The permeants of the terrestrial environment
- Distribution of major terrestrial communities: the biomes

(Applications and Technology)

Resources:
- Conservation of resources
- Mineral resources
- Agriculture and forestry
- Wildlife management
- Aquaculture
- Range management
- Desalination and weather modification
- Land use

Table 4.2 cont'd.

Human migration and its biogeographical consequences
Environmental pollution
The present status of world ecosystems
The continued impoverishment of world ecosystems
The need to maintain biological diversity
The need for conservation

Pollution and Environmental Health:
The cost of pollution
Kinds of pollution
Phases of waste treatment
Strategy of waste management and control
Monitoring pollution
Environmental law
Problem areas

Radiation Ecology:

Remote Sensing as a Tool for Study and Management of Ecosystems:

Perspectives in Microbial Ecology:

Ecology of Space Travel:

Toward an Applied Human Ecology:

Notes:

1. Clearly the tables of contents of each book do not set out the material included in the book with equal detail. Nevertheless, the headings do give a clear picture of the general contents of each in sufficient detail to facilitate comparison. Occasionally, the headings have been abbreviated. In Odum some major headings have been included without the detail provided in the book itself, in order to save space, where they are unnecessary to the comparison. In Krebs the detailed breakdown of each chapter has always been omitted.

2. The close correspondence between Watts and Odum is fairly clear and, even when the books are compared in detail, the distinction between ecology and biogeography made by Watts in his introduction seems unimportant in terms of the substantive content, and it is clearly arguable, on the evidence presented, that the distinction is trivial. If this is so, then why attempt to make it? We are all interested in ecology in general and human ecology in particular. Why not admit it, and cease trying to build insecure fences with rotting posts?

3. The space allocated to the general fields in Watts and Odum is:

	Watts pp.	no. of pages	Odum pp.	no. of pages
Introduction	1-6	6	3-7	5
Ecosystem concept	(3-6)	3	8-36	29
Energy	7-46	29	37-85	49
Biogeochemical Cycles	52-110	29	86-105	20
Limiting Factors	120-188	69	106-139	34
Community Relationships	-		140-161	22
Populations	197-237	41	162-232	71
Species/Individual Relations	-		234-250	17
Ecosystem Development	242-298	57	251-275	26
Mathematical Modelling	-		276-292	17
Major Environments and Biomes	-		293-404	112
Man and Ecosystems	306-368	63	405-467	63
Remote Sensing	-		468-483	16
(Microbial Ecology)	-		484-497	15
(Space Travel Ecology)	-		498-509	12
Applied Human Ecology	-		510-516	17

Some elements which appear to be excluded in Watts in this tabulation are mentioned under other headings but, surely, the major surprise must be the fact that the largest section in Odum concerns the major biomes and habitats which, it might be argued, are of major biogeographical interest, but Watts' consideration is cursory in the extreme. This links into the remarks made in the essay on the new series concerning Ecosystems of the World and the minimal input to that series by geographers.

further textbook setting out yet again much the same material. Those books which have succeeded, such as those of Simmons, ought to be especially valued.

What then of research? It has already been suggested that, in ecology in general, one significant avenue of development has been the exploration of *ecosystem theory and the application of systems analysis*. This has produced an impressive array of theoretical structures which rival the achievements of the population ecologists (van Dyne (ed.), 1969; Patten (ed.), 1971; MacArthur, 1972; Levin (ed.), 1976; and May, 1976), which some would see as competing paradigms, in the Kuhnian sense, or research programmes in the terms of Lakatos' analysis of the activity and structure of scientific investigation (Simberloff, 1980; McIntosh, 1983). This achievement has, however, been largely the accomplishment of ecologists, in both its pure and its applied dimensions. The contribution of geographers, particularly British ones, has been small, even negligible, despite the enthusiasm shown by many geographers in their espousal of systems analysis techniques and, perhaps even more enthusiastically, systems philosophy, as *salvator mundi* in the field of geography (Morgan, 1980).

Nevertheless, despite the lack of rigorous theoretical research by geographers, the concept of the ecosystem underlies much work done in a wide variety of situations by geographers. Many of these have involved situations in which human impact is a significant element in the problem to be tackled. It may, of course, be remarked that this is moving away from the original concept as formulated by Tansley, where it is firmly set in a successional framework, and is directed towards the reconstruction of a situation which would obtain in the absence of anthropogenic influences. But that essentially Clementsian paradigm is now seriously questioned, and the whole development and use of the concept in ecology as well as in geography has been towards emphasising the relationships between biotope and biocoenosis in the abstract, without a specific time reference, still less a developmental or genetic framework of theory (McIntosh, 1983).

In this dimension, as a *concept defining an approach to problem-solving*, numerous examples might be cited. The work of Geertz in Indonesia was a pioneering study which pointed the way for much further work, especially in its demonstration of the usefulness and illumination of a basic pattern of thought (Geertz, 1963). As he writes:

> The ecological approach attempts to achieve a more exact specification of the relations between selected human activities, biological transactions, and physical processes by including them within a single analytical system, an *ecosystem*.

He saw in the concept the means to a more precise specification of the intimate set of relationships involved with a view to understanding both dynamics and change. This he achieved with reference to the two principal types of agricultural ecosystem in Indonesia, namely swidden, of shifting cultivation, whether true shifting agriculture or rotational fallowing, and sawah, or padi rice cultivation. He considered the twin traditional questions, 'How far is culture influenced by environment?', and, 'How far is the environment modified by the influence of man?', to be 'the grossest of questions' to which we 'can give only the grossest of answers: "To a degree, but not completely"'. To appreciate the point he is making is to accept a fundamental paradigm shift, in Kuhnian terms, or, in the framework of Lakatos' thought, a new research programme. The 'gross' questions pre-suppose that there exists, or that there can be certainly reconstructed, an environment which existed without the influence of man; and not only that, but also that it can be described or reconstructed in comparable detail and within the same categories as that which now exists, in order to effect real appreciation of the change that is attributable to man. And, conversely that, if we wish to reverse the question, we can reconstruct a culture which would have existed without environmental influence. In neither case can the prior reconstruction be achieved with any confidence or precision, except in extremely rare instances. Geertz thus proposes and practises a programme which focusses on the relationships, seeks to specify the dynamics of the present system, and then attempts to reconstruct the way in which that dynamics has itself developed through time. Geertz is working with fundamentally different categories. And it is this change which is most significant, and defines the ecosystematic approach in distinction from others. Nor is the paradigm necessarily confined to biogeography; it is applicable in other areas. More than that, in the kind of context in which Geertz applied it, it is an area which geographers are peculiarly qualified to explore, by both training and tradition. Geertz was of course an anthropologist!

The conceptual change is reflected in a number of studies which effectively bridge the physical/human dichotomy so implacably maintained in so much geographical teaching and research. Work on the stability of heavily cultivated islands of closed forest ecosystems in a matrix of savanna in Western Nigeria is one example (Moss and Morgan, 1977), which has been used to point to significant questions concerned with the relations of forest to savanna in West Africa in a more specifically plant-oriented context, thus echoing the perspective of Geertz in a rather different way (Moss, 1982). The general applications of ecosystem study to agriculture/environment studies has been reviewed by Harris in his usual perceptive and concise fashion (Harris, 1978), and to biogeography of plant communities by Harrison (1969). Numerous other examples might be cited, and consultation of the bibliography of almost any major book on biogeography, particularly those concerned to emphasise human activity in relation to the 'living landscape', as it has been termed by Paul Sears (1939 and 1969). As such, ecosystem studies, seen in the broad framework employed by Geertz, have applications in conservation, recreation studies, as well as in agricultural and land use investigations. It is, as has been suggested, that the principal impact of the underlying implications of the ecosystematic approach has been felt in geography (Simmons, 1966 and 1970). Paul Sears has remarked that: 'When the ecologist enters a field or meadow he sees not what is there but what is happening there'. Perhaps some geographers have, through the subtle influence of the concept of the ecosystem on their discipline, in a rudimentary way begun to see the same. Finally, in relation to this influence, it is relevant to point out that increasingly appropriate and sophisticated techniques of analysis are beginning to be developed by geographers to deal with problems of this kind in this perspective. A notable example is the work of Gatrell (1981) on the application of Q-Analysis in this general field (see also Beaumont and Gatrell).

Though the developments outlined in the previous paragraphs are perhaps the most profound and significant impact of developing ecosystem thinking on geography, with the potential for real long-term effects in the field as a whole, in conclusion it is necessary to point to a large area of study in which geographers have played a part, namely the use of the ecosystem as a descriptive tool. Vegetation mapping has been a respectable geographical pastime for many years (Kuchler, 1967). Now that the plant cover of the earth has, apart from a few significant individual locations, like parts of the Amazon Basin, and the remoter reaches of high latitudes and altitudes, been so modified, if not denuded, by man, such maps bear an antiquarian ambience. Nevertheless, the notion of the major biome, which was after all a derivative if the ecosystem concept, bids fair to become the new mapping unit on world or regional maps. The end-papers of the new major series, Ecosystems of the World, furnish world maps of the distribution of the ecosystem which is the concern of each major volume (Goodall (ed.), 1977). The important point seems to be that the focus of attention has been transferred from what is seen to be there to what is happening there; it is a recognition that, even when self-propagated vegetation has been removed, there are potential ecological relations implicit in the location (or biotope?) which it is necessary to recognise as essential properties of the place. The relation between the biotope and the biocoenoses which actually exist in particular places ranges from self-propagated successional communities, through manipulated ecosystems related to them, which are productive in an economic sense, to completely man-made ecosystems, which are the stuff of the modern agricultural industry, in some of which man himself manufactures the biotope (as in factory farming and hydroponics). But the relationships implied by the definition of a particular ecosystem remain a property of the place. It is at this point that distinctly geographical factors begin to attain some importance, notably scale, modification of land properties either deliberately or unintentionally, and so on. Such relationships can be used, where self-propagated plants are a significant component of the vegetative cover, to map properties of the biotope which may be of economic significance to man, as Cole (1970) and her co-workers have so successfully done in the case of mineral deposits. The same principle applies in the case of some interpretation in remote sensing, and indeed Cole has used such techniques in this way. The emphasis implied by the notion of the ecosystem, and enunciated by Geertz, in fact underlies much work of this kind. A similar argument might be mounted in relation to the reconstruction of past environments from various kinds of fossil evidence, including pollen, and even human artefacts. But, in so tracing the influence of ecosystem-type thinking, the content of the concept has been so attentuated as to render it almost trivial. Almost, but not quite, for it might be argued that the importance of Tansley's contribution was not that he invented a concept, but rather that he initiated a new train of thought which has made its mark in many areas of natural science, including geography. He provided a new focus of attention which transferred interest from 'gross' questions to the much more precise ones admitted by the new system to which attention was directed.

Conclusion

The argument has almost come full circle. The new Series already mentioned, in its specification of the Volumes which it will contain when it is complete, is clearly the global offspring of Tansley's major work (Table 4.3).

Table 4.3: Ecosystems described in Ecosystems of the World (Goodall, 1977 onwards)

I TERRESTRIAL ECOSYSTEMS

A. Natural Terrestrial Ecosystems

1. Wet Coastal Ecosystems
2. Dry Coastal Ecosystems
3. Polar and Alpine Tundra
4. Swamp, Bog, Fen and Moor
5. Shrub, Steppe and Cold Desert
6. Coniferous Forest
7. Temperate Deciduous Forest
8. Natural Grassland
9. Heath and Related Shrubland
10. Temperate Broad-Leaved Evergreen Forest
11. Maquis and Chaparral
12. Hot Desert and Arid Shrubland
13. Savannah and Savannah Woodland
14. Seasonal Tropical Forest
15. Equatorial Forest
16. Ecosystems of Disturbed Ground

B. Managed Terrestrial Ecosystems

17. Managed Grassland
18. Field Crop Ecosystems

19. Tree Crop Ecosystems
20. Greenhouse Ecosystems
21. Bioindustrial Ecosystems

II. AQUATIC ECOSYSTEMS

A. Inland Aquatic Ecosystems

22. Rivers and Stream Ecosystems
23. Lake and Reservoir Ecosystems

B. Marine Ecosystems

24. Intertidal and Littoral Ecosystems
25. Ecosystems of Estuaries and Enclosed Seas
26. Coral Reefs
27. Ecosystems of the Continental Shelves
28. Ecosystems of the Deep Ocean

C. Managed Aquatic Ecosystems

29. Managed Aquatic Ecosystems

But a comparison of detailed content in those volumes which have so far been published with that of Tansley's study shows clearly how far research, both theoretically and empirically, has developed since the 1930s. One observation may be made both with respect to 1939 and 1980-83, however, namely that the geographical contribution to both is negligible. Tansley quotes a few geographers. The contributors to the new Series in the published volumes so far who claim or acknowledge a geographical, or even environmental science allegiance, can almost be counted on the fingers of one hand: the geographers in fact can be so counted, and one of them is the Editor of this volume of essays! The many dozens of remaining contributors claim a biological link in research. Yet here we have a field of interest which geographers have not infrequently claimed as their own, or more modestly claimed to be a vital development in their discipline if not their exclusive province.

Perhaps the reasons in both cases are not hard to find, and similar: a phenomenon which might be succinctly described as paradigm isolation. In the 1930s and 1940s the dominant approaches in geography were 'regional' and 'morphogenetic'; ecological approaches were 'relational' and 'populational'; and were rapidly moving along the roads so dictated, which were essentially divergent. Biogeographers were thus torn between loyalty to geography and the necessity to think and move ecologically. In the 1970s and 1980s the dominant approaches in geography are 'spatial analytical', 'systematic physical', 'systematic human', and (still!) 'morphogenetic'; those in biogeography are still 'relational' in that they avoid 'gross' questions as they are inevitably implied by the physical/human dichotomy, seeking more refined questions and more precise answers. The biogeographer is still torn (or at least one is!) between a loyalty to the discipline in which he was trained and the way his interest demands he should go. He frequently finds more in common with those outside the laager of his institutional loyalty than with those within, simply because he shares a general paradigm with the outsiders and is isolated in his allegiance to it amongst those within geography. And even the outsiders are understandably suspicious: with justification no doubt if he is seen to be incubine, in either the medieval or the biological sense of the solecism of the introductory reviewer.

Perhaps, therefore, the ecosystematic approach leads us away from what is seen to be the mainstream of geographical progress, as it is seen by the majority of professional geographers, in which case it might be argued that ecologically minded geographers should change camps, or find another approach more consistent with the prevailing ethos in the field of geography. It is at least worth reflecting as to why, when in the 1960s so much lip service was paid to the use of biological concepts in geography, no major movement in research developed from it, even though the implicit influence on geographical thinking was in some ways not at all insignificant. What cannot be in doubt is the fact that biologically trained ecologists are surely and rightly and competently taking over the ground which geographers have vacated.

REFERENCES

Beaumont, J.R. and Gatrell, A.C. (1982), An Introduction to Q-Analysis, Catmog 34, Geo Abstracts, Norwich.

Bennett, R.J. and Chorley, R.J. (1978), Environmental Systems: Philosophy, Analysis and Control, Methuen, London.

Chapman, R.J. (1931), Animal Ecology with special reference to Insects, McGraw Hill, New York, USA.

Chorley, R.J. (1973), 'Geography as human ecology', in Chorley, R.J. (ed.) (1973), Directions in Geography, Methuen, London, pp. 155-69.

Cody, M.L. and Diamond, J.M. (eds.) (1975), Ecology and the Evolution of Communities, Belknap Press of Harvard University, Cambridge, Mass., USA.

Cole, M.M. (1970), Biogeography in the Service of Man (Inaugural Lecture, Bedford College, University of London).

Cox, C.B., Healey, I.N. and Moore, P.D. (1976), Biogeography: an Ecological and Evolutionary Approach, 2nd edition, Blackwell, Oxford, UK.

Dansereau, P. (1957), Biogeography: an Ecological Perspective, Ronald, New York, USA.

van Dyne, G.M. (ed.) (1969), The Ecosystem Concept in Natural Resource Management, Academic Press, New York, USA.

Gatrell, A.C. (1981), On the Geometry of Man-Environment Relations, with special reference to the Tropics - a Speculation using Q-Analysis, Discussion Paper No. 13, Department of Geography, University of Salford, UK.

Gause, G.F. (1934), The Struggle for Existence, Williams and Wilkins, Baltimore, USA.

Geertz, C. (1963), Agricultural Involution: the Processes of Ecological Change in Indonesia, University of California Press, Berkeley, USA.

Goodall, D.W. (ed.-in-chief) (1977-), Ecosystems of the World (29 vols., Elsevier, Amsterdam, 1977-). (A dozen or so volumes have now been published.)

Haggett, P. (1965), Locational Analysis in Human Geography, Arnold, London, UK.

Harris, D.R. (1978), 'The environmental impact of traditional and modern agricultural systems' in Hawkes, J.G. (ed.) (1978), Conservation and Agriculture, Duckworth, London.

Harrison, C.M. (1969), 'The ecosystem and the community in biogeography' in Cooke, R.U. and Johnson, H.J. (eds.) (1969), Trends in Geography: an Introductory Survey, Pergamon, Oxford, UK.

Hutchinson, G.E. (1948), 'Nitrogen in the biogeochemistry of the atmosphere', Annals of the New York Academy of Sciences, **50**, 221-46.

Krebs, C.J. (1978), Ecology: the Experimental Analysis of Distribution and Abundance, 2nd edition, Harper and Row, New York, USA.

Küchler, A.W. (1967), Vegetation Mapping, Ronald, New York, USA.

Levin, S.A. (ed.) (1976), Ecological Theory and Ecosystem Models, The Institute of Ecology, Indianapolis, USA.

Lindeman, R.L. (1942), 'The trophic-dynamic aspect of ecology', Ecology, **23**, 399-418.

Lotka, A.G. (1925), Elements of Physical Biology, Williams and Wilkins, Baltimore, USA.

MacArthur, R.H. (1972), Geographical Ecology, Harper and Row, New York, USA.

May, R.M. (1976), Theoretical Ecology, Saunders, Philadelphia, USA.

McIntosh, R.T. (1983), 'The background and some current problems of theoretical ecology' in Saarinen, E. (ed.) (1983), Conceptual Issues in Ecology, Reidel, Dordrecht, Netherlands, pp. 1-62.

Morgan, R.K. (1980), The Application of the Systems Approach in Geographical Research, Occas. Publ. No. 10, Department of Geography, University of Birmingham, UK.

Morgan, W.B. and Moss, R.P. (1965), 'Geography and ecology: the concept of the community and its relationship to environment', Ann. Ass. Amer. Geogrs., **55**, 339-50.

Moss, R.P. (1982), Reflections on the Relations between Forest and Savanna in Tropical West Africa, Discussion Paper No. 23, Department of Geography, University of Salford, UK.

Moss, R.P. and Morgan, W.B. (1967), 'The concept of the community: some applications in geographical research', Trans. Inst. Brit. Geogrs., **41**, 21-32.

Moss, R.P. and Morgan, W.B. (1977), 'Soils, plants and farmers in West Africa', in Garlick, J.P. and Keay, R.W.J. (eds.) (1977), Human Ecology in the Tropics, Symposium XVI, Society for the Study of Human Biology, 2nd edition, Taylor and Francis, London, pp. 27-77.

Newbigin, M.I. (1936), Plant and Animal Geography, Methuen, London.

Odum, E.P. (1971), Fundamentals of Ecology, 3rd edition, Saunders, Philadelphia, USA.

Patten, B.C. (ed.) (1971), Systems Analysis and Simulation in Ecology, Academic Press, New York, USA.

Sears, P. (1939), Life and Environment, Columbia University Press, New York, USA.

Shimwell, D.W. (1983), Review of Tivy, J. and O'Hare, G. (1981), Human Impact on Ecosystems (Oliver and Boyd, Edinburgh), J. Ecol., **71** (1), p. 346.

Simberloff, D. (1980), 'A succession of paradigms in ecology: essentialism to materialism and probabilism' in Saarinen, E. (ed.) (1980), Conceptual Issues in Ecology, Reidel, Dordrecht, Netherlands.

Simmons, I.G. (1966), 'Ecology and land use', Trans. Inst. Brit. Geogrs., **38**, 59-72.

Simmons, I.G. (1970), 'Land use ecology as a theme in biogeography', Canadian Geogr., 14, 309-22.

Stoddart, D.R. (1965), 'Geography and the ecological approach: the ecosystem as a geographic principle and method', Geography, **50**, 242-51.

Tansley, A.G. (1935), 'The use and misuse of vegetational terms and concepts', Ecology, **16**, 284-307.

Tansley, A.G. (1939), The British Islands and Their Vegetation, Cambridge University Press, Cambridge, UK (p. 228).

Tivy, J. and O'Hare, G. (1981), Human Impact on Ecosystems, Oliver and Boyd, Edinburgh, UK.

Volterra, V. (1925), 'Variazioni e fluttuarzioni del numero d'individui in specie animali conviventi', Mem. Acad. Lincei, 231-113. A translation is to be found in Chapman, R.N. (1931), Animal Ecology with special Reference to Insects, McGraw Hill, New York, USA.

Watts, D. (1971), Principles of Biogeography: an Introduction to the Functional Mechanisms of Ecosystems, McGraw Hill, London.

Wiener, N. (1948), Cybernetics, or Control and Communication in the Animal and the Machine, Cambridge, Mass., USA.

FURTHER READING

Cox, C.B., Healey, I.N. and Moore, P.D. (1976), Biogeography: an Ecological and Evolutionary Approach, 2nd edition, Blackwell, Oxford, UK.

Krebs, C.J. (1978), Ecology: the Experimental Analysis of Distribution and Abundance, 2nd edition, Harper and Row, New York, USA.

Odum, E.P. (1971), Fundamentals of Ecology, 3rd edition, Saunders, Philadelphia, USA.

Simmons, I.G. (1979), Biogeography: Natural and Cultural, Arnold, London.

Simmons, I.G. (1981), The Ecology of Natural Resources, Arnold, London.

Watts, D. (1971), Principles of Biogeography: an Introduction to the Functional Mechanisms of Ecosystems, McGraw Hill, London.

Chapter 5

VEGETATION ANALYSIS

D. W. Shimwell

Introduction

Vegetation analysis has five main objectives, namely:

(i) an understanding of the plant communities of an area, country or continent;

(ii) how these communities are related to one another;

(iii) how they relate to and express their environments;

(iv) how individual plant species are distributed within these communities; and

(v) how the communities develop and function as organised living systems.

The disciplines of plant geography and plant sociology have traditionally sought to answer most of these questions and, only comparatively recently, have techniques of vegetation anslysis been adopted as a possible study within biogeography. Even then, in most modern biogeographical treatise, the subject is given a scant, often perfunctory treatment. Alternatively, Dansereau (1957) was probably the first biogeographer to give the topic a central focus. The year 1971 saw the publication of three methodological reviews (Shimwell, 1971; Dickinson et al, 1971 and Harrison, 1971) which suggested that methods of vegetation analysis might find a general application in biogeography; this was later emphasised by the recognition of vegetation analysis as one of three major approaches in plant geography by Kellman (1975) and the publication of a short textbook by Randall (1978) on the subject. But by far the greatest advances and most significant contributions to vegetation analysis have been made by practitioners with a training in biological sciences and, thus, the first question that needs to be posed is 'Exactly what aspects of vegetation analysis are of most importance to the biogeographer?' It is impossible to obtain a general consensus of opinion by the analysis of modern textbooks on biogeography or from contributions to biogeographical journals. _Quot homines tot sententiae_. There are those workers who will argue that the whole realm of vegetation study falls within the nature and scope of biogeography. Ideally, this is the true situation; practically, it is impossible within the constraints of even a specialist degree in biogeography. It would seem a more profitable exercise to consider selected aspects of vegetation analysis which introduce the student to basic tenets, properties of vegetation which present themselves for analysis, simple field methods of description and some comprehension of the major world vegetation types. To this end, the following sections discuss the basic approaches to vegetation, describe properties of vegetation which have been used for analysis, outline two popular descriptive methods and review systems of classification for world vegetation types.

Description, Classification and Ordination - Some Basic Premises

The publication in 1973 of Part V of the _Handbook of Vegetation Science: Ordination and Classification of Communities_ under the general editorship of Professor Robert H. Whittaker (1973) marked the culmination of a decade of discussion, dissection and interchange of ideas on how and why vegetation could and should be described, analysed and classified. There had for some years been a growing awareness of the diversity of approaches from even the most fundamental of viewpoints: whether vegetation could be classified into discrete entities or whether vegetation, or 'plants collectively' formed a continuum of variation along one or more major environmental gradients. Moreover, controversy reigned in the debates concerning which properties of vegetation should be measured for analysis. How, if at all, should these properties be weighted? How should the sample measurements be combined and manipulated to present an accurate picture of the characteristics of the vegetation? Finally, why does vegetation need to be analysed and categorised? Should vegetation analysis be a preliminary stage to detailed ecological interpretation or is it an objective in itself? Does it merely serve to foster a deeper understanding of the developments, methodologies and goals of vegetation science, an ever-expanding subject with approaches ranging from traditional plant sociology, as a descriptive science of communities readily observable in nature, to the more abstract approaches that seek a detached understanding derived from a study of quantitative relationships that might be represented as models, diagrams and equations? Several authors had sought to draw together the various traditions by the publication of review articles and monographs (Whittaker, 1962, 1967; McIntosh, 1967), while the few attempts at a comprehensive overview of vegetation science had generally been found wanting because the task proved beyond the competence and expertise of just one (or two) author(s) (e.g. Cain and Castro, 1959; Küchler, 1967; Pielou, 1967; Whittaker, 1970; Shimwell, 1971, _inter alia_). The volume edited by

Whittaker (1973) comprising contributions from nineteen eminent vegetation scientists succeeded where other works had failed and provided a fundamental reference framework and platform for the integration of a variety of related disciplines such as plant geography, palynology, soil science and wildlife management - some of the topics which fall within the broad compass of what is now known as biogeography. But just what aspects of vegetation science are of the greatest importance to modern biogeography? What properties of vegetation do biogeographers need to understand, measure and analyse, and of what value is an understanding of methods of gradient analysis and classification of vegetation? These are the three major topics to which the following review is addressed. But first, some basic concepts need to be elaborated upon.

The meaning of the word vegetation is familiar either in the form of its definition from the Oxford English Dictionary - 'plants collectively' - or in the form of a more complex definition supplied by the ecologist, for example that of Tansley (1939): 'Plants are gregarious beings, because they are mostly fixed in the soil and propagate themselves largely in social masses, either from broadcast seeds (or spores), or vegetatively by means of rhizomes, runners, tubers, bulbs, or corms: sometimes by new shoots ('suckers') arising from the roots. In this way they produce vegetation, as plant growth in the mass is conveniently called, which is actually differentiated into distinguishable units or plant communities'. But, to elaborate, let us begin with the descending definition spiral typical of many sciences. Simply, on a regional basis, vegetation may be regarded as being composed of all the different types of plant communities within the region. The plant community in turn is not merely a random aggregation of plants but an organised complex with a typical floristic composition and morphological structure which have resulted from the interaction of species populations through time.

The term species population is <u>the</u> fundamental vegetation characteristic, and it is the continuous flux of species populations which makes the vegetation and the community so variable. The distribution of each species population is affected by both interspecific and intraspecific factors which are directly related to the genotypic adaptability or phenotypic plasticity of the species. Each species population has a potential optimum size which is affected by internal and external interference so that the population size and space is invariably modified according to the competitive, reproductive and tolerance capacities of the individuals, relative to the magnitude and type of interference. This is the <u>first</u> level of sociological organisation. Secondly, every species population has certain essential requirements of its physical environment or habitat. Several species have pronounced climatic or edaphic requirements and will grow only under such conditions. Every species has a characteristic requirement for a range of external physical environmental conditions for growth, i.e. has a certain ecological amplitude. Different species have different ecological amplitudes and this phenomenon results in variations in the specific composition of vegetation - the <u>second</u> level of sociological organisation.

Studies in vegetation science seek ways of communicating and collating information concerning the two levels of sociological organisation in three contexts, namely, community characteristics, species populations and environmental factors. These sets of information are collected, abstracted and presented according to two broad concepts adopted in vegetation research: classification and ordination. Classification groups together a number of samples representing plant communities into a class or abstract unit on the basis of shared characteristics. Such units of classification are best referred to as community-types, principally to avoid the confusion of nomenclature engendered by the wide variety of classificatory methods. Once the vegetation samples have been classified into a community-type, the community characteristics, species composition and environmental factors may be determined as a basis for relating community-types to environment and landscape. Classification recognises discontinuities in vegetation and environmental characteristics; the alternative approach has the basic tenet that samples from plant communities may be arranged in a continuum of variation in relation to one or more environmental gradients or axes. This process is known as ordination, the basic technique of gradient analysis wherein changes in species populations and community characteristics are related to changes in the environment.

Neither ordination nor classification should be viewed as objectives in themselves. They are simply means for ordering information, and for describing inter-relationships and trends between and within samples that are representative of the plant communities which form vegetation. Classification and ordination are not mutually exclusive approaches to vegetation study. In many ways they are complementary and may be combined effectively to aid the interpretation of relationships between plant communities and their relationships to environment.

Some Basic Properties of Vegetation

The basic methodology of all vegetation analysis involves the collection of community samples, relevés or Aufnahmen which are considered to be representative of, and to reflect the general characteristics of the vegetation under study. Such samples should be large enough in area to represent effectively the composition of the plant community, and they should be homogeneous, efficient and appropriate in the sense that there should be no major trend of change in community structure or composition and that the most important and appropriate information should be collected for analysis. The perception of what homogeneity means and exactly what is the most important and appropriate information for collection has varied widely with the origin, scale, aim and scope of research project and inevitably a universal consensus of opinion is difficult to obtain. Most modern researchers, however, would agree that the three essential components of the sample should be (i) a list of plant species present in the sample area, usually a quadrat or transect, (ii) some indication of the relative importance of these species, and (iii) additional information on community structure, soil and other environmental factors. The most widely used and arguably the most efficient of all types of samples are those based upon the relevé of the Braun-Blanquet school of plant sociology (1932) which, in a variety of modified forms, have been used as a data base for many methods of analysis, ordination and classification, throughout continental Europe and the British Isles. Alternatively, the most detailed investigations of gradient analysis developed in temperate North America have evolved more quantitative sampling methods aimed to produce measurements which express the importance values of each species in a community. Such importance values are commonly derived from five properties of individual species, namely (i) density: the number of individuals of a species per quadrat sample; (ii) cover: the percentage of quadrat sample area above which foliage of a given species occurs; (iii) basal area: the area occupied by cross-sections of stems at 1.3 m above the ground per quadrat sample; (iv) frequency: the percentage of small subquadrats within a larger quadrat, in which a given species occurs; and (v) biomass: the total mass of a species at a given time per quadrat sample. In contrast, studies in tropical environments where problems of time, size, familiarity and access rank high as limiting factors in vegetation analysis or in the description and representation of the broader vegetation zones on a global scale, more simple analyses, using one or more specific features or broad, general characteristics of structure and physiognomy, find favour.

All three approaches rely, to a greater or lesser extent, on the recognition that vegetation possesses two principal properties: floristic composition and physiognomy (Beard, 1973). The former property and the analysis of vegetation by floristic methods involves a detailed knowledge of most of the species of a particular flora and their enumeration in a quadrat sample, simply as a mere record of presence or absence. On the other hand, a vegetation sample may be analysed according to its gross physiognomic characteristics, ignoring actual specific composition. Physiognomy is probably best defined as 'the form and function of vegetation; the appearance of vegetation that results from the life-forms of the predominant plants' (Cain and Castro, 1959), although other authors have distinguished variously between physiognomic and structural, functional and textural properties (Fosberg, 1967; Shimwell, 1971; Barkman, 1979). Whatever the finer points of distinction between such characteristics, the approach to vegetation physiognomy involves, firstly, the selection of useful parameters, secondly, the selection of an appropriate subdivision of the range of values of each parameter and, thirdly, the actual measurement of the parameters. What parameters of vegetation physiognomy present themselves for analysis?

Leaf size. The leaf-size spectrum for the characterisation of a vegetation type was a brain child of Raunkiaer (1910 and 1934), who believed that leaf-size was primarily a measure of climatic effect, or the effect of the aerial environment in general. Raunkiaer's main concern was the structural adaptations of plants to 'the water problem' and this obvious morphological character was viewed as the functional diminution of the transpiring surface as an index of adaptation to prevailing climatic conditions. The six Raunkiaerian leaf-size classes are shown in Table 5.1 along with the major modifications of the original system as proposed by Webb (1959) and Barkman (1979). The actual tedium of Raunkiaer's method of tracing the leaf outline on to centimetre squared paper has been alleviated by a much simpler procedure developed by Cain and Castro (1959). This involves the measurement of the length and maximum breadth of leaf lamina followed by the derivation of the lamina area by the calculation of two-thirds of the rectangular length x breadth area. This straightforward procedure works adequately in most circumstances but not for compound leaves or leaves of tropical trees with drip-tips, measurement of which is generally omitted.

Table 5.1: Leaf-size classes

Leaf class	Size range (mm^2)		
	Raunkiaer (1934)	Webb (1959)	Barkman (1979)
(Bryophyllous)	-	-	4
Leptophyllous	25	-	4-20
Nanophyllous	25-225	-	20-200
Microphyllous	225-2,025	225-2025	200-2,000
(Notophyllous)	-	2,025-4,500	-
Mesophyllous	2,025-18,225	4,500-18,225	2,000-10,000
Macrophyllous	18,225-164,025		10,000-50,000
Megaphyllous	> 164,025		> 50,000

In the methods of Raunkiaer and Webb the class limits are calculated using 25 mm as the upper limit of the leptophyll class and the subsequent multiplication of this figure by 9, e.g. 9 x 25 = nanophyll, 9^2 x 25 = microphyll, etc. Barkman modifies Raunkiaer's (1918) original class sizes.

Leaf consistency, like leaf-size, is a characteristic which is also viewed as being of significance with respect to conservation of moisture by plants. Many authors recognise four main types, namely:

malacophyllous - thin, filmy and easily withering
orthophyllous - normal, glabrous or felty
sclerophyllous - thick, dry and leathery
succulent

The former type are typical of high atmospheric humidity; sclerophyllous leaves indicate aridity of climate or a lack of soil nitrogen while succulent leaves are most common in either arid environments or in association with high salt concentrations in soil and water.

Leaf orientation. The orientation of leaves in space as a potential feature for analysis has found some recent application in the work of Parsons (1976) and Barkman (1979) who recognise its importance in affecting the vertical distribution of light intensity and the evapotranspiration in vegetation. Of the two components of leaf orientation, exposition and inclination, the latter is the more important and is usually measured in degrees in relation to the horizontal plane. It is a most tedious process to measure leaf angles systematically and Barkman (1979) has therefore suggested the assignation of whole plants to one of eleven classes by observation of dominant inclinations. The classes are as follows:

Sphaerical (s) - all inclinations $+90^{o}$ to -90°
Hemisphaerical (hs) - all inclinations from $+90^{o}$ to 0°
Erect (e) $+90^{o}$ to $+60^{o}$
Erecto-patent (ep) $+70^{o}$ to $+30^{o}$
Patent (pa) $+40^{o}$ to $+10^{o}$
Spreading (sp) lower half (ep), upper half (h)
Arcuate (ar) lower half (ep), upper half hanging (-30° to -60°)
Recurved (r) lower half (h), upper half (pe) (-40° to -90°)
Horizontal (h) $+20^{o}$ to -20°
Decumbent (d) 0° to -50°
Pendant (pe) -40° to -90°

Although this classification makes use of a genotypic character, inclination can also be affected phenotypically by such factors as age of leaves, climatic conditions, light intensity and trampling, and all uses of this property should acknowledge these facts.

Growth forms and life forms. The much misunderstood, misinterpreted topic of growth form and life form has encouraged a wealth of literature and a vast number of systems of description, many of which are an end in themselves. The topic is extensively reviewed by Shimwell (1971) and there seems to be little merit in discussing further the relative importance or popularity of the various methods. Growth form is simply the vegetative form of the plant body, those morphological characteristics that determine the general architecture and appearance of a plant; life-form is specifically 'the morphological expression of the adaptation of organisms to their environment', or their epharmony (Barkman, 1979). Of the pure growth-form systems of plant classification that of Dansereau and Arros (1959) is perhaps the most widely used and well-known in the production of 'Danserograms'. Five main types are recognised: erect woody plants, lianas, herbs, bryoids and epiphytes, and these types are subdivided according to four independent criteria: function, in terms of seasonal leaf periodicity; height; form and size of leaves; and leaf texture. The full characteristics of this major contribution to biogeography are shown in Figure 5.1. A second methodology which has had a profound influence on biogeographical research is the life form classification of Raunkiaer (1910, 1934). In the selection of characters to determine life form, Raunkiaer lists three prerequisites: the character must be fundamental in the relationship of the plant to climate; it must be easily recognisable in the field; and it must represent a single aspect of the plant which lends itself to a comparative statistical treatment of the vegetation of different regions. He thus chose, as the basis for characterisation, the adaptation of the plant to survive the unfavourable season, or in other words, the position on the plant of the vegetative buds. Fifteen main types of life form, divided into five broad categories: phanerophytes, chamaephytes, hemicryptophytes, cryptophytes and therophytes, are recognised. Raunkiaer's scheme was later modified by Braun-Blanquet (1932) by the introduction of a series of Latin binomials, and this system was in turn extended by Ellenberg (1956) and elaborated by Ellenberg and Müller-Dombois (1967). Of all the types of life form classification, the most widely applied is that of Ellenberg (see Shimwell, 1971: Appendix I; Box, 1981).

Vertical structure. The vertical arrangement of species into layers or strata has been widely used as a primary descriptive tool in many preliminary vegetation analyses. A cursory examination of any forest of the north temperate zone reveals a stratification composed of five layers: a canopy of trees, an understory of shrubs and saplings, a herb or field layer, a moss or ground layer and the subterranean root systems and soil microflora layer. This structure is constant for most types of forest and woodland with the exception of tropical rain forest which frequently has a layer of emergent crowns above the main canopy. It was tropical forest studies which gave rise to the profile diagram method of analysis. Based on narrow sample strips of forest commonly of the dimensions 60 x 8 m and known as clear felling plots, scale profile diagrams are constructed. Vegetation lower than an arbitrary height is cleared and the following features of the emergent canopy and undergrowth trees documented: location, trunk diameter, total height, height to first main branch, lower limit of crown and crown width. The classification of the tree species in the plots has concentrated upon the height of stems and inevitably there have been several interpretations placed upon what is a tall, medium or low tree. The five main schemes of tree and shrub height classes are compared in Table 5.2.

Horizontal structure. The spatial distribution of individuals of the component species of a vegetation sample confer a pattern on each species and give the vegetation as a whole a horizontal structure. The largest scale of pattern which depends upon the overall morphology of a vegetation type is whether the individual component plants are so spaced as to form a continuous lateral contact. This situation is known as closed vegetation and in stratified vegetation,

Table 5.2: A comparison of tree and shrub height classes of various authors

Metres	Raunkiaer	Küchler	Dansereau	Cain and Castro	Heyligers	Metres
35						
	Mega-	t tall	Tt Trees tall		Tv Very tall trees	
30						30
25				Tall tree stratum	Tt tall	
20	Meso-	m Medium	Tm Trees medium			20
15						
					Tm	
				Intermediate tree stratum		
10						10
8			Tl trees low			
	Micro-	l Low	Ft Shrubs tall	Low tree- High shrub	Tl low Sv very tall shrub	
5						5
3					Tp - pygmy tree	
2		s	Fm	Low shrub		
					St - tall shrub	
1	Nano-					
0		z	Fl		Sm,l,p - shrubs	

at least one or all layers are touching or overlapping. Where there is space between individuals which can be colonised and where the space is not more than twice the diameters of the predominant individuals, the term open vegetation is applied. Sparse vegetation refers to any situation where there is a greater amount of ground space than in the previous case, so that substrate, not vegetation dominates the landscape. For these three types, Fosberg (1967) proposed the term primary structural group, and used them as basic categories in his key to the major vegetation types of the world. Other authors have used more categories for the pattern of both individual species and vegetation generally under the general head of coverage or cover. Dansereau (1951), for example, uses four coverage letters: b-barren, i-discontinuous, p-grouped, c-continuous, while both Heyligers (1965) and Küchler (1967) adopt six categories which are based on actual estimated percentage-cover values.

The methods of analysis used by Heyligers (1965) were devised specifically for the interpretation of vegetation patterns from aerial photographs. Since this early work, a considerable amount of research has been undertaken into the possible uses of remote-sensing techniques in vegetation analysis (Curtis, 1978; Curran, 1980) and a variety of systems of interpretation have been evolved using three main properties of photographs: pattern, texture and, in the case of infrared colour photographs, tone. Good examples of methods of interpretation are given by Lillesand and Kiefer (1979), ranging from simple land-use/land-cover mapping to the identification of forest tree species, using crown shape, size and texture, and the distinction between aquatic vegetation types using the pattern, texture and tone of infrared images. Additionally, developments in remote sensing in the past decade have made possible the estimation of the amount of vegetation in terms of biomass, cover and density through an understanding of the spectral reflectance of different plant species and vegetation types (Curran, 1980).

Abundance, cover and dominance. Horizontal structural patterns are determined by the abundance, coverage and dominance of certain species. At the coarsest scale of analysis, adjectives and initial letters may be used to describe a certain type of pattern for the vegetation type as a whole or for the individual species; at the finest scale, actual measures of cover may be achieved by detailed point-sampling (Greig-Smith, 1964) to give an insight into the quantitative floristic composition of the vegetation. In common usage in plant sociology are scales-of-cover values which, when assigned to the component species of a vegetation type, in some ways reflect their importance or dominance. Several scales have been widely applied, but all differ slightly in cover class delimitation (Shimwell, 1971). Today, the two most common scales of cover-abundance are those of Braun-Blanquet (1932) and Domin (Evans and Dahl, 1955). These scales are shown in Table 5.3 along with the 2m, 2a, 2b modification proposed by Barkman et al (1964) and the ordinal transformation of both scales (Westhoff and Maarel, 1973) for the purposes of species weighting and the use of the data in a variety of objective analytical

Table 5.3: The cover-abundance scales of Braun-Blanquet and Domin and their ordinal transformation

Braun-Blanquet		ordinal transformation	Domin	
			+	one individual, reduced vigour
r	one or few individuals	1	1	rare
+	occasional and less than 5% of total plot area	2	2	sparse
1	abundant and with very low cover, or less abundant but with higher cover; in any case less than 5% cover of total plot area	3	3	4%, frequent
2	very abundant and less than 5% cover, or 5-25% cover of total plot area			
	2m very abundant	4		
	2a 5-12.5% cover, irrespective of number of individuals	5	4	5-10%
	2b 12.5-25% cover, irrespective of number of individuals	6	5	11-25%
3	25-50% cover of total plot area, irrespective of number of individuals	7	6 7	26-33% 34-50%
4	50-75% cover of total plot area, irrespective of number of individuals	8	8	51-75%
5	75-100% cover of total plot area, irrespective of number of individuals	9	9 10	76-90% 91-100%

and synthetic methods. Arising from the assignment of cover values is the property of the dominance of individual species over others in the various strata of the vegetation type. Dominants are basically those species which have the greatest total biomass. Much has been written on the topic of dominance and many misunderstandings have been perpetrated. For the basic aspects of vegetation analysis, it is probably adequate to adopt the definition of Cain and Castro (1959) for the term ecological dominance. In general ecology, the species with the greater foliage cover in each stratum is used as an indication of ecological dominance or, in forestry, the foliage cover or crown class are used as measures of community dominance. Foresters commonly classify the component forest trees as dominant, co-dominant, intermediate and suppressed, based on their crown height with reference to the strata of the forest. Dominance, in this case, refers to whether their crowns are in the superior forest stratum, irrespective of the contribution to stratum cover. Intermediate trees are those whose crowns are partially overtopped by the canopy, and suppressed trees are those forming a stratum below the canopy. Basal area of trees: the cross-sectional area at a height of 1.3 m, is also used as a measure of ecological dominance.

A Choice of Descriptive Methods

The biogeographer is mainly interested in two aspects of vegetation description: the physiognomy of the vegetation, and its floristic composition. Practically, for each stand of vegetation chosen for analysis, a selection of field methods which best describe and facilitate the communication and comprehension of the essential characteristics of the vegetation by other workers, must be carefully made. To this end, the following descriptive methods are recommended. The process of description begins with the delimitation of homogeneous stands of vegetation, using physiognomic and structural features, and proceeds to a method of floristic description to characterise each community-type thus recognised.

A physiognomic descriptive system. Shimwell (1971) and Beard (1973) have categorised two distinct schools of physiognomic descriptive systems, those of British Commonwealth ecologists and those of the North American plant geographers, exemplified by the works of Kuchler (1949, 1967) and Dansereau (1951, 1957). From this latter school, the most appropriate method for use by the biogeographer is that of Dansereau (1951) which combines the vegetation characteristics of life-form, size, function, leaf-size and shape, leaf texture and coverage (Figure 5.1) in a symbolic representation of a typical vegetation profile or a danserogram.

A floristic descriptive method - the relevé. As previously mentioned, perhaps the most efficient of all types of plant community sample are those based upon the relevé of traditional plant sociology which lists the species present in the sample, assesses their relative importance and gives additional data on community structure, soil and other environmental factors. There are many variations on the relevé theme. Figure 5.3 provides an illustration of the type used by the National Vegetation Classification scheme in a survey of the

Figure 5.1: Physiognomic symbols of Dansereau (1951). Data are commonly collected from a belt transect or quadrat and a semi-quantitative estimate of the relative abundance of each plant type in each stratum is made and presented in the danserogram (Figure 5.2)

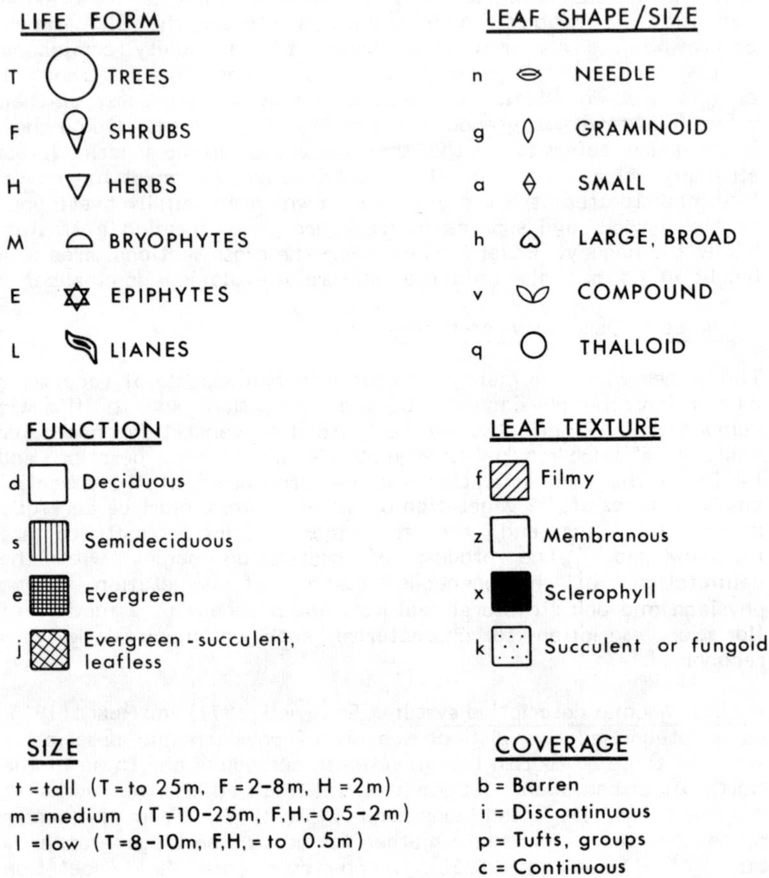

British Isles between 1976 and 1982. All species of vascular plant, bryophytes and macrolichens are listed, and quantitative presence is assessed using the Domin Scale of cover/abundance (see Table 5.3).

Figure 5.2: Examples of Danserograms. A descriptive formula consisting of six letters (one from each of the major categories) is composed for each stratum of the vegetation. Dansereau et al (1966) modified the system of presentation in a later methodological critique but, as the use of this latter type of danserogram has been limited, the former traditional method is to be preferred

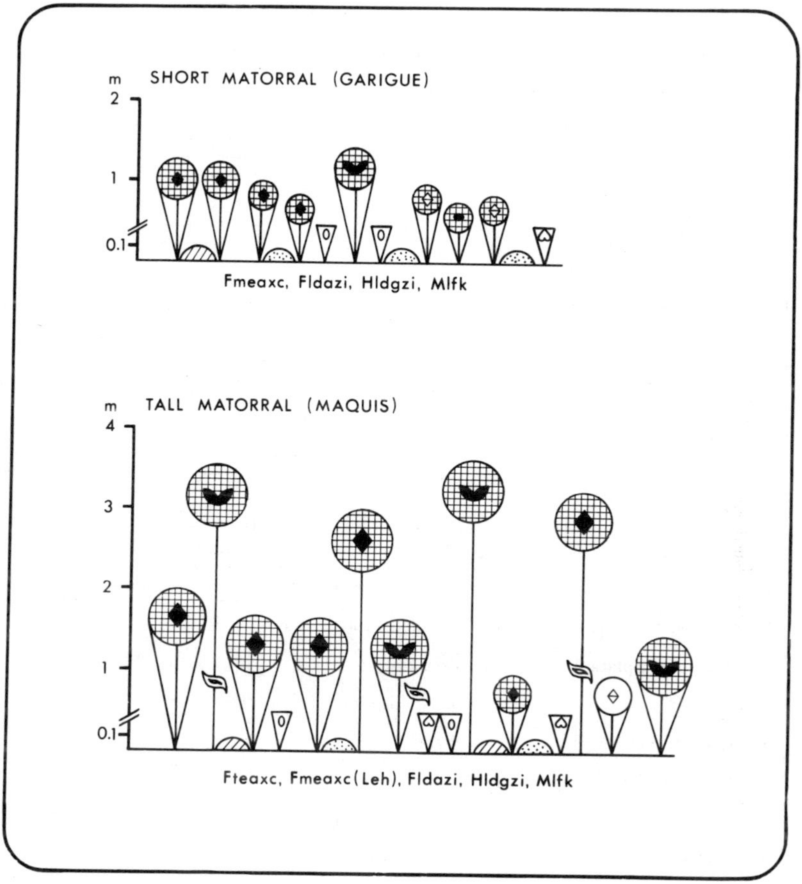

The components of layered vegetation are listed separately for each layer, and a species contributing to the cover of more than one layer is recorded for each layer; bare rock, soil, litter and open water are

Figure 5.3: An NVC Record Sheet - an example of a releve

Location	Grid reference	Region	Author
Saltwells Wood LNR, Dudley	33/933870	Mid	DWS

Site and vegetation description

Species-rich, basiphilous, mixed deciduous woodland, variable canopy with combinations of oak, ash and sycamore; understorey hazel, with evidence of coppicing; ungrazed; grading into Oak-*Holcus mollis* woodland on flatter surfaces.

Date	Sample no.
23/06/82	11
Altitude 110 m	Slope 20°
Aspect 360°	Soil depth 50+ cm
Stand area 80 x 20 m	Sample area 10 m x 10 m

Layers: mean height
20 m 4 m 0.3 m

Layers: cover
80% 60% 60% 20%

Geology
Downtonian Red Marl (Devonian)

Species list:

Canopy
6 Quercus robur
7 Fraxinus excelsior
6 Acer pseudoplatanus

Understorey
7 Corylus avellana
4 Acer pseudoplatanus

Field
5 Hedera helix
3 Brachypodium sylvaticum
6 Rubus fruticosus
4 Endymion non-scriptus
3 Deschampsia caespitosa
2 Dryopteris filix-mas
2 Urtica dioica
2 Festuca gigantea
1 Heracleum sphondylium
3 Mercurialis perennis
2 Galium aparine

Bryophytes
5 Eurynchium praelongum
2 Minium hornum

Tree seedlings
2 Fraxinus excelsior

Soil profile:

Moist, silty-loam brown-earth of Bromyard Series

pH 6.5

entered at the end of each list. Data on the site and vegetation description include notes on community structure, spatial relationships with neighbouring communities, temporal relationships with other communities and the biotic factors which appear to be of importance in influencing community development. The total stand area is recorded and the following quadrat sample areas are in current use:

tree canopies, sparse scrub	50 x 50 m
dense scrub, tall woodland ground layers, mosaics	10 x 10 m
short woodland ground layers, tall herb communities, mosaic components and heaths	4 x 4 m
grasslands, dwarf shrub heaths	2 x 2 m
bryophyte communities	50 x 50 cm

If square samples are impossible, an alternative shape of identical area should be employed. In layered vegetation, the height and cover of each stratum is recorded. Environmental data collection comprises details of altitude, slope angle in degrees, aspect in compass degrees and such detail of soil profile as type, depth, layering and pH. A number of relevés of similar floristic composition may then be tabulated to give an overall impression of the community-type characteristics. Traditionally, these groups are arrived at by a progressive tidying of a sequence of tables (Shimwell, 1971) but recent innovations have enabled the production of summary tables by a variety of computational assortment techniques, notably those of Maarel et al (1978) and Huntley, Huntley and Birks (1981). The end product is a series of constancy tables which give details of the general floristic composition of a group of community-types (Table 5.4). Constancy refers to the percentage presence in the group of relevés of a particular plant species according to the scale of classes: I = 1 - 20%; II = 21 - 40%; III = 41 - 60%; IV = 61 - 80%; V = 81 -100%. Thus, if a species occurs in all ten relevés of a particular community-type it is given a constancy value of V with the appropriate range of cover values appended.

Five Approaches to the Classification of World Vegetation

Formations and Formation-types. The concept of the vegetation formation is one of the oldest in biogeography, being introduced by Grisebach (1838) as a phytogeographical grouping of plants, such as a meadow or forest, 'that has a fixed physiognomic character', and taken up by Warming (1909) as a physiognomic unit that is 'an expression of certain defined conditions of life and is not concerned with floristic differences'. From these early works the plant formation emerged as a key concept defined by the physiognomy or structure of the plants themselves, particularly growth form, related to broad environmental features of total climate and habitat. Hence, Beard (1973) provides the following definition: 'a formation is a major kind of plant community on a given continent, characterised by

Table 5.4: A constancy table for British Reedmace swamps

Community-Type	A	B	C	D	E	F
Number of Relevés	41	14	6	5	11	8
Average Species Number	2	4	7	7	6	4
Characteristic Species						
TYPHA latifolia	V^{6-10}	V^{6-10}	V^{6-10}	V^{5-9}	V^{5-10}	V^{8-10}
Differential Species of Community-types						
LEMNA minor	-	V^{3-10}	-	II^{1}	-	-
L. trisulca	-	II^{4-5}	I^{2}	-	-	-
EQUISETUM fluviatile	I^{1}	-	V^{3-10}	I^{3}	-	-
CAREX rostrata	-	I^{1}	-	V^{3-4}	-	-
EPILOBIUM hirsutum	-	I^{3}	II^{3}	-	V^{1-7}	-
AGROSTIS stolonifera	I^{4}	I^{4}	-	I^{4}	-	V^{3-7}
SOLANUM dulcamara	-	-	-	-	-	IV^{1-7}
Other species of Constancy III+						
MENTHA aquatica	I^{1-5}	II^{2-3}	IV^{3-7}	-	I^{5}	I^{2}
GALIUM palustre	-	I^{3}	III^{3-5}	II^{2-4}	I^{2}	II^{2-4}

physiognomy and a range of environments to which that physiognomy is a response'. Thus, savanna (tropical grassland) and steppe (temperate grassland) are distinct formations although they are dominated by the same graminoid growth forms. This is the usual scale of study to which the concept of the vegetation formation is applied. More detailed local studies have required the recognition of sub-divisions or subformations while broader descriptions of vegetation on a continental or world scale have used the concept of formation-type or formation-class (equivalent to the biome-type of the zoogeographer). The formation-type recognises that there are convergences of physiognomy in the similar major environments of the world and thus has been used as a basis for several world-wide systems of vegetation classification. Perhaps the best known are the traditional systems of Rübel (1930, 1936) and Schimper and von Faber (1935) but, more recently, Ellenberg and Mueller-Dombois (1967), Schmithüsen (1968) and Whittaker (1970) have all produced variations on the theme. Four of the most popular systems of physiognomic classification are compared in Table 5.5.

Formations and formation-groups. A more radical modification of the formation system of classification was produced by Fosberg (1961) and revised and developed for categorisation in the International Biological Programme Check Sheet Guide (Fosberg, 1967). Physiognomy, structure and function are the main vegetation criteria for the recognition of three primary structural groups: closed, open and sparse vegetation. Formation-class is used for the next level, units which are defined by habit and stature, for example, forest, scrub and grass. The third rank is characterised according to whether the dominant vegetation layer is evergreen, deciduous or seasonally dormant; the term formation-group is proposed. Fourthly, the formation category is used and interpreted in its traditional physiognomic sense, with emphasis on dominant growth form, leaf texture and other epharmonic features. The subformation category is used in a similar fashion where further subdivision seems appropriate. Fosberg's system is probably unique in that it actually provides a dichotomous key for the analysis of a particular stand of vegetation to 'a convenient reference point on a continuum'. Table 5.6 reproduces the key to formation classes, while Table 5.7 exemplifies the types of formation-group, formation and subformation within the dwarf scrub formation-class. Similar physiognomic categorisation has been adopted by Specht (1979) for the primary classification of heathlands and Mediterranean scrub vegetation, respectively.

Formation-series. Beard (1944) developed a system of formations for tropical America in which community-types were typified by physiognomy and related to one another and environmental factors in terms of sequences along environmental gradients which were both climatically and edaphically controlled. To such sequences the term formation-series was applied. Beard recognised five formation-series: a lowland seasonal series, controlled by decreasing rainfall: rain forest, deciduous forest, thorn woodland-desert; a montane series, controlled by altitude: lowland rainforest, montane rain forest, elfin woodland, alpine paramo-puna; and three edapic formation-series, comprising swamp, seasonal swamp and dry evergreen formations. On a world scale Whittaker (1970) related the most broadly defined formations to four major climatic gradients. Beginning with tropical rain forest, three formation series or ecoclines proceed along gradients associated with altitude, latitude and in terms of drought, while the fourth describes vegetation from deciduous forest to desert typical of the climate of the North Temperate zone. The formation-series is thus conceived as a broad

Table 5.5: A comparison of four physiognomic classifications (after Beard, 1973)

Schimper and von Faber 1935	Rübel 1930	Schmithüsen 1968	Whittaker 1970
1. Tropical rainforest	1. Rainforest	IA. Evergreen forest	1. Tropical rainforest
2. Subtropical rainforest			
4. Temperate rainforest	3. Laurel-leaved forest		3. Temperate giant rainforest
			4. Montane rainforest
			8. Elfin woodland
6. Needleleaf forest	12. Needle-leaved forest		7. Taiga
			6. Temp. evergreen needleleaf sclerophyll
7. Evergreen hardwood	5. Sclerophyll forest		
3. Monsoon forest	10. Raingreen forest	IB. Deciduous forest	2. Tropical seasonal forest
5. Summergreen deciduous forest	8. Summergreen forest		5. Temperate deciduous forest
8. Savanna woodland		IIA. Evergreen woodland	11. Temperate woodland
		IIB. Deciduous woodland	

9.	Thorn forest and scrub	11.	Raingreen scrub	IIC.	Xeromorphic woodland	9.	Thorn woodland
				IIIC.	Xeromorphic shrubland	10.	Thorn scrub
		9.	Summergreen scrub	IIIB.	Deciduous shrubland	12.	Temperate deciduous heath sclerophyll subalpine-needleleaf sublapine-broadleaf
12.	Heath	7.	Heath	VIC.	Heath (in part)		
		6.	Sclerophyllous scrub	IIIA.	Evergreen shrubland		
		13.	Needle-leaved scrub				
		4.	Laurel-leaved scrub				
10.	Savanna	14.	Hardgrass prairie	IVA.	Savanna	13.	Savanna
11.	Steppe and semidesert			IVB.	Steppe (in part)	14.	Temperate grassland
		20.	Dry desert			19.	Cool-temperate desert scrub
13.	Dry desert			VIIA.	Desert	17.	Tropical desert
						18.	Warm-temperate desert
14.	Tundra	15.	Meadows	IVC.	Meadows	15.	Alpine grasslands
				VID.	Tundra	16.	Tundra
15.	Cold desert	21.	Cold desert			20.	Arctic-alpine desert
		19.	Moss moor	VIE.	Moss moor	21.	Bog
	(edaphic)	2.	Rain scrub			24.	Mangrove swamp
		17.	Marsh			25.	Saltmarsh

Table 5.6: A dichotomous key to the main formation-classes (after Fosberg, 1967)

1	Closed vegetation. Crowns or peripheries of plants mostly touching or overlapping.	2
	Open and sparse vegetation. Crowns or peripheries of plants mostly not touching.	17
2	Trees present.	3
	Trees absent, or nearly so.	7
3	Tree canopy closed.	1A Forests
	Tree crowns mostly not touching.	4
4	Shrub layer closed.	5
	Shrub layer absent or open.	6
5	Tree crowns separated by less than their diameters.	1D Open Forest with closed lower layers
	Trees scattered, separated by more than their crown diameters.	1E Closed scrub with scattered trees
6	Closed layer of dwarf shrubs.	1F Dwarf scrub with scattered trees
	Closed layer of tall grasses, etc.	1I Tall savanna
	Closed layer of short grasses, etc.	1J Low savanna
7	Shrubs and dwarf shrubs present.	8
	Shrubs and dwarf shrubs absent, or nearly so.	12
8	Shrubs present.	9
	Shrubs absent, or nearly so; dwarf shrubs present.	11
9	Shrub layer closed.	1B Scrub
	Shrub layer open or sparse.	10
10	Shrubs separated by less than their diameters	1G Open scrub with closed ground cover
	Shrubs scattered, separated by more than their diamenter	1K Shrub savanna
11	Dwarf shrub layer closed.	1C Dwarf scrub (see Table 5.7)
	Dwarf shrub layer open.	1H Open dwarf scrub with closed ground cover
12	Dominated by plants the leaves of which are adapted to lengthy or permanent submersion or to floating.	13

Dominated by grasses, etc., herbs, bryophytes or lichens whose leaves are not adapted to lengthy or permanent submersion or to floating. 14

13 Plants rooted. 1P Submerged meadows
Plants floating, not rooted on bottom. 1Q Floating meadows

14 Dominated by short and/or tall grasses, etc. 15
Dominated by broad-leaved herbs, bryophytes or lichens. 16

15 Dominated by graminoid plants exceeding 1 m in height.
1L Tall grass
Dominated by graminoid plants less than 1 m tall.
1M Short grass

16 Dominated by broad-leaved herbs.
1N Broad-leaved herb vegetation
Dominated by bryophytes and/or lichens.
1Q Closed bryoid vegetation

17 Plants or tufts of plants not touching, but crowns not separated by more than their diameter; plants, not substratum, dominating landscape. 18
Plants so scattered that substratum dominates landscape. 27

18 Trees present. 19
Trees absent or nearly so. 20

19 Tree layer open; lower layers may be open or sparse.
2A Steppe forest
Trees scattered, not forming a well-defined layer.
2D Steppe savanna

20 Shrubs and dwarf shrubs present. 21
Shrubs and dwarf shrubs absent, or nearly so. 27

21 Shrubs present. 22
Shrubs absent, or nearly so: dwarf shrubs present. 23

22 Shrub layer open. 2B Steppe scrub
Shrubs sparse, not forming a distinct layer.
2E Shrub steppe savanna

23 Dwarf shrub layer open. 2C Dwarf steppe scrub
Dwarf shrubs sparse, not forming a distinct layer.
2F Dwarf shrub steppe savanna

24 Dominated by plants, the leaves of which are adapted to lengthy or permanent submersion or to floating. 25
Dominated by herbaceous vegetation whose leaves are not adapted to lengthy or permanent submersion or to floating. 26

25 Plants rooted. 2I Open submerged meadows
Plants floating, not rooted on bottom.
2J Open floating meadows

26 Dominated by grasses, etc., and/or broad-leaved herbs.
2G Steppe
Dominated by bryophytes and/or lichens.
2H Bryoid steppe

27 Trees present. 3A Desert forest
Trees absent, or nearly so. 28

28 Shrubs and dwarf shrubs present.
3B Desert scrub
Shrubs and dwarf shrubs absent or nearly so. 29

29 Herbaceous plants predominate.
3C Desert herb vegetation
Plants with leaves adapted to lengthy or permanent submergence or to floating predominate.
3D Sparse submerged meadows

---

habitat grouping of the physiognomically-related formations and a simple method of gradient analysis.

Bioclimatic diagrams. Several authors have sought to relate formation-types to major climatic types in the form of bioclimatic diagrams and framework charts.

Ecosystem-types. Several major treatments of vegetation at a global scale use no formal classification system (Walter, 1968) while others prefer the ecosystem-type to the term formation. Perhaps the most notable of the modern approaches to a comprehension of world vegetation types is the series of monographs, *Ecosystems of the World*, edited by Professor D.W. Goodall (1977-). The thirty ecosystems recognised are shown in Table 5.8, where the general departure from simple vegetation analysis and the trend towards a unifying study of biocoenoses may be clearly seen. Within each monograph a series of papers cover the physiognomic, floristic and ecological approaches to description and classification and present a biogeographical analysis in the broadest and most complete sense (Castri *et al*, 1981).

Table 5.7: Dwarf scrub formations and subformations (after Fosberg, 1967)

IC Dwarf scrub (Closed predominantly woody vegetation less than 0.5 m tall)		
1C1 Evergreen dwarf scrub	1 Evergreen orthophyll dwarf scrub	e. Rohdodendron mat (Eastern Himalaya)?
	2 Evergreen broad sclerophyll dwarf scrub	
	(a) Mesophyllous broad sclerophyll dwarf scrub	e. Arctostaphylos uva-ursi mat (Northern temperate region)
	(b) Microphyllous evergreen dwarf scrub	(Without significant peat accumulation) e. Coastal Osteomeles scrub (Miyako Island) Calluna heath without peat (Western Europe)
	(c) Microphyllous evergreen dwarf heath	(With peat accumulation) e. Empetrum heath (Arctic and Subarctic) Loiseleuria heath (Arctic)
	3 Evergreen dwarf shrub bog	(Dwarf shrub with significant peat accumulation, root systems of plants adapted to constant immersion) e. Mountain bogs, more closed phases (Hawaii) Chamaedaphne bog (Eastern North America)
1C2 Deciduous dwarf scrub	1 Deciduous orthophyll dwarf scrub	
	(a) Deciduous orthophyll dwarf scrub	(Without significant peat accumulation) e. Low bush Vaccinium scrub (North temperate and subarctic regions)
	(b) Deciduous orthophyll dwarf heath	(With peat accumulation) e. Vaccinium myrtillus heath (Subarctic regions)

Table 5.8: A classification of World Ecosystems (from Ecosystems of the World, ed. Goodall, D.W.)

I TERRESTRIAL ECOSYSTEMS

A. Natural Terrestrial Ecosystems

1. Wet Coastal Ecosystems
2. Dry Coastal Ecosystems
3. Polar and Alpine Tundra
4. Swamp, Bog, Fen and Moor
5. Shrub, Steppe and Cold Desert
6. Coniferous Forest
7. Temperate Deciduous Forest
8. Natural Grassland
9. Heath and Related Shrublands
10. Temperate Broad-Leaved Evergreen Forest
11. Mediterranean-type Shrublands
12. Hot Desert and Arid Shrubland
13. Savannah and Savannah Woodland
14. Seasonal Tropical Forest
15. Equatorial Forest
16. Wetland Forests
17. Ecosystems of Disturbed Ground

B. Managed Terrestrial Ecosystems

18. Managed Grassland
19. Field Crop Ecosystems
20. Tree Crop Ecosystems
21. Greenhouse Ecosystems
22. Bioindustrial Ecosystems

II. AQUATIC ECOSYSTEMS

A. Inland Aquatic Ecosystems

23. Rivers and Stream Ecosystems
24. Lake and Reservoir Ecosystems

B. Marine Ecosystems

25. Intertidal and Littoral Ecosystems
26. Ecosystems of Estuaries and Enclosed Seas
27. Coral Reefs
28. Ecosystems of the Continental Shelves
29. Ecosystems of the Deep Ocean

C. Managed Aquatic Ecosystems

30. Managed Aquatic Ecosystems

Some Conclusions

Vegetation analysis is primarily a field study which relies upon careful observation, the collection of accurate data and a certain amount of intuitive decision-making by the observer. The end results of the later laboratory analyses are only as good as the primary field data; errors due to incomplete or inconsistent sampling manifest themselves and the whole treatment, whether as a classification or an ordination, must be regarded as an artificial framework, littered with artefacts. The vital stage following analysis is a return to the field to test the validity of the gradients and community-types which have emerged as the most important components of the vegetation variation. Vegetation scientists research such problems at a variety of scales. The autecologist is most commonly concerned with small-scale areal variations in community-types and the erection of a framework on which to hang other sets of environmental data; the plant sociologist tends to operate at the meso-scale, focussing attention on the relationships of community-types of similar floristic and physiognomic characteristics, or on the analysis of all the community-types of a certain region or country; the plant geographer operates at the macro-scale, concentrating on the analysis of vegetation formations and formation-types on a continental or global scale. In reality, there should be scope within a basic biogeography course for studies at all three scales. But, where should the emphasis be placed? What major contribution can the biogeographer make to vegetation analysis?

The publication in 1980 of the World Conservation Strategy, through the agency of the International Union for Conservation of Nature and Natural Resources, itemised a number of priority actions necessary for global conservation. Two statements seem to provide a focus for the biogeographer through work on vegetation analysis.

World Conservation Strategy, Section 12:11:

> Research programmes should cover three broad overlapping areas: inventory - this includes research on the distribution of ecosystems and species in each country ...

World Conservation Strategy, Section 17:9:

> Review of the distribution of protected areas of the land shows that 35 of the 193 biogeographical provinces listed have no national parks or equivalent reserves and a further 38, while having at least one park or reserve, are inadequately covered. Representative samples of the ecosystems in these provinces should be given protection as soon as possible. In addition, all the other provinces need to be examined for the quality of their coverage ...

Surely, it is here that the biogeographer, through the use of rapid vegetation and environmental analysis techniques, can and should make a major contribution to global conservation.

REFERENCES

Barkman, J.J. (1979), 'The investigation of vegetation texture and structure' in Werger, M.J.A. (ed.), The Study of Vegetation, Junk, The Hague, Netherlands, pp. 123-160.

Barkman, J.J., Doing, H. and Segal, S. (1964), 'Kritische Bemer kungen und Vorschläge zur quantitativen Vegetationsanalyse', Acta botanica neerlandica, **13**, 394-419.

Beard, J.S. (1944), 'Climax vegetation in tropical America', Ecology, **25**, 127-158.

Beard, J.S. (1973), 'The Physiognomic Approach' in Whittaker, R.H. (ed.), Ordination and Classification of Communities, Handbook of Vegetation Science, Part V, Junk, The Hague, Netherlands, pp. 355-386.

Box, E.O. (1981), Macroclimate and plant forms: an introduction to predictive modeling in phytogeography, Tasks for Vegetation Science, Volume 1, Junk, The Hague, Netherlands.

Braun-Blanquet, J. (1932), Plant Sociology, the Study of Plant Communities, McGraw-Hill, New York, USA.

Cain, S.A. and Castro, G.M. de O. (1959), Manual of Vegetation Analysis, Harper, New York, USA.

Castri, F. di (1981), 'Mediterranean-type shrublands of the world' in Castri, F. di, Goodall, D.W. and Specht, R.L. (eds.), 'Mediterranean-type Shrublands', Ecosystems of the World, Vol. 11, Elsevier, Amsterdam, Oxford and New York, pp. 1-52.

Castri, F. di, Goodall, D.W. and Specht, R.L. (1981) (eds.), 'Mediterranean-Type Shrublands', Ecosystems of the World, Amsterdam, Oxford and New York.

Curran, P. (1980), 'Multispectral remote sensing of vegetation amount', Progress in Physical Geography, **4**, 315-341.

Curtis, L.F. (1978), 'Remote sensing systems for measuring crops and vegetation', Progress in Physical Geography, **2**, 55-79.

Dansereau, P. (1951), 'Description and recording of vegetation upon a structural basis', Ecology, **32**, 172-229.

Dansereau, P. (1957), Biogeography: an Ecological Perspective, Ronald, New York, USA.

Dansereau, P. and Arros, J. (1959), 'Essais d'application de la dimension structurale en phytoscoiologie. I. Quelques exemplaires européens', Vegetatio, **9**, 48-99.

Dansereau, P., Buell, P.F. and Dagon, R. (1966), 'A universal system for recording vegetation. II. A methodological critique and an experiment', Sarracenia, **10**, 1-64.

Dickinson, G. et al (1971), 'The application of phytosociological techniques to the geographical study of vegetation', Scot. Geog. Mag., **87**, 83-103.

Ellenberg, H. (1956), 'Aufgaben und Methoden der Vegetations Kunde' in Walter, H. (ed.), Einfuhrung in die Phytologie, Vol. IV, Part 1, Ulmer, Stuttgart.

Ellenberg, H. and Mueller-Dombois, D. (1967), 'Tentative physiognomic-ecological classification of plant formations of the earth', Bericht geobot. Inst. ETH, Stiftg. Rübel, **37**, 21-55.

Evans, F.C. and Dahl, E. (1955), 'The vegetational structure of an abandoned field in Southeastern Michigan and its relation to environmental factors', Ecology, **36**, 685-706.

Fosberg, F.R. (1961), 'A classification of vegetation for general purposes', Tropical Ecology, **2**, 1-28.

Fosberg, F.R. (1967), 'A classification of vegetation for general purposes' in Peterken, G.F. (ed.) (1967), Guide to the Check Sheet for I.B.P. Areas, IBP Handbook No. 4, Blackwell, Oxford, UK, pp. 73-120.

Goodall, D.W. (ed.-in-chief) (1977-), Ecosystems of the World (29 vols.), Elsevier, Amsterdam (a dozen or so have now been published).

Greig-Smith, P. (1964), Quantitative Plant Ecology, 2nd edition, Butterworth, London.

Grisebach, A. (1838), 'Ueber den Einfluss des Climas auf die Begranzung der naturlichen Floren', Linnaea, **12**, 159-200.

Harrison, C.M. (1971), 'Recent approaches to the description and analysis of vegetation', Trans. Inst. Brit. Geogrs., **52**, 113-27.

Heyligers, P.C. (1965), 'Structure formulae in vegetation analysis on aerial photographs and in the field', Symposium on Ecological Research in Humid Tropical Vegetation, Sarawak, 1963, pp. 249-254.

Huntley, B., Huntley, J.P. and Birks, H.J.B. (1981), 'PHYTOPAK: A suite of computer programs designed for the handling and analysis of phytosociological data', Vegetatio, **45**, 85-95.

Kellman, M.C. (1975), Plant Geography, Methuen, London.

Küchler, A.W. (1949), 'A physiognomic classification of vegetation', Annals Association American Geographers, **39**, 201-210.

Küchler, A.W. (1949), Vegetation Mapping, Ronald Press, New York, USA.

Lillesand, T.M. and Kiefer, R.W. (1979), Remote Sensing and Image Interpretation, Wiley, New York, USA.

Maarel, van der E., Janssen, J.G.M. and Louppen, J.M.W. (1978), 'Tabord, a program for structuring phytosociological tables', Vegetatio, **38**, 143-156.

McIntosh, R.P. (1967), 'The continuum concept of vegetation', Botanical Reviews, **33**, 130-187.

Parsons, D.J. (1976), 'Vegetation structure in the Mediterranean scrub communities of California and Chile', Journal of Ecology, **64**, 435-447.

Pielou, E.C. (1967), An Introduction to Mathematical Ecology, Wiley Interscience, New York, USA.

Randall, R.E. (1978), Theories and Techniques in Vegetation Analysis, OUP, Oxford, UK.

Raunkiaer, C. (1910), 'Formationsundersøgelse og Formation statistik', Bot. Tidsskr., **30**, 20-132.

Raunkiaer, C. (1934), The Life Forms of Plants and Statistical Plant Geography, Clarendon, Oxford, UK.

Rübel, E. (1930), Pflanzengesellschaften der Erde, Huber, Berne and Berlin.

Rübel, E. (1936), 'Plant communities of the world' in Goodspeed, T.H. (ed.), Essays in Geobotany in Honor of William Albert Setchel, University of California Press, Berkeley, pp. 263-290.

Schimper, A.F.W. and Faber, F.C. von (1935), Pflanzengeographie auf physiologischer Grundlage, 3rd edition, Fisher, Jena.

Schmithusen, J. (1968), Allgemeine Vegetations geographie, 3rd edition, de Gruyter, Berlin.

Shimwell, D.W. (1971), The Description and Classification of Vegetation, Sidgwick and Jackson, London.

Specht, R.L. (1979), 'Heathlands and Related Shrublands: descriptive studies', Ecosystems of the World, Vol. 9A, Elsevier, Amsterdam, Oxford, New York.

Tansley, A.G. (1939), The British Islands and their Vegetation, Cambridge University Press, Cambridge.

Walter, H. (1968), Die Vegetation der Erde in ökophysiologischer Betrachtung, Fisher, Jena.

Warming, E. (1909), Oecology of Plants: An Introduction to the Study of Plant-Communities, Oxford University Press, Oxford, UK.

Webb, L.J. (1959), 'A physiognomic classification of Australian rain-forests', Journal of Ecology, **47**, 551-570.

Westhoff, V. and Maarel, E. van der (1973), 'The Braun-Blanquet Approach' in Whittaker, R.H. (ed.), Ordination and Classification of Communities, Handbook of Vegetation Science, Part V, Junk, The Hague, Netherlands, pp. 619-726.

Whittaker, R.H. (1962), 'Classification of natural communities', Botanical Reviews, **28**, 1-239.

Whittaker, R.H. (1967), 'Gradient analysis of vegetation', Biological Reviews, **42**, 207-264.

Whittaker, R.H. (1970), Communities and Ecosystems, MacMillan, New York, USA.

Whittaker, R.H. (ed.) (1973), Ordination and Classification of Communities, Handbook of Vegetation Science, Part 5, Junk, The Hague, Netherlands.

FURTHER READING

Dickinson, G., Mitchell, J. and Tivy, J. (1971), 'The application of phytosociological techniques to the geographical study of vegetation', Scot. Geogr. Mag., **87**, 83-103.

Fosberg, F.R. (1967), 'A Classification of Vegetation for General Purposes' in Peterken, G.F. (ed.) (1967), Guide to the Check Sheet for IBP Areas, IBP Handbook No. 4, Blackwell, Oxford, UK, pp. 73-120.

Greig-Smith, P. (1964), Quantitative Plant Ecology, 2nd edition, Butterworth, London.

Harrison, C.M. (1971), 'Recent approaches to the description and analysis of vegetation', Trans. Inst. Brit. Geogr., **52**, 113-127.

Harrison, C.M. and Frenkel, R.E. (1974), 'An assessment of the usefulness of phytosociological and numerical classificatory methods for the community biogeographer', J. Biogeog., **1**, 27-56.

Orloci, L. (1978), Multivariate Analysis in Vegetation Research, Junk, The Hague, Netherlands.

Vegetation Analysis

Randall, R.E. (1978), Theories and Techniques in Vegetation Analysis, Oxford University Press, Oxford, UK.

Shimwell, D.W. (1971), The Description and Classification of Vegetation, Sidgwick & Jackson, London.

Whittaker, R.H. (ed.) (1973), Ordination and Classification of Communities, Handbook of Vegetation Science, Part V, Junk, The Hague, Netherlands.

Chapter 6

THE GEOGRAPHY OF ANIMAL COMMUNITIES

R. J. Putman

The Distribution of Animal Species: Limits to Tolerance

One of the most striking features of biogeography is a sense of order or pattern. Animals and plants are not just scattered haphazardly over the globe, as if distributed by some gigantic pepper-shaker, but occur in specific associations in particular contexts. Particular species or species assemblages crop up again and again in one particular context, yet never occur outside that context. In short, we can define for any animal or plant species a set of circumstances in which it will not occur; equally we can recognise a number of environments for which it is characteristic. In this chapter we will attempt to offer some explanation for this predictable distribution of animal species.

One part of the explanation for the observed distribution of particular animal or plant species is of course that any living organism is adapted to a particular set of environmental circumstances and can only survive within a limited range of physical and chemical conditions: its distribution is limited by its environmental tolerances.

The very word 'environment' conjures up an impression of a structural, physical 'stage-set' upon which background, biological processes are acted out: a sort of passive, non-reactive context. Yet this impression is misleading: an organism's relationship with its environment is in fact highly interactive, and there are a vast number of facets of the environment which may affect the performance or survival of any organism attempting to live there. In practice, of course, any organism's environment includes not only the physico-chemical stage-set but also other organisms; living organisms of its own or different species are as much a part of its environment as are more structural, abiotic components, and it is important not to overlook these biotic elements as we shall see later. However, the physico-chemical, abiotic background is perhaps the most obvious and fundamental facet of the environment; abiotic relationships are the primary factors determining whether or not any organism can exist in a certain environment.

Relationships between an organism and this abiotic environment are primarily physiological and there is a profusion of specific facets of the abiotic environment with which the organism must interact. Temperature, light intensity, concentration of oxygen, carbon dioxide, wind, topography, chemical nature of soil or water flow, humidity: these and a host of other characteristics of the abiotic environment may exert an influence on the organisms trying to live there.

Because physiological processes proceed at different rates under different conditions, any one organism has only a limited range of conditions in which it may survive; within those limits is a further restricted range where the organism may operate at maximum efficiency. The very fact of having such tolerance limits restricts markedly the conditions under which an animal or plant can operate efficiently and provides the mechanism whereby influences from the abiotic environment may affect its ecology in such a fundamental sense.

Figure 6.1: A tolerance curve: the efficiency of performance of some hypothetical organism over a range of a given physico-chemical parameter

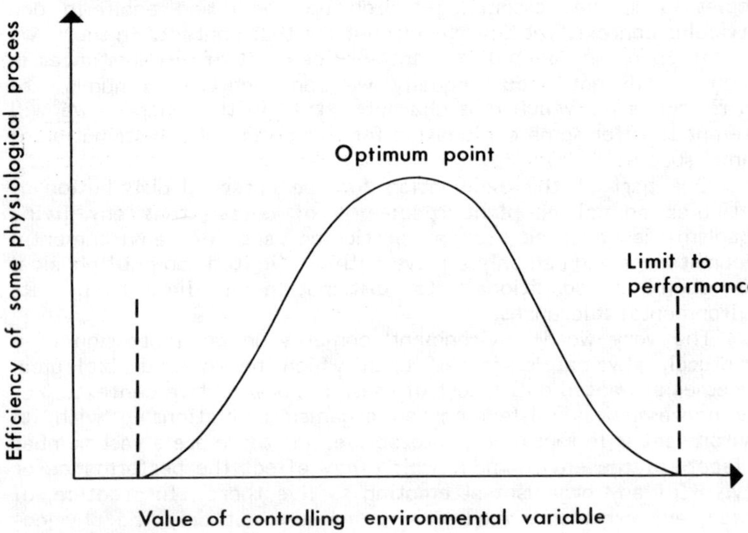

Curves of performance may be drawn for any organism for any particular physiological process, representing its efficiency of operation over a range of a given physico-chemical parameter. Such curves, known as 'tolerance curves' (Shelford, 1913) are typically bell-shaped, with their peaks representing optimal conditions for a particular physiological process, and their tails representing limits of tolerance (Figure 6.1). It is usual to define such limits as a series of 'nesting' inner limits. Thus, within an upper and lower lethal limit at which death occurs, we may recognise a pair of inner limits, the critical maximum and minimum, outside which the organism, though not dead, is ecologically inviable (let us say, a range beyond which it is no longer able to move). Within this we may define yet another, narrower, preferred range, and within this again, an optimum range (Figure 6.2). From consideration of such tolerance curves for individual physiological processes, we may derive tolerance limits for the organism as a whole.

Figure 6.2: Nesting limits of tolerance

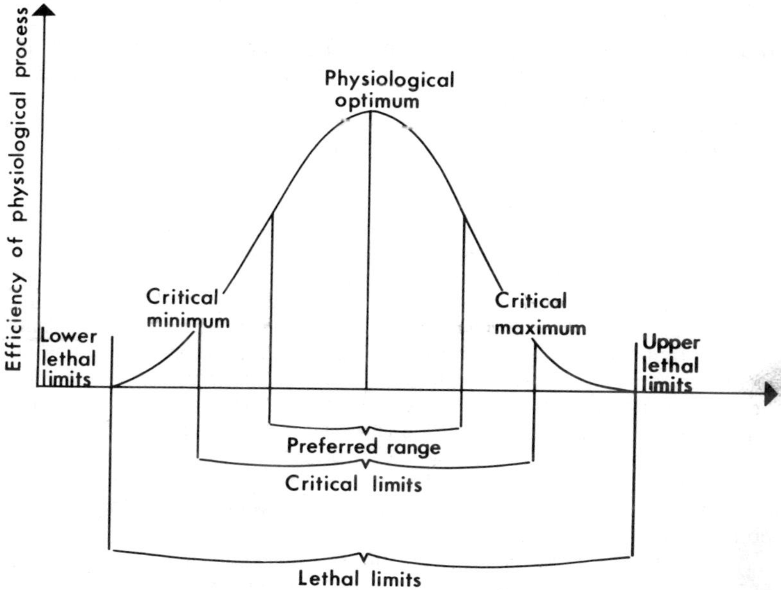

Some individuals and species have narrow, high-peaked tolerance curves for any particular environmental variable; such are described as steno-topic organisms. (The prefix steno- (narrow) may also be applied to specific responses, thus: steno-thermic, steno-haline: narrow tolerance to temperature, salinity, respectively.) In other organisms (or in the same organism for a different environmental variable) curves may be broader and flatter. (Broad tolerance curves are given the prefix eury, thus, eurythermic, euryhaline, etc.)

There is considerable interaction between the effects of different environmental variables, so that it is misleading to consider responses to any one abiotic factor in strict isolation. Tolerance to temperature for instance, in many organisms, is intricately bound up with tolerance to relative humidity, because the physiological processes affecting temperature regulation are themselves controlled by water availability. In more general terms, one may frequently observe interdependence of tolerances to pairs or groups of environmental variables which affect the same physiological process. Thus, organisms near the limits of their range in one environmental variable, or those which have 'chosen' to become physiological specialists, in order to exploit an extreme environment, become less able to tolerate variation in conditions in other abiotic parameters as well as that one actually under stress. This reduced tolerance to all abiotic variables by organisms at extremes of tolerance to one variable is clearly a factor of extreme importance in conservational exercises. For such organisms, variability in environmental conditions may become as important a limitation as the actual values of the abiotic variables themselves.

Such physiological limits to tolerance of a whole range of abiotic parameters clearly have a profound influence on the distributions of animals and plants, and the efficiency of their operation once they are there. Indeed, it is probably fair to say that much of the observed distribution of living organisms is determined by such abiotic considerations. (More strictly, it is fair to say that abiotic factors, through limits to tolerance, may determine where an organism may not occur, but in practice will not dictate where such an organism will occur. For within those systems where abiotic conditions might permit any organism to exist, its actual occurrence may then be limited by other factors: relationships with other organisms, through competition or predation, for example, may exclude an organism from many systems in which physiologically it could survive.)

However, such tolerance curves are not, as is sometimes presumed, immutable. Their position within the environmental gradient, and their breadth, may be shifted to a certain extent by genetic or evolutionary change, or by physiological or behavioural changes during the organism's lifetime. Thus, an organism may adjust over time to a slight deviation in environmental conditions or may even evolve to exploit extremes of conditions.

Expansion of Tolerance Limits

Organisms may adjust very slightly their range of operation with respect to some environmental parameter, or suite of parameters, through acclimation. Prolonged exposure to conditions slightly to one side of the optimum (but usually still within the 'preferred range') results in displacement of the tolerance curve to produce a new optimum at the ambient value. This process of acclimation involves changes in enzyme systems (for it is the restricted range of conditions under which enzymes operate efficiently which imposes the tolerance limits in the first place), but the details of such changes are uncertain. Such acclimation may be completed within a relatively short time: as little as 24 hours for most small animals. Over a longer time-scale such changes may become genetically 'fixed' as the organism adapts evolutionarily to the new environment. But such adaptation must be at some cost. In general the broader the tolerance, the lower the overall efficiency at optimum. There appears to be something in the nature of a trade-off: a jack of all trades can be master of none, while organisms that 'choose' to specialise and become extremely efficient over a narrow range of conditions, lose the ability to operate at all under other conditions.

A more fundamental mechanism for overcoming the restrictions placed upon an organism by physiological tolerance limits may be developed in assuming greater 'responsibility' for controlling its own internal environment, divorcing it to a greater or lesser extent from external conditions, in the evolutionary development of sophisticated mechanisms of homeostasis. By maintaining their own internal state in some way independent of external conditions, such homeostatic organisms are capable of tolerating a far wider range of values of environmental variables. There are, still, overall limits beyond which homeostatic mechanisms are ineffective in controlling body state, or beyond which they become too energetically expensive to maintain, but the development of some degree of homeostasis does permit at least partial release from the immediate control of environmental conditions.

Such homeostatic control may be physiological or behavioural. Thus many animals have a degree of homiothermy: control over body temperature. Control of body temperature may be achieved through physiological control of heat production and heat loss by those animals which are capable of independent heat production as metabolic heat (endotherms); those animals dependent on the environment as heat source and sink (ectotherms), must perforce regulate their temperature purely by behavioural means, but are nonetheless able to achieve a remarkably sophisticated control of body temperature; homiothermy is by no means, as is widely misbelieved, restricted to endotherms. The same sort of control may be exercised by different organisms over internal salt-concentration (through a variety of mechanisms of osmo-regulation).

Such homeostasis is a primary mechanism in expansion of environmental tolerance and is widely used by a variety of organisms. But, in practice, even if it does enable organisms to expand their tolerance range, it does not enable them to escape entirely from the imposition of environmental limits; the organism remains limited by the restricted range over which homeostasis can be maintained. In effect, a homeostatic organism expands its tolerances to become eurytopic.

A final alternative mechanism for overcoming the restrictions of physiological limits to tolerance is to 'opt out when times are hard'. That is to say, any organism may colonise and indeed do extremely well in environments which are strictly well beyond its true range, if it is able to become dormant during those periods when conditions become too severe. Conditions which would be well outside tolerance limits, if not lethal limits, were the organism active, may be accommodated by inactivity. The organism enters a phase of dormancy, or suspended animation, and in this state may withstand a far wider range of conditions. Extreme examples may be cited of Amoebae encysting when a temporary pond dries up, or even more complicated organisms such as *Chirocephalus* whose eggs may remain dormant in the mud base of temporary (and irregularly flooded) ponds for many years. Many desert amphibians also bury themselves under ground between rains, and become dormant.

Even under less extreme conditions, seasonal dormancy may be a valid way of colonising an environment otherwise beyond the organism's tolerance ranges. Many insects enter a state of *diapause* in unfavourable weather, where metabolic rates are reduced to one-tenth of normal; such insects show increased ability to withstand cold and desiccation. Endothermic animals, too, i.e. those which regulate their own body temperature by physiological means relatively independently of prevailing conditions, may enter a state of dormancy when environmental temperatures exceed the range within which they can effectively regulate, entering a state of *torpor*. In such a state they become functionally ectothermic, allowing body temperatures to change with environmental temperature. For many ectotherms, cold temperatures *directly* curtail activity and induce dormancy; true *torpor* is the endothermic equivalent of this. The more complex phenomena of *hibernation* (winter dormancy) and *aestivation* (summer dormancy) are triggered instead by intermediate stimuli (changes in photoperiod, for example) so that the animal can prepare for the dormant period in advance by building up food reserves.

Many physiological changes accompany dormancy. The onset of hibernation in mammals is anticipated by the accumulation of a specific type of fat, with a low melting point, that will not harden and cause stiffness at low temperature. Heartbeat is reduced (hibernating ground squirrels have heart rates of 7-10 per minute by comparison with normal active rates of 200-400 per minute (Svihla *et al*, 1951); with a reduced rate of blood flow, blood chemistry must change to prevent clotting. In ectothermic cold diapause water is chemically bound, or reduced, to prevent freezing, and metabolism drops to near zero.

None of the various mechanisms just described for adjustment of tolerance limits permits complete escape from their restrictions.

They may result in expansion of the range over which the organism may function, or a gross shift in that range, but they do not entirely remove overall limitations. Thus, however, an organism may attempt, in evolutionary terms, to 'play the system', physiological limits to tolerance still play a major role in determining that organism's geographical distribution. More strictly, as we have already noted, it is fairer to say that abiotic factors of the environment, through limits to tolerance, may determine where an organism may not occur; for in practice such considerations cannot dictate where such an organism will occur. For within those systems where abiotic conditions might permit any organism to exist, its actual occurrence and distribution may then be limited by other factors, primarily those due to interaction with other living organisms around it. Competition or predation may restrict the ecological range of an organism to only a small part of its potential physiological range. When at the limit of that physiological range (where it will be performing suboptimally), an animal or plant may be excluded through competition with another species which is better adapted to those particular conditions (for which they fall nearer to the centre of its preferred range) (Figure 6.3). An animal may be excluded from an area because its food species is absent, and thus, although abiotic factors are ideal for its survival, it cannot support itself. Brimstone butterflies, Gonepteryx rhamni, for example, are limited in their colonisation of climatically suitable sites in Great Britain by the presence of buckthorn, the caterpillar food plant Rhamnus catharticus (Ford, 1947). A host of biotic interactions in any particular set of circumstances may ultimately be superimposed upon the distribution limits set by physiological tolerance, in determining where any organism will or will not occur.

World Biomes

The abiotic factors which we may consider to affect living organisms are, broadly speaking, characteristic of a relatively large regional area. Different geographical areas may be classified in terms of a set of physico-chemical characteristics: soil-type, geological formation, climate, etc. Because of the different, and quite specific physiological tolerance limits of different animal and plant species, each of these climatic and geochemical subdivisions is characterised, whenever it occurs, by a predictable and characteristic ecological community adapted for that particular range of conditions. We call these bioclimatic assemblies biomes. Whittaker (1975) defines a biome as a grouping of terrestrial ecosystems that are similar in vegetation structure or physiognomy, in the major features of the environment to which this structure is a response, and in some characteristics of their animal communities. We have space only to review the more important of these biome types here, but for detail the reader is referred to Whittaker (1975).

In structural terms, there are six major physiognomic types on land: forest, grassland, woodland, shrubland, semi-desert shrub and true desert. These intergrade in various directions with one another

Figure 6.3: Examples of how the ecological range may be different from the physiological range due to competition (after Barbour et al, 1980)

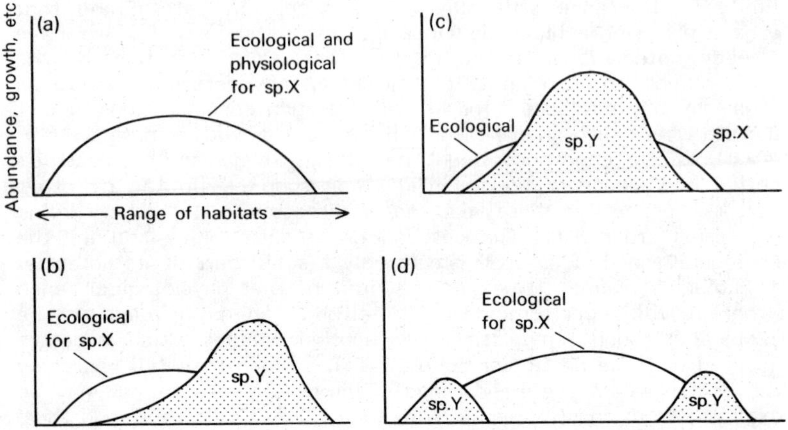

In (a), species X grows alone and laboratory experiments determine its full physiological range. In nature, it faces competition from other species which displace it from habitats it could grow in alone. In (b) its ecological range is shifted far to the left of its physiological range by competition from one side. In (c), species X competes poorly and is displaced from the middle of its range; it appears to have two ecological optima. In (d), species X competes poorly at the range extremes, with the result that the range is shortened (from Walter, H., 1973, after Barbour, Burk and Pitts, 1980).

and with certain other formation-types. Each of these structural types occurs in so wide a range of environments that more than one biome-type is defined within it on the basis of major differences in climate. Thus alpine meadow, temperate grassland and tropical savannah are separate biome-types although all are dominated by grasses or grasslike plants; rain forests or cloud forests are separate biome-types from other forests. Although vegetational terms are used to define these various biome-types, it should be noted that this is purely for identification purposes only. By definition the biome embraces the whole of the community characteristic of a given bioclimatic region.

Combination of community-type with environment thus allows us to recognise amongst forests: tropical rain forest (forests of the humid tropics where rainfall is abundant and distributed throughout the year), tropical seasonal forests (of tropical climates which have pronounced wet and dry seasons), temperate rain forests, temperate deciduous forests and temperate coniferous forests. The taiga, or subarctic/alpine needle-leaved forests are usually separated from temperate evergreen forests as a distinct biome type of colder climates.

Tropical broadleaved woodlands replace tropical seasonal forests towards drier climates; temperate woodlands form a distinct but rather varied biome-type: communities of small trees, characteristic once again of rather drier climates than those of true forests. Grasslands are also more characteristic of somewhat drier climates. Temperate grasslands occur in great areas of moderately dry continental climates in North America and Eurasia; savannahs are tropical grasslands usually with, but sometimes without, scattered trees or shrubs. Savannahs are most extensive in Africa but also occur in Australia and South America. Both savannahs and temperate grasslands are subject to regular periodic fires which affect the structure of the communities and permit them to extend into climates that might otherwise support forest. Alpine grasslands are the principal communities in montane areas or to the far north, in either case beyond the timberline. Where they occur to the north, they grade with the tundra, treeless arctic plains with varied vegetation which may consist primarily of grasslike species, or may become extensively scrubby.

The tundra vegetation tends towards dominance by dwarf-shrubs and scrub in drier environments. Further to the south, such environments are characterised by the various scrubland biomes of first cooler, then warm semi-desert. These in turn grade into the true desert biome. True deserts are primarily sub-tropical, but extend, in areas of severe aridity into temperate zones. Distinct biome types are recognised of warm-temperate, and cool-temperate and sub-tropical desert. Arctic, alpine deserts at the cold extremity of climate are deserts created more by cold than by aridity: landscapes of low plant cover, dominated by snow, ice or rock.

The sequence from forest through woodland, grassland and shrubland to desert follows a climatic gradient of decreasing rainfall or humidity. The relationship of all the different biome-types

Figure 6.4: A pattern of world formation-types in relation to climatic humidity and temperature (after Whittaker, 1970)

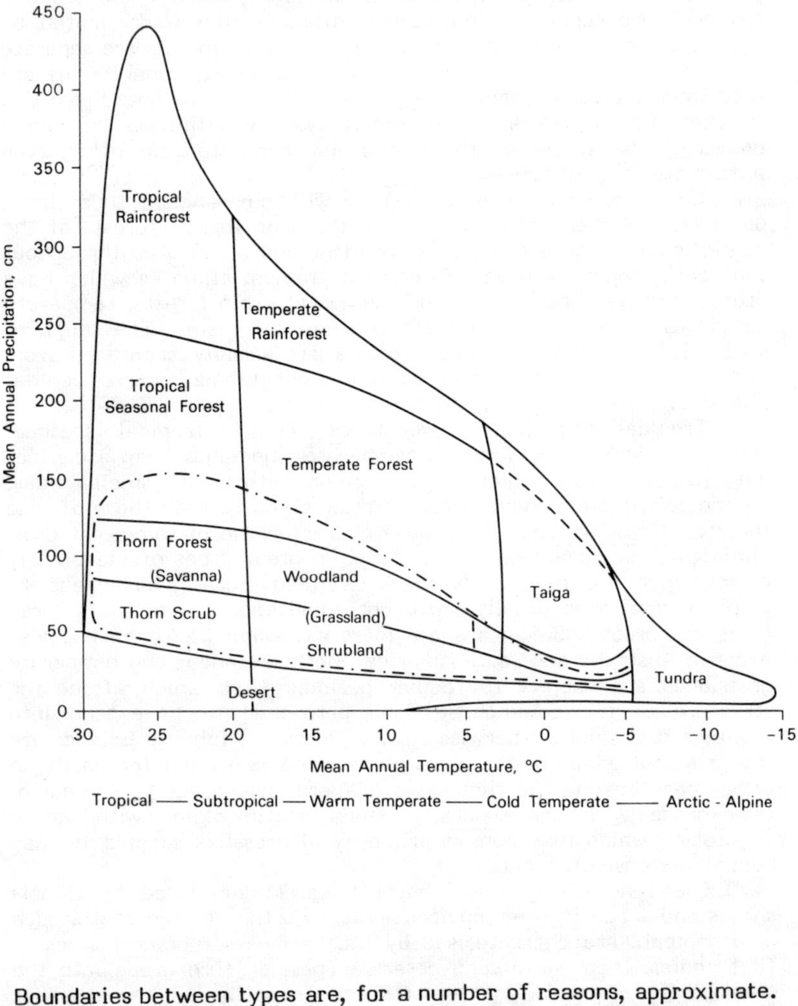

Boundaries between types are, for a number of reasons, approximate. In climates between forest and desert, maritime versus continental climate, soil effects, and fire effects can shift the balance between woodland, shrubland, and grassland types. The dot-and-dash line encloses a wide range of environments in which either grassland, or one of the types dominated by woody plants, may form the prevailing vegetation in different areas (from Whittaker, 1970).

considered here to each other and to environmental conditions are elegantly summarised by Whittaker, in an earlier work (1970), as in Figure 6.4.

A number of other biome-types can however be identified which do not fall into this classification. These are the formations which occur only locally, characteristic of peculiar conditions which crop up, so to speak unpredictably. Hydric communities are adapted to wet soils and we may recognise a number of biome types: sphagnum bog, tropical and temperate freshwater swamps, the more saline mangrove swamps of the tropics, and saltmarshes of temperate regions. Mangroves and saltmarshes are characteristic of coastal regions where we may identify further biome types: the littoral biomes of rocky and sandy shores and mud-flats. Inland, we may also recognise distinctive biomes amongst fresh water formations in lakes, rivers and streams. Our aim here, however, is merely to introduce the diversity of such biomes and to account for the fact that such 'biotic packages' may exist as distinct bioclimatic types, purely because of the different physiological tolerances of different species.

Species Diversity

In considering the distribution of animals between the different world biomes, and the reasons underlying the pattern of such distribution, there is one other important factor we should consider. Not only do the types of animals - the actual species - which occur in different climatic zones differ, in relation to different environmental tolerances, but the actual number of animal species that are present is also different in different biomes: that is the areas differ not just in species composition but in the richness and diversity of those species. The number of animal species physiologically-adapted to the arctic tundra is far lower than the number of species which might be found in tropical rain forest, or on a coral reef.

This difference in diversity between different biomes was noted by many early naturalists: both total species number for the region as a whole, and diversity of individual communities increase dramatically from tundra to tropics. So far in our discussions we have considered only those factors affecting the distribution of individual species. Why is there this overall difference in total diversity between the different climatic regions? What is it, in fact, that determines the observed species diversity of any system?

The earliest hypothesis attempting to account for differences in species diversity in different systems merely postulated that diversity was a simple function of time. All animal communities tend to diversify with time, therefore older communities are more diverse than young ones. Such a concept embraces increases in diversity due to (a) immigration to an area, (b) specialisation of the existing species and (c) evolution of new species. All of these processes may contribute in time to a greater diversity within the community. It accounts for increased diversity during ecological succession and also for differences between tropical, temperate and tundra communities, if we may assume that tropical ecosystems are considerably older than these others. But it offers no real explanation for the mechanism involved: it is an observation rather than a true explanation, and in truth is also really a statement of the obvious. Species richness can only increase through immigration to an area or through evolution of new species, and both do take time. The hypothesis does not explain what allows the community to become more diverse in the first place.

An alternative hypothesis claims that diversity may be a function of the biological productivity of a region - that, in our context, the tropics are more diverse than the arctic because they are more highly productive. There is no doubt that very frequently an increase in species diversity, a community, or between communities, is associated with an increase in primary productivity (e.g. Cody, 1974; Brown and Davidson, 1977). But equally such a relationship is not invariable and an increase in productivity may in practice be accompanied by either an increase or a decrease in diversity (e.g. Yount, 1956; Abramsky, 1978). We could review a host of other ideas and hypotheses, but perhaps the most plausible explanation for observed differences in the species richness of particular communities or, in our context, biogeographic regions, is that put forward by Slobodkin and Sanders (1968).

As noted earlier, any organism may adjust its tolerance limits to adapt itself so that it may colonise extremes of environmental conditions; but such adaptation to extremes is at considerable cost: an animal can only survive in really harsh conditions by becoming a physiological specialist. Specialisation to such extremes is at the expense of flexibility: of lability in tolerance of a wider range of conditions. An organism can afford to specialise in this way only if conditions are effectively constant. It has been proposed by Slobodkin and Sanders (1968) that this may be the key to understanding the different species richness of different biomes. Because organisms can adapt to harsh conditions only when these are constant, harsh and unpredictable environments cannot be colonised by any animal or plant. Severe but constant environments, or those which are variable but subject to predictable change like the arctic tundra, may be colonised by physiological specialisation, but because of the necessary commitment involved in such specialisation, these areas are normally relatively species-poor. As severity decreases or predictability increases, the specialisation required to colonise environments declines, and more and more species can utilise any particular environment. Ultimately, in favourable and predictable environments, a whole host of organisms may survive (and, as severity decreases, so too, predictability in fact becomes less important). Thus, diversity is seen to increase as environments become more favourable and/or more predictable: and we observe a dramatic increase in the diversity of animal species from tundra to tropics. Movement from the tundra to the tropics will indeed be a movement from environmental severity to an environment which is more favourable and stable.

These various considerations help us to understand how many animal species may occur in a given biome and which particular animal species, because of their specific environmental tolerances, may be found there. But, our deliberations so far have been concerned primarily with latitudinal zonation (our subdivision of the globe into gross climatic regions is essentially a latitudinal classification); we have ignored completely equally apparent differences in animal species composition, even within a constant climatic zone, with longitudinal change.

The Different Zoogeographic Regions of the World

Because geographical barriers and limits to free movement prevent the distribution of any one organism to fill all instances worldwide of the niche for which it may be adapted, equivalent niches in different regions are occupied by different 'local' organisms which have evolved in that region to fulfil that particular ecological role. Thus while in equivalent communities across the globe the same 'jobs' may be available, the vacancies are not always filled by the same organism in every region. Certain species are more cosmopolitan, more widespread in their distribution than others, but many of the animal species of the different regions of the globe are peculiar to one particular region. Geographical barriers, present and past ocean currents, deserts and mountains, and resultant isolation, have led in each case to the development of a distinct 'local' fauna.

The most obvious illustration of this is the amazing development of marsupial fauna in Australia. Precise ecological analogues were developed for many eutherian species, so that the ecological jobs within the community are filled in the same way, but with a fauna unique to that particular region.

But this is only an extreme example of a more general phenomenon. The degree of isolation (and the length of time for which that isolation was maintained) influences the degree of distinctiveness of the fauna. In addition, a constant isolation in geographic or geological terms has different relative effects on different relative size-scales of organism (dependent upon their dispersal ability), so that boundaries between faunal assemblies are not always so sharp, or may differ for different taxonomic groups. Nonetheless it is apparent that there are a number of distinct faunal regions of the globe. It is customary to identify six main regions.

The Palearctic region is the northerly part of the Old World. The region includes Eurasia, north of the tropics, and the Mediterranean coast of Africa. The climate is on the whole temperate but extends north into tundra and taiga zones.

The Nearctic region covers the whole of North America, and extends South to the middle of Mexico. It includes Greenland to the east and the Aleutian Islands to the west. The range of climate and vegetation in this region resembles that of the Palearctic. Although a number of animal species are unique to this region, with few exceptions the families involved are also represented in the Palearctic. Many authors thus combine these two regions into a single category, the Holarctic.

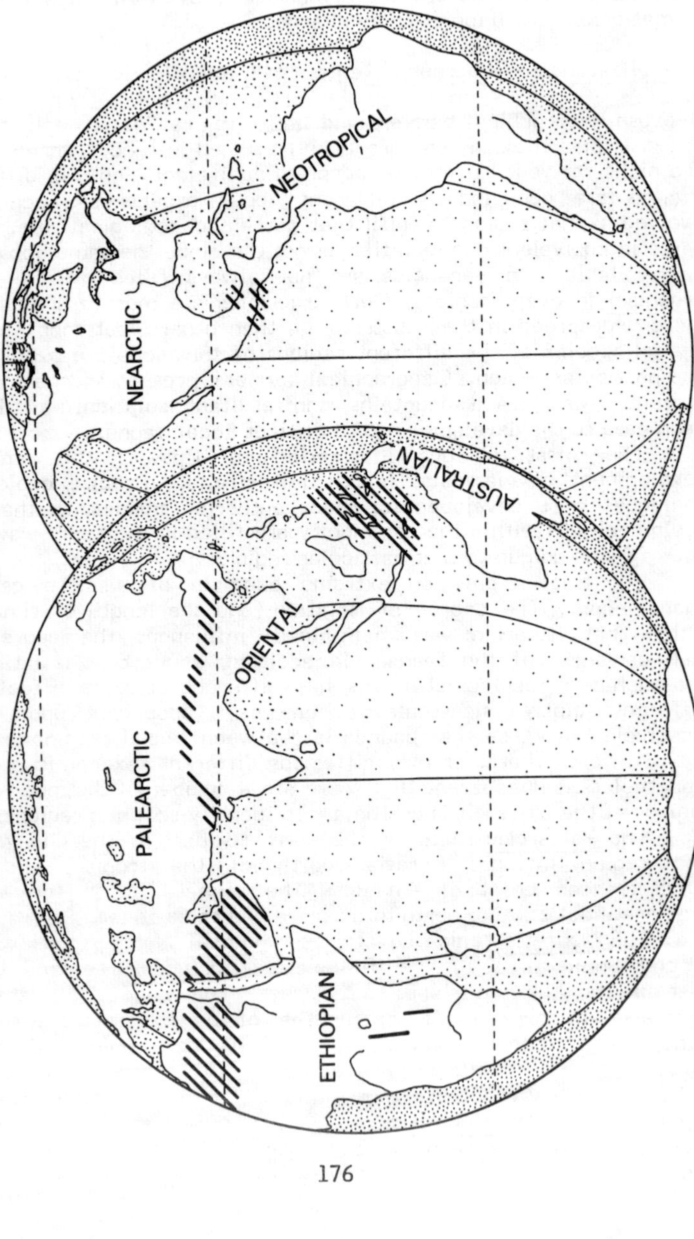

Figure 6.5: The six continental faunal regions (after Wallace 1876); diagonal hatching shows approximate boundaries and transition areas (after Darlington, 1957)

The Ethiopian region embraces the major part of the old world tropics: essentially the whole of Africa (except the northern corner) with part of Southern Arabia. The Ethiopian region shares with the more northerly Palearctic families of dormice, jerboas, coneys and wild horses. But it differs markedly from that region in being without moles, beavers, bears and camels. It has twelve unique families of mammals and a vast number of endemic species, birds, reptiles and amphibians.

The Oriental region abuts the Ethiopian and shares with it eight of its mammal families (including old world monkeys, lorises, apes, pangolins, elephant, rhinoceros and chevrotains). The region includes India, Indochina, South China and Malaya, with the westerly islands of the Malay archipelago. Once again, the climate is essentially tropical. The fauna resembles that of the Ethiopian region but it is not so varied and is not so rich in endemic families.

The Australian region: Australia, Tasmania, New Guinea and associated islands. A region with a unique fauna as we have already noted with the dominant mammals being marsupial.

The Neotropical region: Like the Australian region the Neotropics are almost completely isolated from other faunal regions by large ocean barriers. The region consists basically of the whole of South America, but includes also most of Mexico and the West Indies. Its nearest neighbours are the Nearctic to which it is joined by a narrow isthmus, and the Australian region. The region is mostly of tropical climate and has a distinctive and varied fauna. Many of the commonly occurring mammal and bird families have no representatives in the Neotropics, while the area has the highest number of endemic families of any faunal zone.

(Our treatment and description of the main faunal regions and their characteristics is necessarily sketchy. Darlington (1957) notes 'It is impossible to describe the regional faunas adequately in a few pages. Wallace (1876) devoted nearly 500 pages to them'. The reader unfamiliar with the details of these regions should consult a specialist zoogeographical atlas, such as Darlington (1957) or Crowther (1962).)

The boundaries of these various faunal zones are, as might be expected, by no means rigid. Many animal families, many individual species, occur in more than one zone; only the Neotropical and Australian regions, bounded by extensive oceans, are anywhere near distinctive or in any sense spatially discrete. Each region owes such uniqueness as it may possess to its periods of isolation. As land-bridges between the major continents formed and broke, as oceans and mountains rose and fell during geological history, so each region became successively joined to or isolated from first one, then another of its neighbours. Recent or present connections are represented by shared species, connections in the past followed by isolation: by shared families which have then independently undergone their own adaptive radiations. Markedly discontinuous animal distributions of the present day (such as tapirs, which occur only in Malaysia and in South America) show where land connections may have existed in the past - or may be relics of a much wider distribution in prehistory, extending over the whole area between the two, currently relict, populations.

Thus, we find transitions and discontinuities between the regions because, although they are described in terms of current faunal distributions, that present-day distribution is the result of evolutionary and geological change, and distribution in times past. Further, the situation is in continuous flux, as present day faunas continue to disperse and colonise areas open to them by courtesy of present day bridges and barriers between the regions. Perhaps the clearest illustration of this is 'Wallacea': the area of transition between the Oriental and Australian regions. Between these two zones of the Old World tropics is an enormous area where the two faunas mix. The main land masses of the two regions are distinct, but they are separated from each other by several island chains, the islands of the Malay archipelago, with varying mixtures of the two faunal types. Early zoogeographers tried to define the boundaries of the Oriental and Australian regions. The famous English naturalist Alfred Russel Wallace (who based his theories of evolution on sixteen years' study of the faunas of this region) first drew the dividing line between Bali and Lombok, between Borneo and the Celebes and between the Philippines and the Moluccas (Figure 6.6). Later authors debated the position of this line and some years after, another line was drawn between the two regions because it was thought to divide the two faunas more accurately than Wallace's line. This new line (Weber's Line) was based mainly on observations of the mollusc and mammal faunas of the area, and runs between the Moluccas and the Celebes (now Sulawesi), between the Kei Islands and Timor (Figure 6.6).

Although based on present-day faunal observations, Wallace's Line is drawn coincidentally along the geological limit of what used to be a large land mass joined to Malaya. Weber's Line more or less marks the western limit of what was at one time the extent of the Australian continent. It is clear that what is to the west of Wallace's Line is Oriental; to the east of Weber's Line the fauna is clearly Australian. But between the two are a number of islands of intermediate fauna. In despair of ever being able to draw a single line to divide the two regions, many zoogeographers have suggested keeping both Wallace's and Weber's lines and recognising the region between them for what it is, a transition zone: Wallacea.

In an earlier paragraph we considered differences in species diversity between different biomes and possible explanations for these differences in species richness. In the same sense, what dictates, or limits, the diversity of animal species which may develop in any one zoogeographic region? To some extent, as we have seen, the 'uniqueness' of the fauna and its diversity, depend upon the degree of isolation of that particular region from adjoining zoogeographic zones and the length of time for which it has remained isolated. Once barriers are removed, creatures from other regions may invade to fill any remaining gaps in the species assembly. But these are proximate answers only. What ultimately determines the number of species that may develop, through evolution or colonisation?

Figure 6.6: Wallacea and the transition between the Oriental and Australian faunal regions (after George, 1962)

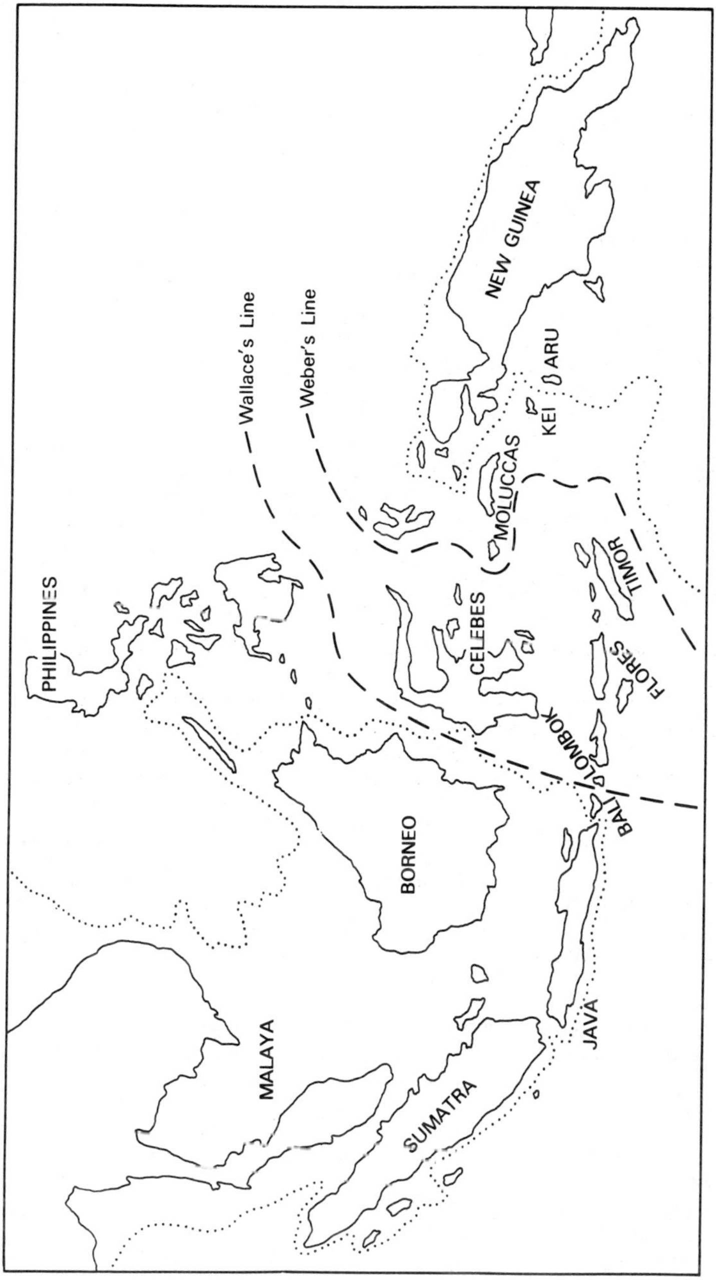

Global diversity

Clearly, the actual number of species in existence in our world and their geographic distribution are a complex function of local evolutionary selection pressures and adaptive radiations, coupled with biogeographical diversions and barriers. The total diversity of any geographic region, or of a single taxonomic group with that region, is a function of local selection pressures and radiations, of niche availability and needs; because biogeographical barriers and limits to free movement prevent the distribution of any one organism to fill all instances of the niches worldwide for which it may be adapted, such radiations and developments can be repeated and replicated in each separate biogeographical 'continent', as we have already noted. Global 'diversity', in this case species number, is thus far greater even than the number of potential niches, for equivalent niches in different regions are occupied by different 'local' organisms.

However, any particular environment must have a limit to the number of potential niches available. For a time, new species can be accommodated by progressive 'niche-narrowing' of existing and established members of the community: but sooner or later the environment becomes saturated, and no new species can be accommodated. Only when the environment changes markedly or some new biological development opens up new environments or new potential niches, can the radiation recommence.

Thus we may envisage increases in the global diversity of animal and plant species as a series of discrete stepwise expansions. Gould (1979) has suggested that such expansions are likely to sigmoid, pointing out that the process of speciation in a new environment is exactly analagous to that of the growth of animal and plant in virgin resources. Thus increase in the number of species is slow at first (rates of speciation are as fast as they will ever be, but the founder species are few in number, so the development of new species from them is slow). As the numbers increase, in geometric progression, we enter an explosive phase of development, but this increase cannot continue indefinitely and, as the environment nears saturation, the rate of increase declines, and the 'population' levels out. Such a sigmoid pattern may be repeated over and over; as, for whatever reason, a new environment becomes available for colonisation, so the process is regenerated (Fig. 6.7).

But as at each stage the existing environment approaches saturation, so the number of species must tend towards an upper limit (until a major new environment is opened). With no new niches to be exploited, a new species can evolve only at the expense of the extinction of an old one. As the environment approaches saturation so it becomes progressively more and more difficult to establish a new species; species number and diversity become a constant. As the Red Queen advised Alice in 'Alice Through the Looking Glass': 'From here on it takes all the running you can do just to stay in the same place' (Carroll, 1886). Such an ecological phenomenon has been formalised by van Valen (1975) as 'The Red Queen Hypothesis' (e.g. Van Valen, 1975; Stenseth, 1980). But what determines saturation

Figure 6.7: Increase in total number of species in an area with time, through evolutionary development. A series of sigmoid curves is generated as, after each 'explosion' in response to environmental change, species number approaches saturation

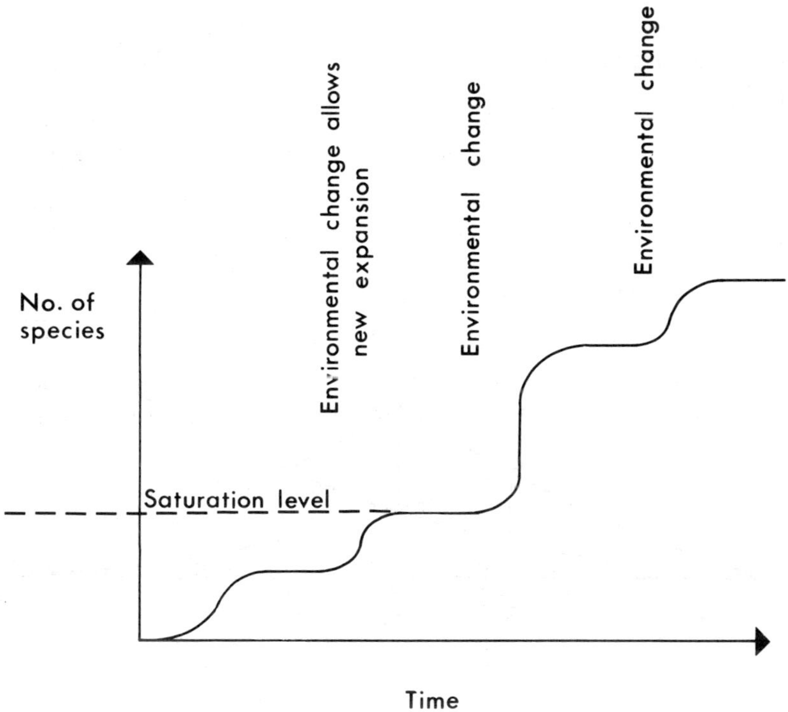

level? Clearly the number of different species which can be developed (number of different niches which can be supported) depends upon a number of factors, notably available area of the region and habitat heterogeneity.

The Theory of Island Biogeography: Species Number and Island Size

The size, as area, of a zoogeographical region or ecological habitat has a profound influence on species number developed, and not solely through its role in determining the potential number of niches which

Figure 6.8: Relative number of species on small, distant islands (S_1) and large, close islands (S_2) predicted by the MacArthur-Wilson equilibrium model. The number of species on small, close islands and large, distant islands is intermediate (after MacArthur and Wilson, 1967)

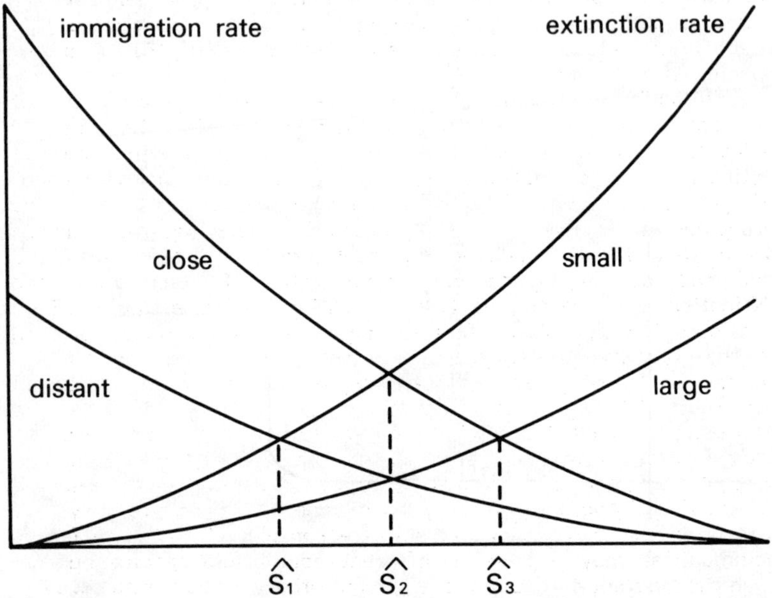

may be available.

At saturation, we have suggested that the species number of any geographical area must remain constant: the establishment of a new species, through evolution or immigration, can only be accomplished at the expense of the extinction of some other species. This same pattern of rapid growth to saturation, and maintained equilibrium between establishment and extinction is as true for the establishment of organisms within a 'present-day' community, by colonisation, as it is for the development of species on a regional scale through evolution. Both zoogeographic regions and the communities within them exist in balance between successive colonisations and extinctions.

In an analysis of the way in which rates of colonisation and extinction might influence the diversity (species number) on islands, MacArthur and Wilson (1967) came to the conclusion that both colonisation and extinction rates were functions of the number of species already on the island. The rate of immigration of new species to the island decreases as the number of species on the island increases (as more and more of the potential mainland colonists are found on the island, fewer of the new arrivals constitute new species); the rate of extinction likewise increases with the number of species on the island.

In addition, rates of colonisation were shown to be a function of the distance of the island from the nearest mainland (or 'seeding' point from which migrants might reach the island), while rates of extinction were a function of island size. With the relative number of species on a given island expressed as an equilibrium between the processes of colonisation and extinction, species number can be shown to differ on islands of different size, and distance from the mainland, as varying factors affect rates of colonisation and extinction and produce different equilibria (Figure 6.8). Such considerations may also affect the equilibrium diversity expressed by any community. Although it may not represent a physical island surrounded by seas, even a mainland ecosystem is in effect an ecological 'island', in that it represents a particular and discrete area of its type, surrounded by a 'sea' of systems of different types. It, too, may be defined in terms of its area, or distance from the nearest 'mainland' of a similar system. Thus, for these mainland communities, too, rates of colonisation and extinction may vary with 'island size' or 'distance from the mainland', and different communities may express a different equilibrium species number even though their saturation levels might otherwise be the same.

MacArthur and Wilson's initial theory is greatly over-simplified: as noted by Gilbert (1980), for example, it is essentially stochastic. In consequence it is applicable only for those species which can be considered to be non-interactive. In addition, MacArthur and Wilson make the further assumptions (i) that immigration is independent of island size, and (ii) that, while immigration rates decrease with increasing distance from the mainland, extinction rates remain the same. Yet it has been suggested by Osman (1977) that larger islands will receive more colonisers because they present a larger 'catchment surface' to the migrant species. Brown and Kodric-Brown (1977) propose that extinction rates will be unlikely to be independent of distance from the mainland. The arrival of individuals belonging to species already represented on the island will reduce rates of extinction. This 'rescue effect' will be lower for more distant islands because of the reduced rate of arrivals of individuals, and thus extinction rates will increase with increasing distance. Nevertheless, such modifications merely affect the shape of colonisation and extinction curves and do not affect significantly the basic conclusions.

Williamson (1981) concludes, however, that 'the MacArthur and Wilson theory is true, to the extent that there is an equilibrium between immigration and extinction, but is essentially trivial'. He argues that the major factors influencing species abundance relationships are biotic factors such as heterogeneity of the environment.

The Structure of Animal Communities

Not all the species present in any one geographic region or biome occur in every ecological community within that area. Such communities are formed of selected 'subsets' of the overall potential species pool. The title of this essay is 'The geography of animal communities' and it seems fitting that we should conclude with some consideration of the factors determining the structure and species composition of *individual* communities.

One of the most curious features of the animal communities of different regions is the surprising constancy of design shown by communities of similar function. There is a remarkable constancy of trophic design, of trophic structure, in communities of similar type wherever they may occur, but in addition there appears surprising constancy of the more detailed structure within this. There is a striking similarity not only of overall form, but also of niche-form and structure: of 'the way the jobs are partitioned out'.

In a study of coral reef fish communities, Gladfelter *et al* (1980) found almost constant species composition on comparable reefs in the Western Atlantic. Still more remarkable, the species present on reefs in the Central Pacific were *also* very much the same: Gladfelter *et al* demonstrate, in brief, that particular reef structures supported almost constant communities.

Even where these communities are in different biogeographic regions and thus cannot share exactly the same species, there remains a striking similarity of niche form and structure. The same *jobs* are there, merely held by different organisms. Cody (1974, 1975), for example, has shown that the potential vacancies for insectivorous birds in the equivalent systems of Californian chaparral, Chilean matorall and South African macchia are divided up in precisely the same way. These various grassland systems may be considered ecological analogues within three different continents. Each is colonised by taxonomically distinct groups of insectivorous birds, yet the way the resources are divided between them shows this striking constancy of pattern, suggesting almost a unique efficient design for the community. Similar results have been shown for assemblages of lizards in montane areas of Chile and California by Fuentes (1976), although Lawton (1982) *failed* to find such constancy among the phytophagous insects of bracken stands in England and America.

The classic example of constancy of trophic structure within a community comes from the work of Simberloff and Wilson (1969) who eliminated the fauna from several very small mangrove islets in the Florida Keys, and then monitored their recolonisation by terrestrial arthropods. In all cases the total number of species on an island returned to approximately its original value, although the actual

Table 6.1: Evidence for stability of trophic structure

Island	Trophic classes																	
	H		S		D		W		A		C		P		?		Total	
E1	9	(7)	1	(0)	3	(2)	0	(0)	3	(0)	2	(1)	2	(1)	0	(0)	20	(11)
E2	11	(15)	2	(2)	2	(1)	2	(2)	7	(4)	9	(4)	3	(0)	0	(1)	36	(29)
E3	7	(10)	1	(2)	3	(2)	2	(0)	5	(6)	3	(4)	2	(2)	0	(0)	23	(26)
ST2	7	(6)	1	(1)	2	(1)	1	(0)	6	(5)	5	(4)	2	(1)	1	(0)	25	(18)
E7	9	(10)	1	(0)	2	(1)	1	(2)	5	(3)	4	(8)	1	(2)	0	(1)	23	(27)
E9	12	(7)	1	(0)	1	(1)	2	(2)	6	(5)	13	(10)	2	(3)	0	(1)	37	(29)
Total	55	(55)	7	(5)	13	(8)	8	(6)	32	(23)	36	(31)	12	(9)	1	(3)	164	(140)

The table is after Heatwole and Levins (1972). The islands are labelled in Simberloff and Wilson's (1969) original notation, and on each the fauna is classified into the trophic groups: herbivore (H); scavenger (S); detritus feeder (D); wood borer (W); ant (A); predator (C); parasite (P); class undetermined (?). For each trophic class, the first figures are the number of species before defaunation, and the figures in parentheses are the corresponding numbers after recolonisation. The total number of different species encountered in the study was 231 (the simple sum 164 + 140 counts some species more than once). From May, R.M. (1981) 'Theoretical Ecology'.

species constituting the total were usually altogether different. Heatwole and Levins (1972) reanalysed this data in terms of trophic organisation, listing for each island the number of species in each of the trophic categories herbivores, scavengers, detritus feeders, wood borers, ants, predators and parasites (Table 6.1). Their results show that, in terms of trophic structure, the pattern is one of striking stability and constancy. On the other hand, in terms of the detailed taxonomic composition of the community of arthropod species on a particular island, there is great variability.

Species number and species composition of such communities might be considered largely a matter of random chance: the chance colonisations and extinctions of ecological islands of different size determining species richness of the community and which actual species occur. Clearly, however, there are constraints. Any community is built upon its primary producers: the autotrophic organisms (usually green plants) which sythesise the organic matter used by the rest of the community; the species array within the plants thus influences much of the structure and species composition of the community which develops upon this level. For any set of environmental conditions (climate, temperature, wind, moisture, soil-type, etc.) there is a characteristic set of dominant plant species which may occupy an area. These, truly, are limited primarily by the physiological limits to tolerance with which we began this essay. Associated with this particular species spectrum of plants, will be those herbivorous animals whose particular food species are represented. Associated with them will be the carnivores which prey upon those particular herbivores and so on. Once the selection of primary plants is made, the rest follows automatically, and it is surprising how constant species associations are, at least in the lower trophic stages. (Higher up the trophic scale such relationships are not so rigid: higher consumers are able to exploit a wider variety of food species, and thus tend to be much less community-specific.) At each stage, the selection of animal and plant species is of course still restricted and controlled by abiotic tolerances: the whole is the complete complex of abiotic restrictions and biotic interactions introduced in earlier paragraphs.

But such a picture is deceptively oversimplified. Although it is true that there are well established species associations between both animal and plant species and animals and other animals, it is also true that these frequently have alternatives. What decides which of these alternatives will be represented in any particular community? From a study of bird assemblies in a number of different islands around New Guinea, Diamond (1975) proposed a number of 'assembly rules' for such communities (once again we may take rules derived for true islands and apply them to mainland communities as 'ecological islands'). Diamond asserts that:

> If one considers all the combinations that can be formed from a group of related species, only certain ones of these combinations exist in nature. Permissible combinations resist invaders that would transform them into forbidden

> combinations. A combination that is stable on a large or species-rich island may be unstable on a small or species-poor island. On a small or species-poor island, a combination may resist invaders that would be incorporated on a larger or more species-rich island. Some pairs of species never coexist, either by themselves or as a part of a larger combination. Some pairs of species that form an unstable combination by themselves may form part of a stable larger combination. Conversely, some combinations that are composed entirely of stable subcombinations are themselves unstable.

Examining data from 147 species of land birds distributed in various combinations over 50 islands in the Bismarck Archipelago near New Guinea, Diamond argues that much of the explanation for these assembly rules has to do with competition for resources and with harvesting of resources by permitted combinations so as to minimise the unutilised resources available to support potential invaders.

He claims that 'communities are assembled through selection of colonists, adjustment of their abundances, and compression of their niches, so as to match the combined resource consumption curve of all the colonists to the resource production curve of the island'. However, a more recent analysis of Diamond's data by Connor and Simberloff (1979) demonstrates that such conclusions are unjustified. They show that most of Diamond's observed distributions might be expected if the species were distributed randomly on the islands, and that thus they offer no support for such 'assembly rules'. Indeed they conclude (Connor and Simberloff, 1979):

> Every assembly rule is either a tautological consequence of the definitions employed, a trivial logical deduction from the stated circumstances, or a pattern which would largely be expected were species distributed randomly on the islands.

Another possible approach to gaining some insight into the 'rules' of community organisations may come through analysis of the details of niche structure and organisation. Limits to the design of individual niches in size or shape, limits to similarities of adjacent niches in the same communities, limits to overlap and to the method of 'packing' of niches within the community may have profound implications for the way communities are structured.

The community however is a complex entity. For many years ecologists have considered populations, and have concentrated their efforts on unravelling the dynamics of those populations themselves. 'Community ecology', says Pianka (1980), 'is in its infancy: it will be some time before all its rules are discovered'.

REFERENCES

Abramsky, Z. (1978), 'Small mammal community ecology: changes in species diversity in response to manipulated productivity', Oecologia, **34**, 113-24.

Barbour, M.G., Burk, J.H. and Pitts, W.D. (1980), Terrestrial Plant Ecology, Benjamin/Cummings Publishing Co Inc, Menlo Park, California, USA.

Brown, J.K. and Davidson, D.W. (1977), 'Competition between seed-eating rodents and ants in desert ecosystem', Science, **196**, 800-82.

Brown, J.K. and Kodric-Brown, A. (1977), 'Turnover rates in insular biogeography: effect of immigration on extinction', Ecology, **58**, 445-9.

Cody, M.L. (1974), Competition and the Structure of Bird Communities, Princeton University Press, NJ, USA.

Cody, M.L. (1975), 'Towards a theory of continental species diversity' in Cody, M.L. and Diamond, J.M. (eds.), Ecology and Evolution of Communities, Harvard University Press, Harvard, USA, pp. 214-57.

Conner, E.F. and Simberloff, D. (1979), 'The assembly of species communities: chance or competition?', Ecology, **60**, 1132-40.

Darlington, P.J. (1957), Zoogeography: the Geographical Distribution of Animals, Wiley, New York, USA.

Diamond, J.M. (1975), 'Assembly of species communities' in Cody, M.L. and Diamond, J.M. (eds.), Ecology and Evolution of Communities, Harvard University Press, Harvard, USA, pp. 342-444.

Ford, E.B. (1845), Butterflies, Collins New Naturalist Series, London.

Fuentes, E.R. (1976), 'Ecological convergence of lizard communities in Chile and California', Ecology, **57**, 3-17.

George, W. (1962), Animal Geography, Heinemann, London, UK.

Gilbert, F.S. (1980), 'The equilibrium theory of island biogeography: fact or fiction?', J. Biogeog., **7**, 209-35.

Gladfelter, W.B., Ogden, J.C. and Gladfelter, E.M. (1980), 'Similarity and diversity among coral reef fish communities: a comparison between tropical western Atlantic (Virgin Islands) and tropical central Pacific (Marshall Islands) patch reefs', Ecology, **61**, 1156-68.

Gould, S.J. (1979), 'Is the Cambrian Explosion a sigmoid fraud?' in Gould, S.J., Ever since Darwin: Reflections in Natural History, Norton, New York, USA.

Haefner, P.A. (1960), 'The effect of low dissolved oxygen concentrations on temperature-salinity tolerance of the sand shrimp *Crangon septemspinosa*', *Physiol. Zool.*, **43**, 30-7.

Heatwole, H. and Levins, R. (1972), 'Trophic structure, stability and faunal changes during recolonisation', *Ecology*, **53**, 531-4.

Lawton, J.H. (1982), 'Vacant niches and unsaturated communities: a comparison of bracken herbivores at sites on two continents', *J. Anim. Ecol.*, **51**, 573-96.

MacArthur, R.H. and Wilson, E.O. (1967), *The Theory of Island Biogeography*, Princeton University Press, NJ, USA.

May, R.M. (ed.) (1981), *Theoretical Ecology: Principles and Applications* (2nd edition), Blackwell Scientific Publications, Oxford, UK.

Osman, R.W. (1977), 'The establishment and development of a marine epifaunal community', *Ecol. Monogr.*, **47**, 37-63.

Shelford, V.E. (1913), *Animal Communities in Temperate America*, University of Chicago Press, Chicago, USA.

Simberloff, D.S. and Wilson, E.O. (1969), 'Experimental zoogeography of islands: the colonisation of empty islands', *Ecology*, **50**, 278-96.

Slobodkin, L.B. and Sanders, H.L. (1969), 'On the contribution of environmental predictability to species diversity' in *Brookhaven Symposia in Biologica*, **22**, 82-95.

Stenseth, N.C. (1979), 'Where have all the species gone? On the nature of extinction and the Red Queen Hypothesis', *Oikos*, **33**, 196-227.

Svihla, A., Bowman, H.R. and Ritenour, R. (1951), 'Prolongation of clotting time in dormant estivating mammals', *Science*, **114**, 298-9.

Van Valen, L. (1977), 'The Red Queen', *Amer. Nat.*, **111**, 809-10.

Wallace, A.R. (1876), *The Geographical Distribution of Animals*, Macmillan, London.

Walter, H. (1973), *Vegetation of the Earth in Relation to Climate and the Ecophysiological Conditions*, Springer-Verlag, New York, USA.

Whittaker, R.H. (1970), *Communities and Ecosystems*, Macmillan, London.

Whittaker, R.H. (1975), *Communities and Ecosystems*, 2nd edition, Macmillan, London.

Yount, J.L. (1956), 'Factors that control species numbers in Silver Springs, Florida', Limnol. Oceanogr., 1, 286-95.

FURTHER READING

Cody, M.L. and Diamond, J.M. (eds.) (1975), Ecology and Evolution of Communities, Harvard University Press, Harvard, USA.

Collier, B.D., Cox, G.W., Johnson, A.W. and Miller, P.C. (1974), Dynamic Ecology, Prentice-Hall, NJ, USA.

Darlington, P.J. (1957), Zoogeography: the Geographical Distribution of Animals, Wiley, New York, USA.

George, W. (1962), Animal Geography, Heinemann, London.

Pianka, E.R. (1978), Evolutionary Ecology, Harper and Row, London.

Putman, R.J. and Wratten, S.D. (1984), Principles of Ecology, Croom-Helm, England.

Whittaker, R.H. (1975), Communities and Ecosystems, Macmillan, London.

Chapter 7

SOILS IN ECOSYSTEMS

R. T. Smith

SOILS, SYSTEMS AND SCALE

Biogeography and Soils

Biogeography has a number of current meanings. For many it represents a concern for biological distributions, a meaning generated early on this century and happily perpetuated by biologists. Yet this narrow, inventorial image no longer provides a true picture of the subject as both Watts (1978) and Simmons (1979a, 1979b) have commented. Though concerned to explain biological distributions in terms of environmental factors, biogeography is bound up with the intriguing question of man's relationship with ecological systems. Broad and integrative in concept, the name, like many others, is one of convenience, and soils have traditionally been included as part of biogeography teaching. This is a logical position but it must also be recognised that there are other valid functional approaches to the study of soils. Nevertheless, in representing the interface between lithosphere and atmosphere, soil forms the very foundation of the terrestrial biosphere. It represents the slow fermentation of geological formations and is the intermediary between the atmospheric fluxes and the essentially long-term business of landform evolution. Although ever a question of degree, soil has an intrinsic biological character and origin, brought about through the medium of water, and it is therefore more than simply weathered rock or sediment.

It is possible to define soil in different ways according to one's viewpoint and three main types of definition, or elements within all-embracing definitions, are encountered. There are those which are concerned with the functions of soil from its support of higher plants and soil-dwelling animals to its non-agricultural uses; those which are substance-orientated and set forth the properties of the soil medium and, finally, those which are system-orientated and which specify the formative influences of the environment upon soils (Brady, 1974). In biogeography, soils are viewed in relation to the various factors which have created their diverse forms and potentialities

(Cruickshank, 1972; Foth and Schafer, 1980; Bridges and Davidson, 1982). Inevitably, this generates a need to study soil properties and processes, so that the study of soils by biogeographers should form an ecological basis both for the study of land utilisation and for the evaluation of land resources (see Taylor, Chapter 9 herein).

Soil Systems

> The principal object of natural philosophy is not the form of the material elements but the composite thing, the totality of the form, independently of which those elements have no existence. (Aristotle)

The Concept. As the biogeographical approach to soils is concerned with environmental interactions, it clearly falls within a systems framework. The theory of general systems recognises that all phenomena, and even ideas, are basically connective and the significance of any individual phenomenon depends on its relationships with others (von Bertalanffy, 1950; Stoddart, 1965; Chorley and Kennedy, 1971; Dury, 1981; Wilson, 1981a). The recent emergence of system-based studies and the current popularity of multidisciplinary fields such as ecology reflect a rise to maturity above the intellectual debris of compartmentalism and specialisation, necessary though the latter may have been. This represents a re-emergence of the holism which characterised classical conceptions of nature. Yet attractive though systems ideas may be in broad outline, their detailed examination in relation to various aspects of soils reveals just how complex the relationships are, and their resolution in quantitative terms, except on the most generalised lines, would appear to be a long way off (Jenny, 1941; Trudgill, 1977; Huggett, 1982). Emphasis here, nevertheless, will be on dynamic aspects of the soil system rather than on form and classification which have tended to dominate - even to plague - the teaching of soils in past decades.

Soil as a Dynamic System. Whether we are dealing with a natural soil or an artificial medium, internal biological and biochemical processes cease to operate under extremes of drought and frost (Russell, 1973; Richards, 1974). In the majority of soils a state of perpetual flux exists with nutrients being added and removed in the manner set out in Figure 7.1. Losses resulting from harvesting crops and gains arising from the addition of agricultural chemicals must also be included. From an ecosystem standpoint, the physiognomy of this system - the values which may be assigned the various fluxes - is second only in importance to the provision of sunlight as a basis for plant growth and soil development. It will be appreciated that the structure of the vegetation and the pattern of plant growth throughout the year will be important factors in retaining nutrients within the system. Although exchanges of energy and nutrients take place within the soil, and sometimes considerable morphological

Figure 7.1: A simple model of a soil and vegetation nutrient system (after Trudgill, 1977)

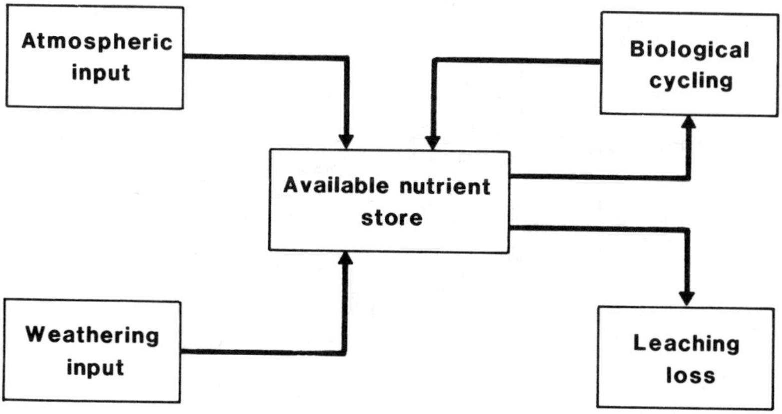

changes arise from seasonal moisture fluctuations, little perceptible change may occur to the morphology or chemistry of the soil from one year to the next. In this respect, many soils can, in the short term, be described as being in a steady-state or in dynamic equilibrium (Trudgill, 1977; Huggett, 1982). But clearly, as an open system, soil must yield in time to the vicissitudes of weathering and geomorphic activities (Figure 7.2). In the long term, many processes, including accretion, truncation, weathering and leaching, act to change both the dimensions and internal horizonation of soil profiles. Such changes, which can be generated quite rapidly by man's influence, may also trigger fundamental changes in pedogenesis (Figure 7.3).

Factors of Soil Formation. The formation of soils depends on nested sets of environmental systems which have varying roles at different terrestrial scales. On the largest scale, soil formation is determined by the climatic template and the temperature and moisture limits it sets for plant formations. However, the role, let alone the relative importance of a particular factor in pedogenesis, is difficult to assess except by allowing that factor to vary while as many others as possible are kept constant (Birkeland, 1974). Various authors have selected attributes such as clay content, depth of carbonate layer, nitrogen or organic matter content (Jenny, 1935) and demonstrated their variation as the climatic parameter is altered (Jenny, 1941). The principal effect is that of temperature on rates of chemical processes, which are accelerated 2-3 times for every $10^{o}C$ rise: the

Figure 7.2: Elements in a soil-landscape system (adapted from Statham, 1977)

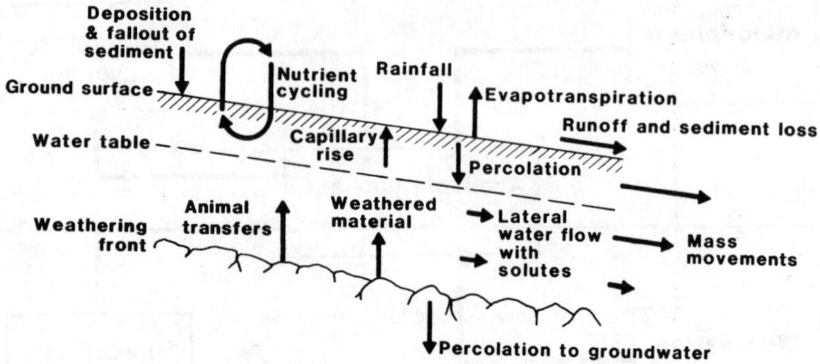

Figure 7.3: The principal factors in soil development

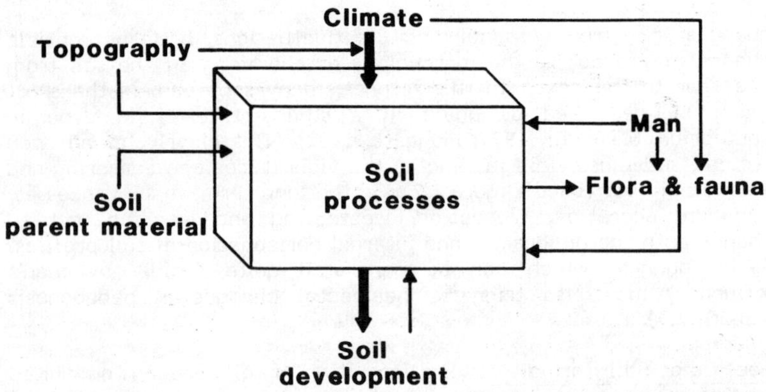

Van t'Hoff rule (see Figure 7.4, with implications for the process of hydrolysis, and also Carroll, 1970; Loughnan, 1969 and Trudgill, 1982).

Figure 7.4: Ionisation of water with temperature

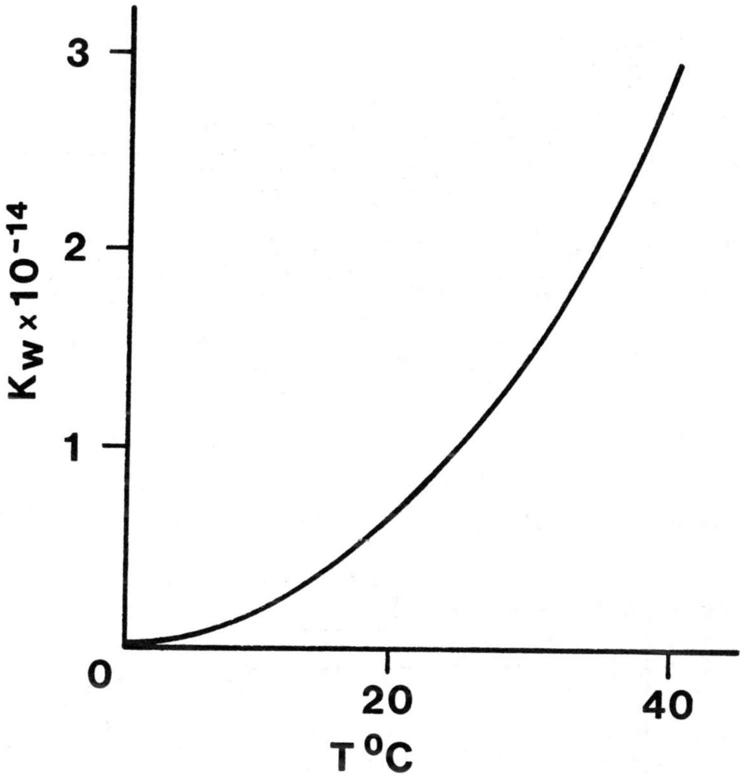

Climate was, of course, the basis for the earlier zonal soil classification scheme (Duchaufour, 1970; Brady, 1974; FitzPatrick, 1980), the main weaknesses of which, were not so much the local exceptions to climatic primacy arising from soil chemistry or geomorphic stability but the assumption that existing soils were necessarily in equilibrium with the prevailing climate. This earlier scheme is, nevertheless, still valid in defining 'soil-process regions', for the actual morphology of soils often indicates development under antecedent conditions of climate and vegetation cover. It is appropriate to note here that a quasi-equilibrium in soils takes very different lengths of time to be achieved under different climates (Figure 7.5 and see Duchaufour, 1978).

On a more local scale, soil variations depend on further templates: geological, altitudinal, geomorphic or hydrological, according to circumstances. It should be noted that in Figure 7.3, the designation 'parent material' incorporates the geological factor, while

Figure 7.5: Time requirement for the formation of diagnostic horizons (from Birkeland, 1974)

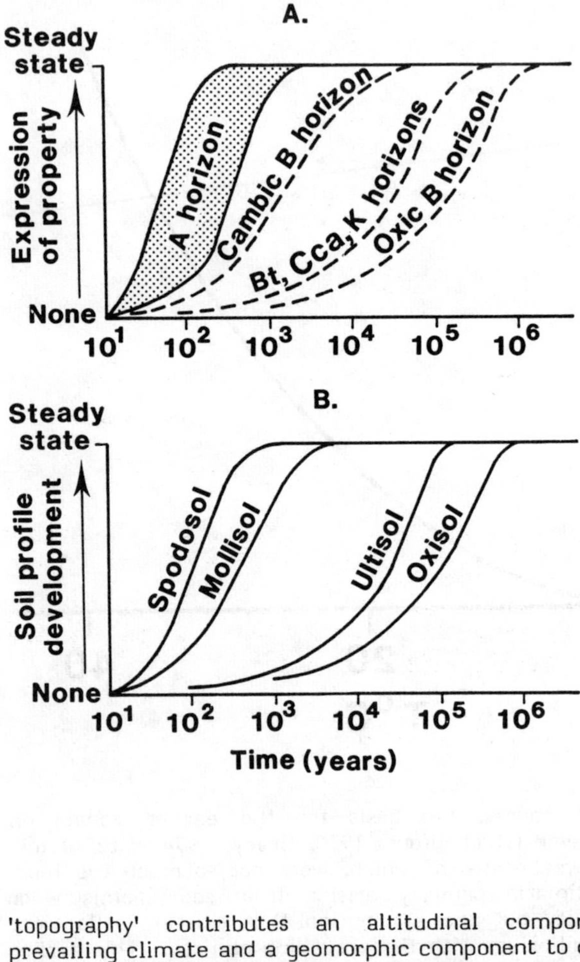

'topography' contributes an altitudinal component to modify prevailing climate and a geomorphic component to determine surface processes and water regime. Accordingly, in an area of uniform geology, whether humid or arid, topography will largely determine the pattern of soils (Birkeland, 1974; Smith and Atkinson, 1975; Conacher and Dalrymple, 1977; FitzPatrick, 1980; Gerrard, 1981).

The influence of biotic factors on soils is manifested on several scales. On the broader latitudinal and altitudinal scales, an association is evident between climatic stress and physiological response as reflected in vegetation type (Levitt, 1980). Soil variation accompanies these changes. On the local scale, patterns of soil and vegetation are dependent on details of terrain and a variety of cultural and historical factors.

Study of large-scale variations inevitably masks the individualistic influence on soil processes of particular plants. In consequence, the latter must be studied within a limited area where environmental variation is minimal. Furthermore, the influence of vegetation upon soils is not one of immediate impact but is progressive over varying lengths of time. While vegetation is a dependent variable in the first instance, the accomplishment of succession, however partial, enables the plant kingdom to win a more independent status for itself and ultimately to impart particular characteristics to soils. Classically, succession involves increased variety of species and complexity of niches, increased nutrient utilisation and biomass even though pioneer species are eliminated in the process (Odum, 1969). As succession advances, soil stability increases and, before examining the individual impact of particular plants or communities, the stabilising influence of vegetation would appear as an essential prerequisite for allowing soils to acquire their characteristic features. The reverse generally holds when disturbance occurs: ecological momentum is lost, nutrients are wasted and the system degenerates to a more primitive level if not to complete bankruptcy.

Soils and Man. Man's capacity to influence soils goes back as far as his ability to alter his vegetated surroundings and many would prefer to argue, following Jenny (1941), that his role should not be considered separately from the other state factors. One is justified, however, in opposing this view, owing to the eventual scale of Man's modification of his environment and the realisation that, unlike the actions of other animal populations, the ecological effects of man's use of fire, even on a periodic basis, were out of all proportion to population size. A further way in which man has been able to influence vegetation is by his control, selectively or otherwise, of other animal populations. The extent of late-Pleistocene faunal extinctions is itself remarkable (Dorst, 1970) and, in the context of man, it is worth considering the possible significance of a variety of controls, in particular of herbivores, on the timing and rate of forest expansion in early Holocene environments (Smith, 1982).

Nevertheless, it is the disturbance of soils by agriculture which has led to the most profound changes, often for the good of the soil in terms of productive use (Pape, 1970). Not only has agriculture imposed characteristics upon soil profiles but, in a wide range of contexts, it has created micro-landforms. These range from field divisions themselves: ditches, banks, hedges, fences, walls, to such features as terraces, lynchets and systems of ridged and sunken fields. The latter reflect a range of adaptive responses such as seeking available moisture, controlling soil drainage, avoiding frost, clearing stones and stabilising slopes. In addition, the size, shape and integration of field systems are a function of technological, cultural and social requirements.

Agriculture has been practised virtually throughout the Holocene period in tropical and temperate areas alike. Thus, around 7,000 BC, while grain crops were being cultivated in Mexico and the

Near East, New Guinea farmers were draining wetlands for growing taro and other vegetatively-propagated plants (Golson, 1977). The capacity of landscape processes to destroy evidence of former intensification of land use has been recognised, while man himself has played no small part in environmental degradation (Hyams, 1952; Hardin, 1968). Such destruction, however, takes place not only on a physical but also on a human level. For example, the lowland Mayas practised three kinds of agriculture: shifting, terrace and drained-field; only the first was to survive the eventual collapse of the Mayan culture (Turner and Harrison, 1981). There is surely a parallel here with the demise of natural ecosystems when they too become degraded by man's activities (Dimbleby, 1962).

Soil Variability. Soils have been surveyed and mapped in different parts of the world over the last 50 years, employing a variety of methods from the contiguous field-to-field approach to point-sampling on the basis of grid squares. Varied use has also been made of available air photography (Beckett and Bie, 1975; Bridges, 1982). In reality, soils form a continuum of changing properties across the landscape rather than presenting easily compartmentalised units. This presents a conceptual problem when attempting to classify them. Indeed, it is precisely the issue of homogeneous-unit versus continuum which, in the investigation not only of soil types but also of plant communities, has served both to divide and confuse (Shimwell, 1971; Kershaw, 1973). On whatever scale soils are considered, they usually present spatial mosaics rather than simple zoned distributions, with fully characterised profiles occupying only a proportion, and often a minority proportion, of a given survey area (Beckett and Webster, 1971; Bie and Beckett, 1971; Hodgson, 1978). How often, for instance, does the student learn from field work that 'textbook' soil profiles are the exception rather than the rule? Soil mapping units are thus statements of probability while soil boundaries merely reflect the changing predominance or proportionality of particular or modal profiles. At best they represent lines across which the rate of change in one or more soil characteristics is more marked or more consistent than elsewhere. It will be evident that the only effective way to consider soils is as individuals, and for this reason, the pedon (Brady, 1974; USDA, 1975) or pedounit (FitzPatrick, 1983) has become the basic unit of study for the spatial analysis of soils. Soil classes, as mapped, thus comprise collections of individuals and, as with the delineation of vegetation communities, they will each tend to possess different amounts of internal variability.

Ecosystems and Soils

As our planet comprises one vast ecosystem, arbitrarily demarcated portions of it are merely ecological subsystems. Nevertheless, in practice, we study ecosystems on different scales from the biome to the individual stand of vegetation. Indeed, in order to derive data which may be extrapolated to the large-scale, we first require detailed studies of well-defined communities and such has been a principal objective of the International Biological Programme. In turn, planetary budgets can be estimated from an aggregation of the various compartments of data collected. By defining ecosystems as interactions of energy and nutrients between biotic and abiotic realms (Odum, 1971), we can see how soils comprise an essential nexus between biological systems and the physical landscapes which support them. But, if soils are part of ecosystems, it is also possible to conceive of an ecosystem existing within the soil (Sheals, 1969; Phillipson, 1971; Richards, 1974; Szabo, 1974). Although soil organisms are merely a component of a larger system, so important are they to the support of plant growth that the study of their micro-environment has become a specialist field in itself (Burges and Raw, 1967; Alexander, 1977).

Such considerations highlight the essential macrocosmic-microcosmic nature of environmental systems (Odum, 1969). Indeed, this would appear to be a pervading principle from the cosmic to the atomic scale and, together with the insights that ecosystematic study provides for the future management of the environment and the challenges which are offered in the field of theoretical modelling, this might be considered as the ultimate philosophical justification for a systems approach. Furthermore, it is normal to find that problems manifesting at one scale find their explanation at another; rates of decomposition, for example, are dependent on micro- and ultimately upon macro-climatic considerations. We may, in the process of investigating an hypothesis, move from the community scale to the physiological and finally to the biochemical scale. We may attempt to solve a problem of soil morphology firstly with reference to clay content and thereafter in terms of minerology and surface chemistry (Greenland and Hayes, 1978, 1981).

It will be seen that there can be no single concept of soils in ecosystems; there are simply differences in scale which confine our attention to particular landscape systems as they relate to soil. Soil is nothing if it is not part of an interactive system. In the following, we shall be looking at different approaches, spatial and temporal, to the study of soil within ecosystems.

SOILS AND CONTEMPORARY ECOSYSTEMS

Plant-Soil Relationships

Environmental Controls. As we have seen, vegetation is dependent on the physical nature of terrain, though in the course of succession, later arrivals are dependent upon the habitat prepared by preceding plants. Species are adapted to a variety of habitats, some having wider ranges of tolerance than others (Kershaw, 1973; Kellman, 1978). For example, while the common heath or ling (*Calluna vulgaris*) is a cosmopolitan plant, bell heather (*Erica cinerea*) inhabits the wetter end of the heathland spectrum (Specht, 1981). Particular species thus show adaptations to aquatic or bog habitats, to sand dunes or salt marsh. Figure 7.6 presents two examples of

Figure 7.6: Relationships of vegetation with terrain (from Hunt, 1972)

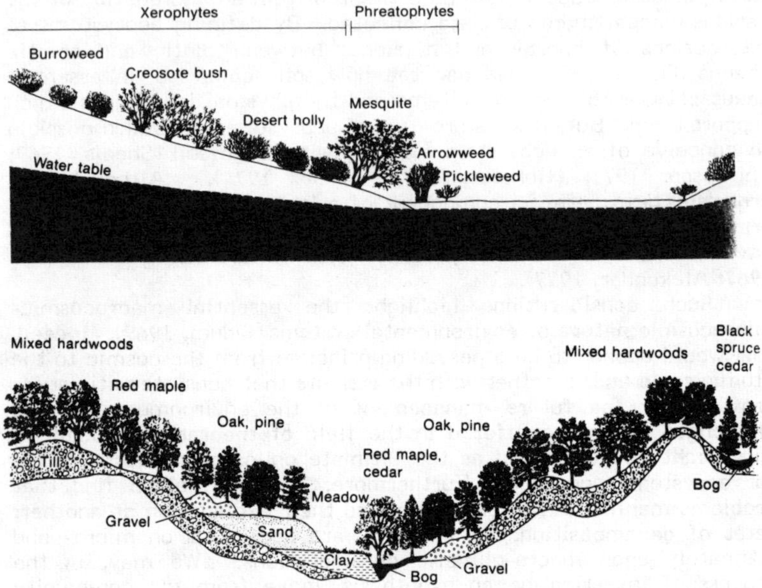

toposequences from the USA. The first comprises a true ecological gradient while the second is largely controlled by change in soil parent material. The upper drawing, a transect across Death Valley, California, reflects the availability and quality of water for plant growth. Capillary transfer from a high water-table makes the centre of the valley too saline for the growth of flowering plants. In arid situations where ground water is deeper, xerophytes will be more widespread and will extend towards the margins of playas (Hunt, 1966). The lower illustration is of a glaciated valley in Connecticut with woodland composition varying in relation to soil texture, drainage and chemistry (Lunt, 1948).

A well-known example of chemical control is associated with the presence or absence of lime. Plants are thus designated calcicoles or calcifuges according to whether they are adapted to nutrition and metabolism under calcium-rich or calcium-poor habitats. Although contrasts arise from differences in soil pH, the distinction is nevertheless a broad one, and it is very unclear as to what particular edaphic stresses individual plants are responding. For example, calcareous sites are not infrequently prone to drought, while high Ca^{2+}, high HCO_3^- and low available Fe, P and K are, individually or collectively, responsible. (Rorison, 1969; Woolhouse, 1981; Etherington, 1982). Particular species may even exhibit edaphic ecotypes capable of exploiting different soil conditions as, for example, the common bent grass (Agrostis tenuis). Species of Eucalyptus (gum) in Australia have also exploited particular soil habitats (FitzPatrick, 1980; Groves, 1981).

Plants and Soil Development. On the other hand, in a given area, the impact which plants have upon the soil depends initially on the available flora (see Figure 7.3) and conditions of drainage appropriate to the development of soil horizons. The diversity and structure of the community and its nutrient status will be of prime importance, as will the length of time that the particular assemblage of plants has occupied a site. Although it is a common observation that soils reflect the vegetation growing on them, it must be realised that such observations are more commonly made in man-dominated ecosystems in which a single dominant species imposes its signature upon the soil undiluted by contrary influences. The distinctive soil-vegetation associations of the mid-latitude grasslands and Boreal Forests are similarly attributable to a homogeneity deriving from major ecological stresses, although variations do occur, notably through ground water movements. By contrast, in the Equatorial Rain Forest it is difficult to distinguish soil variations attributable to particular plants or assemblages except in relation to parent material. Indeed, a greater variety of species thus occurs on clay-rich soils than on sandy ones, further evidence that diversity is fundamentally a function of ecological stress (Holling, 1973; Longman and Jenik, 1974; Goodman, 1975; UNESCO, 1978). Plants which grow on sandy soils do, moreover, have higher levels of phenolic substances in their leaves which are less appetising to grazer and decomposer alike. Even in the tropical rain forest, these conditions generate podzolic-type profiles (McKey et al, 1978).

Individual plants have come to be associated with particular soil-forming attributes. In the Canadian prairies, podzolisation occurs under aspen groves (Populus tremuloides) while in New Zealand 'egg-cup' podzols are associated with individual kauri pines (Agathis australis) (FitzPatrick, 1980). Among studies of the localised effects of plants on soil conditions the results of Grubb, Green and Merryfield (1969), working on ling (Calluna vulgaris), and of Zinke (1962), on redwood (Sequoia gigantea) and Lodgepole pine (Pinus contorta), are of particular interest; the latter is incorporated in Figure 7.7. Other genera such as oak (Quercus) and bracken (Pteridium) are, by contrast, associated with a richer nutrient cycle and will tend to be associated with a higher base status, cation exchange capacity and pH irrespective of parent material (Duvigneaud and Denaeyer-De Smet, 1970; Mitchell, 1973; Armson, 1977; Smith, 1977; Etherington, 1982).

Soil Organic Matter Dynamics and the Nutrient Cycle. Nutrient utilisation by plant assemblages is crucial to soil processes because

Figure 7.7: Gradients of pH and nitrogen (% by weight) of surface mineral soil under a specimen of Pinus contorta. Wind frequencies are indicated (after Zinke, 1962)

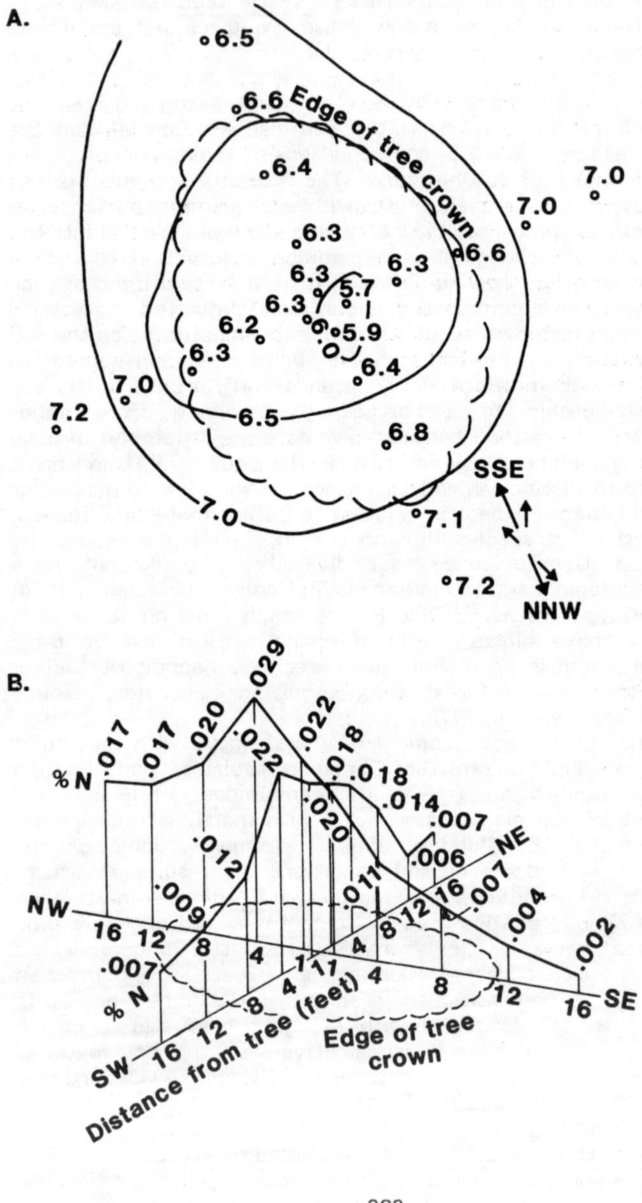

the palatability of litter affects the rate at which organic matter is returned to the soil as mineralised substance. Climatic and edaphic factors retard or accelerate microbiological decay (Parkinson, et al, 1971) while in grassland and sclerophyll communities fires supplement the process (Kozlowski and Ahlgren, 1974). Although burning accelerates mineralisation to the productive advantage of vegetation in cool regions, it does lead to air-borne loss of minerals (Humphries, 1966; Evans and Allen, 1971; Gimingham, 1972; Raison, 1979), and unless ground-level vegetation recovers quickly, leaching losses may be considerable (Nye and Greenland, 1960; Rowe and Hagel, 1973; Debano and Conrad, 1978). Decomposition of organic materials depends otherwise on a favourable balance between temperature and moisture levels. The content of organic matter which characterises particular soils is thus a function of the productivity of the habitat, the existence of any checks to decomposition and the stability of the humic compounds synthesised (Kononova, 1966; Schnitzer, 1977). Figure 7.8A shows that for soil organic-matter levels to remain constant, the annual rate of addition of materials on or in the soil (labelled production) must be matched by equal amounts of decomposition and mineralisation.

Figure 7.8: Equilibrium (A) and disequilibrium (B) in the soil organic matter store

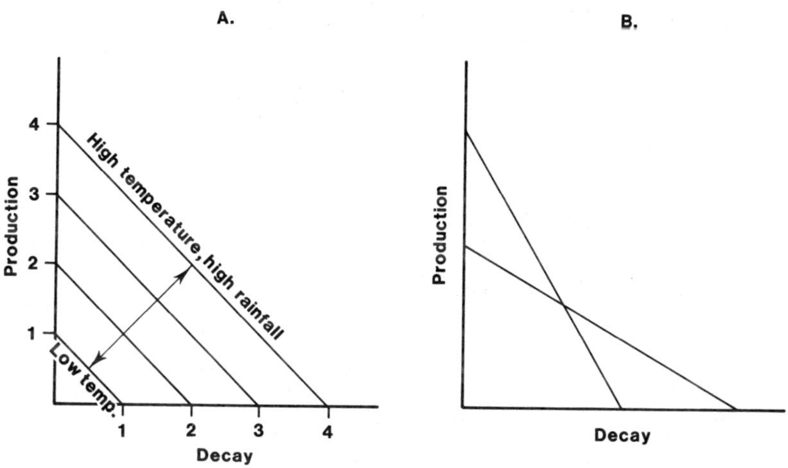

Organic matter will accumulate (Figure 7.8B) if biological activity is inhibited, as under intensely acid or waterlogged conditions, or be depleted if production fails to keep pace with mineralisation, as can happen under continuous cropping (Allison, 1973; Jenkinson, 1981).

Nutrient Cycling in Selected Ecosystems

Although very different amounts of net primary production are associated with different ecosystems (Odum, 1971; Eyre, 1978, Cannell, 1982), the significant comparisons lie (1) in the total quantities of nutrients participating and (2) in the proportions of total nutrients held within various compartments. In the Boreal Forests, growth of the predominantly coniferous vegetation and its acidophile ground flora leads to a surface accumulation of organic matter which, unless recently burned over, displays a characteristic set of layers according to its degree of alteration (Waksman, 1938; Duchaufour, 1982). In such circumstances, nutrients become locked up, permanently in the case of muskeg bog, but more often are subject to slow release (Kendrick, 1959). It is significant that mycorrhizal development on the roots of spruce (as with countless other species growing under acid conditions) renders Boreal Forest peculiarly well equipped to exploit surface organic layers (mor), the main decomposers of which are fungi. Numerical data on nutrient cycling in Boreal Forests are scarce (Greer, 1974) but figures provided by Larsen (1980) confirm the dominance of the soil as a nutrient store and the fact that uptake of calcium and nitrogen is more than matched by inputs and releases. However, the large amounts of calcium released by decomposition in mature spruce forest, together with unspecified amounts weathered and leached, point to calcium depletion of the ecosystem over time.

Table 7.1: Actual content and proportions of nitrogen in the above- and below-ground biomass of prairie grassland (Osage site, Risser _et al_, 1981)

Components	gN/m^2 (percentages of total in parenthesis)					
	1970		1971		1972	
Above ground live	0.62	(0.24)	1.75	(0.62)	1.68	(0.68)
Standing dead	2.43	(0.94)	4.31	(1.54)	3.35	(1.36)
Litter	1.11	(0.16)	1.23	(0.43)	1.60	(0.64)
Crown	1.77	(0.67)	0.79	(0.27)	1.21	(0.48)
Root	3.62	(1.38)	4.31	(1.54)	2.34	(1.15)
Soil 0-20 cm	250.25	(96.65)	265.85	(95.54)	235.30	(95.10)

Comparisons of coniferous with deciduous woodland ecosystems (Figure 7.9) consistently show the latter produce larger fluxes of nutrients from soil to foliage to soil than the former. A disproportionate amount of the flux in the case of coniferous forests

Figure 7.9: Biogeochemical cycling of four macronutrients in hardwood and softwood forests (figures are kg/ha/yr) (after Duvigneaud and Denaeyer-De Smet, 1970)

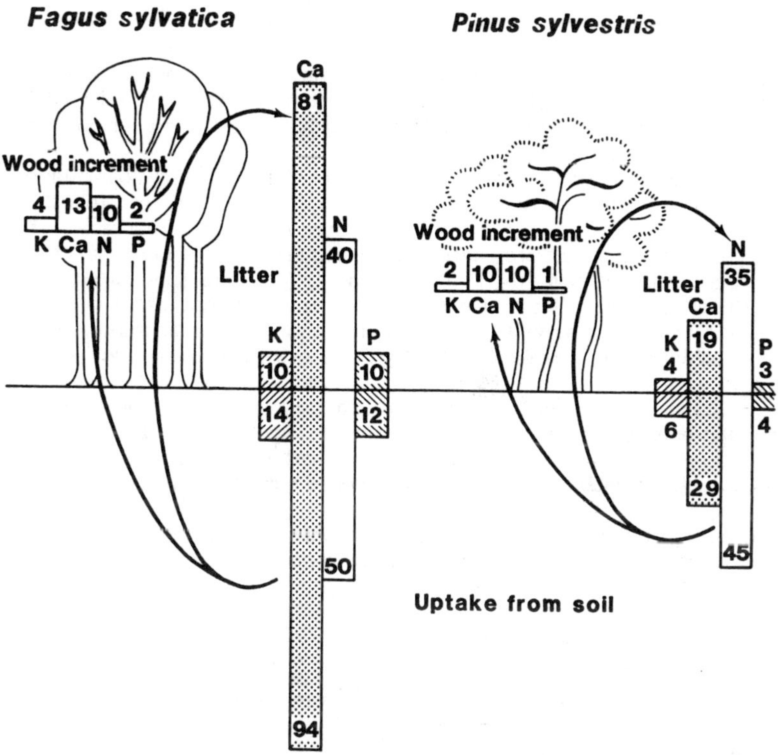

is attributable to the ground flora (Ovington, 1965). More detailed discussion is available in Cole, et al (1968), Duvigneaud and Denaeyer-De Smet (1970), Armson (1977) and Trudgill (1977).

In the prairie ecosystem (Coupland, 1979; Risser et al, 1981) the highly organic soil likewise constitutes a vast store of nitrogen and other nutrients (Table 7.1 and Figure 7.10). The humus is highly polymerised and complexed with clay minerals. Characterised by limited leaching and clays with a high cation exchange capacity, it is understandable that these soils form a basis for sustained agricultural production.

A limited number of studies have been undertaken in heathlands (Gimingham, 1972; Groves, 1981) and again reveal a low level of nutrients in the growing plants compared with the surface soil. Vegetation composition, here as elsewhere, is variable so that comparisons are best made in relation to a single species such as Calluna vulgaris. Grazing pressure and management practices will lead to variations as will the age of the stand, so any detailed studies of heathlands should take the full cycle of growth into account (Table 7.2).

In the Tropical Rain Forest much larger relative and absolute quantities of nutrients are held in the foliage (Table 7.3), the basis for this being the uninterrupted hot, wet climate, the resulting continuous growth and the complex layered structure of the forest.

Figure 7.10: Estimates of nitrogen flows and storage in a prairie ecosystem (g/m^2/yr). Approximations in parenthesis (after Dahlman et al, 1969)

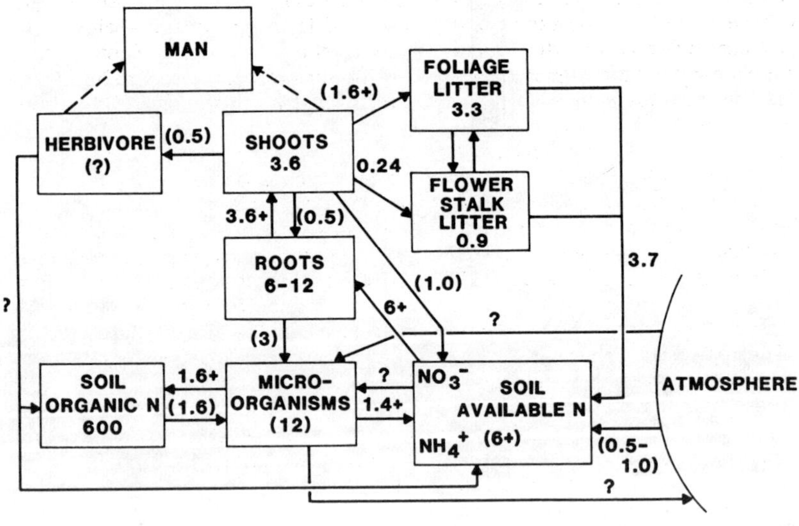

Table 7.2: Calcium and phosphorus balance sheets for 12-year old *Calluna* heathland (after Chapman, 1967)

Data compartment	Kg/ha Ca	P
Vegetation	33.0	4
Litter	15.2	4
Total above ground	48.2	8
Soil 0-20 cm	229	37
Proportion of above ground biomass and litter lost on firing	26%	
Remaining (from fire)	35.7	6
Lost (from fire)	12.5	2
Input from precipitation/yr	4.7	0.01
Input (12 yr)	56	0.12
Balance (gains - losses)	+43.5	-2.1

Additional explanations for this are the speed with which plant debris is consumed and mineralised (Madge, 1965; Lee and Wood, 1971; Mohr *et al*, 1972) and the limited cation exchange capacities of most soils under tropical rain forest. The relatively great age of many of these soils, coupled with high weathering rates, conspire to produce deep but generally infertile soils despite the luxuriance of the vegetation (Richards, 1957; Burringh, 1970; Duchaufour, 1978). Indeed, when these forests are cleared and the ground is prepared for agriculture by burning, much of the organic matter - the vital contribution to the soil's cation exchange capacity - is lost, ensuring that soil exhaustion is the inevitable long-term outcome of shifting agriculture (Nye and Greenland, 1960).

Nutrient Budgets. We should note that investigations of the relative mobility of ions (concentration in solution in relation to abundance in local rocks), based either on wells or stream water, took place long before detailed studies of nutrient flows within particular ecosystems (Carroll, 1970; Engelstad, 1970). Of the latter studies, an increasing proportion incorporate data on nutrient gains via precipitation and losses due to cropping, burning or from drainage water (Table 7.4). These figures, incomplete though they often are in terms of defining the whole system, enable an assessment to be made of the overall mass balance or nutrient budget which is as near as one can get to appraising the chemical efficiency of an ecosystem. While the amount and composition of rainfall are, in principle, straightforward measures, representative data on losses are much more difficult to achieve. In studies of drainage (or leaching) losses, lysimetry may prove expedient in relation to local conditions but there can be little doubt that a small catchment furnishes a superior research design, a noteworthy example being provided by the Hubbard Brook watershed in New England (Likens *et al*, 1977; Bormann and Likens, 1979).

Table 7.3: Selected data for moist tropical forests and for a temperate Douglas fir forest

Table 7.3: Selected data for moist tropical forests and for a temperate Douglas fir forest

	Tropical								Temperate			
Components	Panama (Golley *et al*, 1975)				Ghana (Nye, 1961)				Douglas Fir (Cole *et al*, 1968)			
	N	P	Ca	K	N	P	Ca	K	N	P	Ca	K
STORES (g/m^2)												
Vegetation	123	15	379	306	201	13	263	90	32.6	6.7	34.2	22.7
Litter	13	1.4	32	4	3.5	0.1	4.5	1	17.5	2.6	13.7	3.2
Soil	-	2.2	2217	35	460	1.2	258	65	281	388	74.1	23.4
	(0-30 cm)				(0-25 cm)				(0-60 cm)			
FLUXES ($g/m^2/yr$)												
Vegetation to litter	23	1.0	28	18	25	1.4	32	29	1.64	0.06	1.85	1.58
Rainfall input	1	0.1	2.9	0.9	1.5	0.04	1.2	1.8	1.1	Tr	0.28	0.08

Most work in the latter field has been concerned with forested catchments (Edmunds, 1980) and it will be noted in Table 7.4 that variations occur in the amounts of substances dissolved in rainfall, a major source being that of industry, e.g. Sollins and McCorison (1981) but compare east and west USA. Rainfall is responsible for certain fluxes within ecosystems, notably for the leaching of foliar nutrients in throughfall (Nye, 1961; Abel and Lavender, 1972). When it becomes soil water, it is responsible for weathering and for leaching of materials beyond the rooting zone. Weathering action is, furthermore, dependent on the acidity of the soil solution and local conditions of redox (White, 1979; Rowell, 1981; Duchaufour, 1982). Losses of mineral nutrients in Table 7.4 reflect the latter processes together with the chemistry of rock formations. High Ca values for the Tennessee data, for example, derive from dolomitic rocks. In the case of forest stands, harvesting of the trees at maturity (Table 7.5) and accelerated runoff and leaching losses associated with activities such as logging and site preparation (Figure 7.11), should strictly be accounted for in the long term when calculating the overall nutrient budget (Boyle and Ek, 1972; Sollins and McCorison, 1981). Table 7.5 shows distinctively high values for 100-year-old oak, but soil conditions do play a part in determining the amounts of nutrients taken up by the vegetation.

Table 7.4: Nutrient budgets for selected elements and forest ecosystems. Figures in parenthesis are values for uptake. Sources: (1) Ulrich and Mayer, 1972; (2) Fredrickson, 1975; (3) Cole et al, 1968; (4) Woodwell and Whittaker, 1968; (5) Likens et al, 1977 (inorganic N only); (6) Henderson and Harris, 1975, Henderson et al, 1978; (7) Swank and Douglass, 1975.

	Ecosystem	Kg/ha/yr Total Nitrogen			Calcium			Potassium		
		Input	Output	Net	Input	Output	Net	Input	Output	Net
(1)	Beech (Solling, Germany)	23.9	6.2	17.7	12.4	14.1	-1.7	2.0	1.6	+0.4
(2)	Douglas Fir (Cascades, Oregon)	0.09 (23)	0.38	+0.52	2.33 (45)	50.32	-47.99	0.11 (23)	2.25	-2.14
(3)	Douglas Fir (Oregon)	1.1	0.6 (38)	+0.5	2.8	4.5 (24)	-1.7	0.8	1.0 (29)	-0.2
(4)	Mixed Deciduous (Brookhaven)	-	-	-	3.3	8.0	-4.7	2.4	3.3	-0.9
(5)	Deciduous (Hubbard Brook, New Hampshire)	6.5	3.9 (66)	+2.6	2.2	13.7 (53)	-11.5	0.9	1.9 (53)	-1.0
(6)	Oak-Hickory (Walker Branch, Tennessee)	13.0	3.1 (57)	+9.9	12.0	148.0 (100)	-136.0	3.0	7.0 (46)	-4.0
(7)	Oak-Hickory (Coweeta, North Carolina)	8.8	3.2 (56)	+5.6	4.8	7.7 (75)	-2.9	2.1	5.6 (62)	-3.5

Figure 7.11: Changes in ionic concentrations in stream water following clear felling (after Likens et al, 1970)

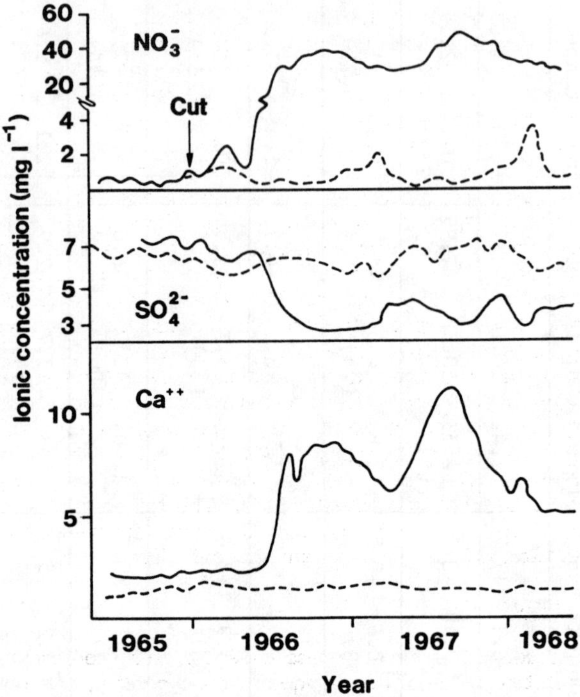

Table 7.5: Estimates of amounts of nutrients (kg/ha) lost in logging

	Species and age (yr)	Kg/ha N	P	K	Ca
(1)	Oak 115-160	386	17.5	219	769
(2)	Mixed hardwoods 45-50	120.3	12.1	60.2	129.6
(3)	Douglas fir 36	125	19	96	117
(4)	Scots pine 55	161	14	98	210
(5)	Loblolly pine 40	138	11.4	96	121
(6)	Jack pine 65	84.5	4.8	52.2	90.6

Nevertheless, the amounts lost by clear felling and removal of boles represent only modest amounts on a per-annum basis.

From studies of agricultural and moorland ecosystems (Crisp, 1966; Frissell, 1978), it also appears that the harvesting of crops and livestock is but a drop in the ocean of nutrients circulating in ecosystems as a whole. In agricultural ecosystems the main characteristic is greater wastage associated with enhanced leaching, runoff and sediment movement. This is the price paid for running an artificial ecosystem at a given level of productivity within the limitations of soil and growing season (Spedding et al, 1981).

In studies of this kind there are often difficulties of evaluating the contribution of the biotic system to the overall nutrient balance (Table 7.4), and the ideal situation would be to employ local comparison sites or watersheds (Likens et al, 1977). Since geology outweighs other factors influencing losses, it is worth considering alternative approaches which sidestep the large background effect of soil chemistry. In a study of lowland heath, Chapman (1967) estimated the nutrient balance over a 12-year period between burns (Table 7.2), and noted that appreciable amounts of potassium, calcium and magnesium were being lost from the litter layer each year and that these losses increase the older the stand. Although no data on drainage losses are given, it will be appreciated that these are dependent on rainfall as much as on parent material. Furthermore, if an upland heath were considered in comparison, although leaching losses would increase, the total mass of nutrient received in precipitation would be greater and likewise the mass of nutrients stored in peat. Such realisations favour a less-than-absolute approach to nutrient budget estimates. The latter may also be more realistic if the aim is to identify balances between the operation of the biotic system and management controls.

Environmental Pollution

Catchment-based studies are nevertheless essential for monitoring the effects of vegetation or land-use change on water quality, and for analysis of the factors responsible for release of a range of chemical and biological materials into rivers subsequently used for water supplies (Cooke, 1976; Harriman, 1978; Walsh, 1980; McDonald and Kay, 1982). Some substances are potentially toxic and, in this respect, the heavy metals differ greatly in their adsorptive behaviour, biological transmission and environmental mobility (Nriagu, 1979; Royal Commission, 1979; Waldron, 1980). If, for example, we compare lead, zinc and cadmium, the latter two have a high physiological and environmental mobility and are more highly concentrated in consumer organisms than lead which is relatively immobile (Hughes et al, 1980). Such considerations will determine the speed with which the environment can be contaminated, or decontaminated (Haan and Zwerman, 1976; Graham-Bryce, 1981; Jones and Jarvis, 1981). In the latter respect, lead is strongly adsorbed by soil organic matter while zinc and cadmium are leached from the soil, albeit with the aid of chelating agents.

A widespread and alarming problem is that of acid rain, the ecological consequences of which are rapidly becoming evident. Acidification of lakes and loss or reduction of fish populations are undisputed while the effects on vegetation, soils and human health remain as urgent research problems (Lewis and Grant, 1979; Johnson *et al*, 1981; Smith, 1981). Areas of acidic rocks are particularly vulnerable since their soils have a very limited capacity to neutralise acids and, ironically, these areas normally have acid-tolerant vegetation which itself promotes rather than checks soil acidification. It would seem likely that afforestation of grasslands will lead independently to increasingly acidic river water. For those pressing for international legislation on the control of atmospheric pollutants it is therefore unfortunate that acidification may be attributable to a variable balance of factors in different areas.

Considerations of this kind show the importance not only of describing the present characteristics of ecosystems but of studying changes through time.

SOILS AND CHANGING ECOSYSTEMS

Soils in the Quaternary

Soils almost invariably alter in appearance with depth. These changes are attributable to sedimentary as well as to pedological processes, both of which may generate features which persist for thousands of years (Yaalon, 1971, 1982; FitzPatrick, 1983). Noteworthy among such features are fossil ice wedges and flame structures associated with freeze-and-thaw; iron pans may subsequently develop along interfaces between depositional units in sand; subsoil structure may pseudomorph the original parent material. Many soil profiles therefore contain a record of history as well as of current processes, thereby greatly increasing the scope of their study.

Within the Quaternary period soil formation has been profoundly affected by climatic change. It was not only in higher latitudes, overridden by ice or buried by aeolian deposits, that changes were forced upon the landscape but also in low latitudes as the limits of rainforest and savanna altered, as deserts expanded and as montane belts shifted up and down slope (Flenley, 1979a, 1979b). Aside from soils severely truncated or restricted in age to the Holocene, it would therefore seem that extensive areas of soils have formed for at least as long under glacial climates as under interglacial conditions. Ancient soils are associated with the oldest geomorphic surfaces but, although soils acquire characteristics as a function of age, notably in relation to the chemistry and minerology of the clay fraction (Birkeland, 1974; Brady, 1974; Brewer, 1981), later events will often frustrate this relationship.

It is not surprising that cyclical relationships between soils and landscape development provide a framework for pedological studies of the Quaternary (Butler, 1959; Yaalon, 1971, 1982; Ruhe, 1975; Gerrard, 1981). Periods of stable soil development took place during periods of complete vegetation cover while surface processes were more active when plant cover was reduced by drought or cold (Walker, 1962a, 1962b; Vita Finzi, 1969; Bintliff, 1981). Mechanical weathering and surface transport in glacial and arid phases are therefore to be contrasted with *in situ* weathering and solutional losses during fully-vegetated phases. The study of lake sediments can also provide an indirect record of pedological progression and retrogression within a single catchment area (Mackereth, 1965; Nriagu et al, 1979; Birks and Birks, 1980).

Such landscape changes create a complex soil stratigraphy, particularly in conjunction with river valleys, and the appearance of the various layers can often lead to problems of interpretation. A soil which has long been buried, for example, is likely to reflect the conditions of its burial as opposed to its pre-burial stage of development and may therefore give a misleading view of previous weathering history. The gumbotil problem in the USA is a good example (Hunt, 1972; Birkeland, 1974) while the identification of loessal soil parent materials in Great Britain is of a similar kind and depends on the nature of the rock type and drainage of the terrain on which these aeolian materials were laid down (Smalley, 1975; Catt, 1979; Smith, 1983).

Given a framework for the long-period changes we can now consider some of the characteristics of shorter-term trends in pedogenesis.

Trends in Soil Development

Sequence and Stability. There have been numerous studies of soil development in relation to plant succession on surfaces ranging from volcanic ash to sand dunes (Olsen, 1958), including surfaces of varying age in a given locality (Vreeken, 1975). The latter, known as chronosequences, can be particularly valuable in providing insights into changes characteristic of different lengths of time. Californian mudflows were studied by Dickson and Crocker (1953/4) while similar studies have been undertaken on Michigan lakeshore terraces (Franzmeier and Whiteside, 1963) and on glacier moraine in Alaska (Crocker and Major, 1955) and the Yukon (Jacobson and Birks, 1980). Although the latter provide templates for demarcating stages in soil development, few are ideal in experimental design, notably on account of variations in parent material and assumptions about man's activities over the range of study sites. Nevertheless there is broad agreement that leaching in unconsolidated materials initially causes a decline in the pH value while organic matter and nitrogen contents rise sharply. In the Yukon example, soil pH falls from 8 to near 6 in 200-year-old soils while nitrogen levels rise from near zero to 0.7% in the same period.

These studies have helped pedologists to appreciate the sequential nature of certain basic processes. For example, although chelation can take place in the presence of calcium carbonate, the latter so stimulates the activity of soil organisms that it must be eliminated before active podzolisation can occur. Furthermore, the movement of clay and humus will normally post-date iron translocation. Whereas certain initial processes are accomplished relatively rapidly, others, including the weathering of silicate minerals, occur much more slowly. The rate of change in soil properties also decelerates after the initial stages of pedogenesis when the ecologically-dominant species establish an equilibrium level of organic matter. An asymptotic maximum level of nutrients stored in the soil will be determined at this stage, which provides a basis for defining the steady-state. This latter is also founded upon the fact that soils, as they develop, eventually acquire considerable in-built resistance to change. For example, as labile phosphate is taken up by plants, immobilised forms become available to maintain an equilibrium. So it is with other adsorbed ions, these processes being maintained so long as weathering can sustain the demands of the system. This can be described as the 'buffering capacity' and most soils possess such a characteristic in different ways and to differing degrees. Soils with high lime or clay content will be buffered against the consequences of leaching, and organic soils will supply cations at lower pH values than will normal mineral soils. It is for such reasons that only after persistent cropping will micronutrient deficiencies or structural problems arise (Debano, 1969; Hudson, 1971; Morgan, 1981).

Progressive and Retrogressive Trends. Change in the early stages of pedogenesis is invariably progressive, yet such trends apply also in other contexts. Ovington (1965), for example, illustrates the role of oak woodland in raising pH, cation exchange capacity and available bases in soils, subsequent to removal of coniferous species. Work on bracken reported by Mitchell (1973) and Smith (1977) suggests a similar capability in this fern-geophyte. Again, the contributions of Dimbleby (1952), Floate (1970) and Maltby (1975) represent different types of study of the impact on soil conditions of upland sward improvement. It is clear that a reversal of modern soil trends is possible provided that the ecosystem is strictly regulated, in this case its grazing regime. Many other instances could be quoted of lands reclaimed for productive use with the aid of modern technology (Cairns *et al*, 1977; Holdgate and Woodman, 1978).

Retrogressive trends in pedogenesis are commonplace, and useful if emotive terms such as degradation and deterioration have been adopted. These imply that we know the original circumstances from which the modern soil is derived: often we do but we may often be less sure. We must also be cautious about type-casting particular soil processes as degradational. For example, while podzolisation frequently comes to mind in this context, it may, under particular climatic and edaphic conditions, be the dominant factor in pedogenesis and eventually bring about soil characteristics which are, when translated into biological productivity, superior to the initial state of the system. Thus, Franzmeier and Whiteside (1963) associate the development of mature forest on their earliest surface with a maximally differentiated podzol, having greater moisture retentiveness and increased cation exchange capacity.

The term degradation is mostly applied when man is identified as the key factor, the assumption being that a particular direction in soil development would either not otherwise have taken place or not have proceeded at the rate which is apparent (Dimbleby, 1962; Bidwell and Hole, 1965; Yaalon and Yaron, 1966; Bridges, 1978;

Davidson, 1982). A number of examples where man has either initiated or reinforced soil trends deserve mention and include salinisation and alkali-solonetz formation (Jacobson and Adams, 1958; Allison, 1964), the genesis of acid sulphate soils from sulphidic muds (Bloomfield and Coulter, 1973), the rubefaction of soils in the context of sclerophyllous vegetation (Duchaufour, 1970) and the deflation and oxidation of organic terrain (Godwin, 1978). They also include podzolisation following clearance of temperate forest and with the planting of conifers (Dimbleby, 1962; Hamilton, 1965; Page, 1968; Kittredge, 1976) and soil losses from limestone terrain leading to increasingly calcareous habitats and even to total soil loss (Hyams, 1952; Carter and Dale, 1974; Smith, 1975, 1983). Indeed, accelerated erosion has occurred since time immemorial so that both historical and contemporary examples of the latter are legion.

The complexity of certain degradative processes is illustrated by reference to a grazing system. In a temperate, maritime upland environment, an inactive nutrient cycle associated with less nutritive and less palatable herbage promotes soil acidity. This acidity and the rank growth of vegetation often associated with it, induce a fall in environmental redox. As a result, gleying is encouraged as pasture quality deteriorates (Rowell, 1981; Armstrong, 1982). The gleying in turn induces floristic changes, including the growth of rush (*Juncus*) and even bog moss (*Sphagnum*). This example of positive feedback would appear to identify threshold conditions as potentially important in determining and reinforcing pedological change.

Transition or Discontinuity? Our present knowledge of soils has identified certain pathways in pedogenesis but has yet failed to tackle the way in which the balance of pedological processes may shift in time. *In situ* profile changes, which might be thought to have been accomplished in an orderly, continuous sequence, may in fact have been generated along a series of discontinuities in the manner of shocks followed by adjustments. This is, of course, how soils react to floristic change or to catastrophic changes such as truncation or burial. So, when forest removal has taken place, a variety of dire consequences have occurred including soil loss, intensified leaching and rise of ground water. In the latter respect, both hill peat formation (Taylor, 1973; Moore, 1975) and lateritisation (McFarlane, 1976) can be largely attributed to deforestation and subsequent hydrological change (see Figure 7.12). Again, with a fall in pH, biological thresholds will be crossed and eventually the structures of clay minerals will begin to collapse (Greenland and Hayes, 1978). Rates of change will, therefore, rarely be constant in time, and major pedological changes may be concentrated within short periods owing to the operation of critical thresholds and multiplier effects. For example, positive feedback applies both to the intensification of acidity (Trudgill, 1977, 1982) and in the case of surface gleying (Crampton, 1965; Ugolini and Mann, 1979; Taylor and Smith, 1980). Abrupt changes in the state factors determining the soil system understandably cause sudden soil responses but it would seem that slow change over a prolonged period can similarly generate future change of a sudden nature. This is the principle which Thom (1975) and others (Renfrew, 1978; Wilson, 1981) have identified as operating elsewhere and which is susceptible to mathematical resolution within catastrophe theory.

Finally, it is pertinent to consider the link that exists between climatic changes and those, some climate-dependent, others not, occurring in vegetation and soils. Obviously there can be no instant relationships because of fundamental response times (otherwise termed lag or retardation) involved in all environmental changes. In the first place, the improvement in climate itself after the last glacial maximum was delayed because huge amounts of radiant energy were needed cumulatively to melt the large masses of ice prior to sea-level rise and ocean expansion. Rapid changes of climate involve a time lag for the immigration of species while the eventual composition of vegetation depends on competitive strategy rather than simple climatic causation. It has been shown, furthermore, that the introduction of new species into the forests of the eastern USA has been a continuous process during the Holocene (Davis, 1976). Therefore, although the notion of climax vegetation may be helpful in providing a conceptual point of reference within a continuum of change, it must now be viewed as essentially illusory. Soil is likely to respond to both vegetational and climatic shifts such that no matter what lengths of time are being considered, it is as likely to be adjusting *to* them as in a supposed equilibrium *with* them.

Ecosystems and Soils Futures

If we look ahead to consider the future, we are involved not only with natural science but with the applied fields of resource management and conservation as well as with straight futurology (Fowles, 1978). The discussion must also be concerned with man as much as with the physical bases of ecological systems. The future of the earth's soil and biological resources may be a foregone conclusion to those who prophesy doom, yet we must continue to believe that a productive future is achievable.

The Problem of Soil Exhaustion. Soil profiles can be likened to factories in which different products have been manufactured. Evidence of former activities may remain while new equipment, extensions and demolition occur from time to time. These vital factories must remain in business and it is clearly our continuing duty to see that more and more profiles do not have to decline and go out of production owing to mismanagement or bad planning. Thousands of hectares of agricultural land are lost per annum, through a lack of awareness of the systemic relations between land use, soil and climate, and thousands more due to the spread of urban areas. Much of our ignorance arises from a pitiful inability to reform existing practices, responsibility for the latter often being in the hands of governments or agencies (local or international) rather than individuals (Morgan, 1981).

Figure 7.12: Pathways to peat formation in a moorland ecosystem (after Taylor and Smith, 1980)

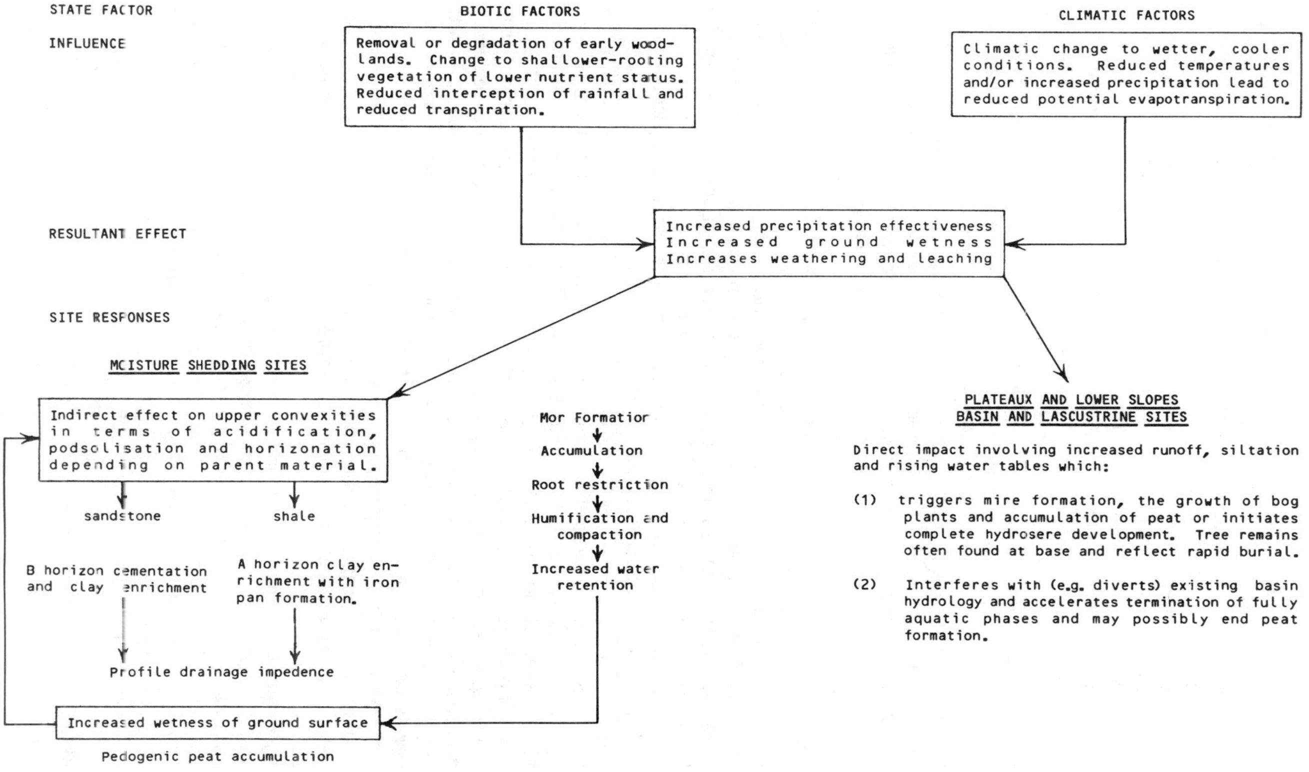

Very few land-use systems can truly be said to have long-term ecological stability, let alone merits, and this applies to Third World and developed countries alike. In this respect we might mention the production of cash crops in tropical Africa and the upland economy of the United Kingdom, for neither are ecologically sound. On the other hand, pious conservationist rhetoric will not solve these problems. The real need in education is for those studying the natural sciences to become more fully informed about the economic, social and political milieu within which current land-use practices are enmeshed, and to be aware that the resource-base upon which mankind has been dependent in the past will not continue indefinitely (Eyre, 1978; Manners, 1981). We must not only introduce conservation measures to buttress existing practices but also be willing to reappraise our whole philosophy concerning the destiny of land-usage. The latter involves asking questions not just about the quantity of food we can produce but increasingly about the quality of the food we eat and the ways in which our health depends upon this.

Future Research. It might be concluded that to serve the future needs of mankind, research should now be directed exclusively towards the study of contemporary land and soil problems. Yet it has to be accepted that a knowledge of the past, whether in terms of agrarian history or palaeoecology, has been valuable in coming to terms with our present world. This, in turn is a necessary precursor to predicting, planning and, we hope, to regulating the future. In the latter respect contributions in the field of theoretical modelling, albeit of single soil or vegetation characteristics, will for a time at least be of great potential value in a period of rapid change (Kline, 1973; Burgess and Sharpe, 1981; Kirkby, 1981; Huggett, 1982). It is always tempting in this kind of discussion to imagine that pure research is doomed and that evaluative surveys and agricultural extension services are all that is truly justifiable. Yet for each macroscale investigation there can be many valuable inputs concerned with such attributes as soil structure, soil organisms or the persistence of residues which ultimately must be resolved for successful future management. Environmental acidification is an example of a problem requiring examination on a soil profile, river basin, national and international scale. We require to know how susceptible individual soils are to this problem, while if international agreements cannot eventually be achieved, biological productivity would appear to be very seriously jeopardised in the years ahead. Just as problems of soil stability, if unchecked, graduate to becoming matters affecting whole landscapes, so it is with pollution problems at the present time. Nor does it help to fall back on that well-worn tactic of claiming that ecosystems are essentially resilient and that we merely require time for adjustments to take place (Hare, 1981). In fact many environmental problems emerge as serious issues only after a prolonged gestation period comparable to the nature of discontinuous change in soils discussed in the previous section.

The Question of Technology. When addressing these problems and the general questions of human starvation and malnutrition, it is tempting to indulge in the belief that technology will contribute new intensive methods of food and fibre production from oil-based protein to hydroponics, thus increasingly emancipating us from the soil. Though such developments may indeed help bale us out of future shortages, the problem with most schemes of this kind is that they consume energy, notably from fossil fuels. We are therefore involved with the matter of future prolongation of supplies and people's ability to afford them as prices are forced upwards. Indeed, it is worth applying the underlying principles of cost and availability to the raw materials upon which intensive, mechanised agriculture depends, and it is quickly realised that these have considerable in-built assumptions about continuing national wealth and the availability of both raw materials and energy to process them (Schumacher, 1973; Eyre, 1978; Foth, 1982). Also, one of the fallacies of estimations of population and food production based on technological capability (Hare, 1981) is that they fail to account for land losses and reduced productivity occasioned by the introduction and persistence of soil-exhausting forms of agriculture.

True Ecological Harmony. It is maintained, in contrast, that an ecological approach to agriculture would fail to sustain yields and be costly in terms of labour. Nevertheless, advocacy of ecological and biodynamic methods (Koepf et al, 1976; USDA, 1980) is more than ever justified if we are truly to regard soil as a renewable resource. The increasing scale of farm and forestry enterprises and, more especially, the increasing size of fields and the supremacy of chemical controls bring inevitable ecological change in the interests of efficiency. This, of course, means economic efficiency at an ecological price. Nor should questions such as the ecological role of organic husbandry be unconnected in our minds from aspects of our urbanised existence. Economic circumstances have, in many parts of the world, led to the depopulation of rural areas. Is it not so inconceivable that, in the face of acute urban unemployment, rural areas could again provide work within an ecologically-sound and more labour-intensive farming system? It must be abundantly clear that the question of unemployment more than any soil problem threatens to destroy human health and national morale. It is clear that just as the need for land reform lies at the root of countless present-day social and economic problems so also, in the broadest sense, does it underlie ecological concerns.

Ultimately, it will be essential to relate the use of soils to the needs of society, whether in terms of food, fibre, wildlife or indeed employment, and to treat all soils remaining in production as renewable resources (Naveh, 1982). Soil is not simply a substance, nor indeed a profile, but it is land, it is space. The ecosystem of which soil is a part thus embraces not only the conservation of nature but of the whole of mankind.

REFERENCES

Abel, A. and Lavender, D. (1972), 'Nutrient cycling in throughfall and litterfall in 450 year-old Douglas fir stands', in Research on coniferous forest ecosystems, eds. Franklin, J.F. et al, pp. 133-43, USDA Forest Service, Portland, Oregon, USA.

Alexander, M. (1977), Introduction to soil microbiology, 2nd edition, Wiley.

Allison, F.E. (1973), Soil organic matter and its role in crop production, Elsevier, Amsterdam.

Allison, L.E. (1964), 'Salinity in relation to irrigation', Advances in Agronomy, 16, 139-80.

Armson, K.A. (1977), Forest soils: properties and processes, Univ. of Toronto Press, Toronto, Canada.

Armstrong, W. (1982), 'Waterlogged soils', in Environment and plant ecology, by Etherington, J.R., Ch. 10, pp. 290-330, John Wiley and Sons, Chichester, UK.

Beckett, P.H.T. and Bie, S.W. (1975), 'Reconnaissance for soil survey 1. Pre-survey estimates of the density of soil boundaries necessary to produce pure mapping units', J. Soil Sci., **26**, 144-54.

Beckett, P.H.T. and Webster, R. (1971), 'Soil variability: a review', Soil and fertilisers, **34**, 7-15.

Bidwell, D.W. and Hole, F.D. (1965), 'Man as a factor in soil formation', Soil Science, **99**, 65-72.

Bie, S.W. and Beckett, P.H.T. (1971), 'Quality control in soil survey. Introduction: 1. The choice of mapping unit', J. Soil Sci., **22**, 32-49.

Bintliff, J. (1981), 'Archaeology and the Holocene evolution of coastal plains in the Aegean and circum Mediterranean', in Environmental aspects of coasts and islands, eds. Brothwell, D. and Dimbleby, G.W., British Archaeological Reports, International Series No. 94, pp. 11-31.

Birkeland, P.W. (1974), Pedology, weathering and geomorphological research, Oxford University Press, New York, USA.

Birks, H.J.B. and Birks, H.H. (1980), Quaternary Palaeoecology, Edward Arnold, London.

Bloomfield, C. and Coulter, J.K. (1973), 'Genesis and management of acid sulphate soils', Advances in Agronomy, **25**, 265-326.

Borman, F.H. and Likens, G.E. (1979), Pattern and process in a forested ecosystem, Springer-Verlag, New York, USA.

Boyle, J.R. and Ek, A.R. (1972), 'An evaluation of some effects of bole and branch pulpwood harvesting on site macronutrients', Can. J. For. Res., **2**, 407-12.

Brady, N.C. (1974), The Nature and Properties of Soils, 8th edition, Collier-Macmillan, London.

Brewer, R. (1981), Fabric and Mineral Analysis of Soils, 2nd edition, Wiley, Chichester, UK.

Bridges, E.M. (1978), 'Interaction of soil and mankind in Britain', J. Soil Sci., **29**, 125-39.

Bridges, E.M. (1982), 'Techniques of modern soil survey', in Principles and Applications of Soil Geography, eds. Bridges, E.M. and Davidson, D.A., Longman, London.

Bridges, E.M. and Davidson, D.A. (eds.) (1982), Principles and Applications of Soil Geography, Longman, London.

Burges, A. and Raw, F. (eds.) (1967), Soil Biology, Academic Press, London.

Burgess, R.L. and Sharpe, D.M. (1981), Forest Island Dynamics in Man-dominated Landscapes, Springer-Verlag, New York, USA.

Burringh, P. (1970), Introduction to the Study of Soils in Tropical and Sub-tropical regions, Centre for Agricultural Publication and Documentation, Waageningen, Netherlands.

Butler, B.E. (1959), Periodic Phenomena in Landscape as a Basis for Soil Studies, CSIRO, Soil Publ. 14.

Cairns, J. et al (1977), Recovery and Restoration of Damaged Ecosystems, University of Virginia Press, Charlottesville, USA.

Cannell, M.G.R. (1982), World Forest Biomass and Primary Production Data, Academic Press, London.

Carroll, D.M. (1970), Rock Weathering, Plenum, New York, USA.

Carter, V.G. and Dale, T. (1974), Topsoil and Civilisation, University of Oklahoma Press, 2nd edition, USA.

Catt, J.A. (1979), 'Soils and Quaternary geology in Britain', J. Soil Sci., **30**, 607-42.

Chapman, S.B. (1967), 'Nutrient budgets for a dry heath ecosystem in the south of England', J. Ecol., **55**, 677-89.

Chorley, R.J. and Kennedy, B. (eds.) (1971), Physical Geography: a Systems Approach, Prentice-Hall, London.

Cole, D.W., Gessel, S.P. and Dice, S.F. (1968), Primary Productivity and Mineral Cycling in Natural Ecosystems, Univ. Maine Press, USA.

Conacher, A.J. and Dalrymple, J.B. (1977), 'The nine-unit land surface model: an approach to pedogeomorphic research', Geoderma, **18**, 1-154.

Cooke, G.W. (1976), 'A review of the effects of agriculture on the chemical composition and quality of surface and underground water', MAFF Tech. Bull., 32, HMSO, London.

Coupland, R.T. (1979), Grassland Ecosystems of the World, IBP 18, Cambridge University Press, UK.

Crampton, C.B. (1965), 'Vegetation, aspect and time as factors of gleying in podzols of South Wales', J. Soil Sci., **16**, 210.

Crisp, D.T. (1966), 'Input and output of minerals for an area of Pennine moorland: the importance of precipitation, drainage, peat erosion and animals', J. Appl. Ecol., **3**, 327-48.

Crocker, R.L. and Major, J. (1955), 'Soil development in relation to vegetation and surface age at Glacier Bay, Alaska', J. Ecol., **43**, 427-48.

Dahlman, R.C. et al (1969), 'The nitrogen economy of grassland and dune soils', in Biology and Ecology of Nitrogen, National Academy of Sciences, Washington DC, USA, pp. 54-82.

Davidson, D.A. (1982), 'Soils and man in the past', in Principles and Applications of Soil Geography, eds. Bridges E.M. and Davidson, D.A., Longman, London, pp. 1-27.

Davis, M.B. (1976), 'Pleistocene biogeography of temperate deciduous forests', Geoscience and Man, **13**, 13-26.

Debano, L.F. (1969), 'Water repellent soils: a world-wide concern in management of soil and vegetation', Agricultural Science Review, **7**, 11-8.

Debano, L.F. and Conrad, C.E. (1978), 'The effect of fire on nutrients in a chaparral ecosystem', Ecology, **59**, 489-97.

Dickson, B.A. and Crocker, R.L. (1953/4), 'A chronosequence of soils and vegetation near Mount Shasta, California', J. Soil Sci., **4**, 123-41, 142-5; 5, 1-19.

Dimbleby, G.W. (1952), 'Soil regeneration on the north east Yorkshire Moors', J. Ecol., **40**, 331-41.

Dimbleby, G.W. (1962), The Development of British Heathlands and their Soils, Oxford For. Mem. No. 23, Clarendon Press, Oxford, UK.

Dorst, J. (1970), Before Nature Dies (trans. from French by Sherman, C.D.), Collins, London.

Duchaufour, P. (1978), Ecological Atlas of Soils of the World, trans., Masson (Paris), New York, USA.

Duchaufour, P. (1982), Pedology: Pedogenesis and Classification, Allen and Unwin, London.

Dury, G. (1981), An Introduction to Environmental Systems, Heinemann, London.

Duvigneaud, P. and Denaeyer-De Smet, S. (1970), 'Biological cycling of minerals in temperate deciduous forests', in Analysis of Temperate Forest Ecosystems, ed. Reichle, D.E., Springer-Verlag, New York, pp. 199-225.

Edmunds, R.L. (1980), Analysis of Coniferous Forest Ecosystems in the Western United States, US/IBP Series No. 14, Hutchinson Ross, Stroudsburg, USA.

Englestad, O.P. (ed.) (1970), Nutrient Mobility in Soils: Accumulation and Losses, Soil Sci. Soc. Am. Spec. Publ. No. 4, USA.

Etherington, J.R. (1982), Environment and Plant Ecology, (2nd edition), John Wiley and Sons, Chichester, UK.

Evans, C.C. and Allen, S.E. (1971), 'Nutrient losses in smoke produced during heather burning', Oikos, **22**, 149-54.

Eyre, S.R. (1976), The Real Wealth of Nations, Edward Arnold, London.

FitzPatrick, E.A. (1983), Soils; their Formation, Classification and Distribution, Longman, London.

Flenley, J. (1979a), The Tropical Rain Forest: a Geological History, Butterworth, London.

Flenley, J. (1979b), 'The late Quaternary vegetational history of the equatorial mountains', Progress in Physical Geography, **3**, 488-509.

Floate, M.J.S. (1970), 'Mineralisation of nitrogen and phosphorus from organic materials of plant and animal origin', J. Brit. Grassl. Soc., **25**, 295-302.

Foth, H.D. (1982), 'Soil resources and food: a global view', in Principles and Applications of Soil Geography, eds. Bridges, E.M. and Davidson, D.A., Longman, London, pp. 256-74.

Foth, H.D. and Schafer, J.W. (1980), Soil Geography and Land Use, Wiley, New York, USA.

Fowles, J. (ed.) (1978), Handbook of Futures Research, Greenwood Press, Westport, Connecticut, USA.

Franzmeier, D.P. and Whiteside, E.P. (1963), 'A chronosequence of podzols in northern Michigan', Michigan State Univ. Agric. Expt. Stn. Quart. Bull., **46**(1), 2-57.

Frederickson, R.L. (1975), 'Nitrogen, phosphorus and particulate matter budgets of five coniferous forest ecosystems in the Western Cascades Range, Oregon', PhD thesis, Oregon State Univ., Corvallis.

Frissell, M.J. (ed.) (1978), Cycling of Mineral Nutrients in Agricultural Ecosystems, Elsevier, Amsterdam, Netherlands.

Gerrard, A.J. (1981), Soils and Landforms: an Integration of Geomorphology and Pedology, Allen and Unwin, London.

Gimingham, C.H. (1972), Ecology of Heathlands, Chapman and Hall, London.

Godwin, H. (1978), Fenland: its Ancient Past and Uncertain Future, Cambridge University Press, Cambridge, UK.

Golley, F.B. et al (1975), Mineral Cycling in a Tropical Moist Forest Ecosystem, Univ. Georgia Press, Athens, USA.

Golson, J. (1977), 'No room at the top: agricultural intensification in the New Guinea Highlands', in Sunda and Sahul: Prehistoric Study in Southeast Asia, Melanesia and Australia, eds. Allen, J., Golson, J. and Jones, R., Academic Press, London.

Goodman, D. (1975), 'The theory of diversity-stability relations in ecology', Quart. Rev. Biol., **3**, 237-60.

Graham-Bryce, I.J. (1981), 'The behaviour of pesticides in soil', in The Chemistry of Soil Processes, eds. Greenland, D.J. and Hayes, M.H.B., John Wiley and Sons, Chichester, UK.

Greenland, D.J. and Hayes, M.H.B. (eds.) (1978), The Chemistry of Soil Constituents, John Wiley and Sons, Chichester, UK.

Greenland, D.J. and Hayes, M.H.B. (eds.) (1981), The Chemistry of Soil Processes, John Wiley and Sons, Chichester, UK.

Greer, C.C. et al (1974), 'Nutrient cycling in 37 and 450 year old Douglas fir ecosystems', in Integrated Research in the Coniferous Forest Biome, ed. Waring, R.H. and Edmonds, R.L., Univ. Washington Coll. Forest Resources, Conif. For. Biome Bull. 5, pp. 21-34, Washington, USA.

Groves, R.H. (ed.) (1981), *Australian Vegetation*, Cambridge University Press, Cambridge, UK.

Groves, R.H. (1981), 'Nutrient cycling in heathlands', in Goodall, D.W. (ed.) *Ecosystems of the World 9B: Heathlands and Related Shrublands: Analytical Studies*, Elsevier, Amsterdam, Netherlands, pp. 151-63.

Grubb, P.J. *et al* (1969), 'The ecology of chalk heath: its relevance to the Calcicole-Calcifuge and soil acidification problems', *J. Ecol.*, **57**: 175-212.

Haan, F.A.M. and Zwerman, P.J. (1976), 'Pollution of soil', in *Soil Chemistry*, eds. Bolt, G.H. and Bruggenwert, M.G.M., Elsevier, Amsterdam, Netherlands, pp. 192-271.

Hamilton, C.D. (1965), 'Soil changes under *Pinus radiata*', *Australian Forestry*, **29**(4): 48-73.

Hardin, G. (1968), 'The tragedy of the Commons', *Science*, **162**, 1243-8.

Hare, F.K. (1981), 'The planetary environment: fragile or sturdy?', *Geogr. J.*, **145**, 379-95.

Harriman, R. (1978), 'Nutrient leaching from fertilised forest watersheds', *J. Appl. Ecol.*, **15**, 933-42.

Henderson, G.S. and Harris, W.F. (1975), 'An ecosystem approach to the characterisation of the nitrogen cycle in a deciduous forest watershed', in *Forest Soils and Forest Land Management*, ed. Bernier, B. and Winget, C.H., Quebec, Canada, Laval Univ. Press, pp. 179-93.

Henderson, G.S. *et al* (1978), 'Nutrient budgets of Appalachian and Cascade region watersheds: a comparison', *Forest Science*, **24**, 385-97.

Hodgson, J.M. (1978), *Soil Sampling and Soil Description*, Oxford University Press, Oxford, UK.

Holdgate, M.W. and Woodman, M.J. (eds.) (1978), *The Breakdown and Restoration of Ecosystems*, Plenum Press, New York, USA.

Holling, C.S. (1973), 'Resilience and stability of ecological systems', *Am. Rev. Ecol. and Systematics*, **4**, 1-23.

Hudson, N.W. (1971), *Soil Conservation*, Batsford, London.

Huggett, R.J. (1982), 'Models and spatial patterns of soils', in *Principles and Applications of Soil Geography*, eds. Bridges, E.M. and Davidson, D.A., Longman, London, pp. 132-70.

Hughes, M.K., Lepp, N.W. and Phipps, D.A. (1980), 'Aerial heavy metal pollution and terrestrial ecosystems', Advances in Ecol. Research, **11**, 218-327.

Humphries, F.R. (1966), 'Some effects of fire on plant nutrients', in The Effects of Fire on Forest Communities, N.S.W. Forestry Commission Technical Paper No. 13, 53-9.

Hunt, C.B. (1966), Plant Ecology of Death Valley, California, US, Geol. Survey Prof. Paper 509, San Francisco, USA.

Hunt, C.B. (1972), Geology of Soils, W.H. Freeman & Co, San Francisco, USA.

Hyams, E. (1952), Soil and Civilisation, Thames and Hudson, London.

Jacobson, G.L. and Birks, H.J.B. (1980), 'Soil development on recent end moraines of the Klutlan Glacier, Yukon Territory, Canada', Quaternary Research, **14**, 87-100.

Jacobson, T. and Adams, R.M. (1958), 'Salt and silt in ancient Mesopotamian agriculture', Science, **128**, 1251-8.

Jenkinson, D.S. (1981), 'The fate of plant and animal residues in soil', in The Chemistry of Soil Processes, eds. Greenland, D.J. and Hayes, M.H.B., John Wiley and Sons, Chichester, UK.

Jenny, H. (1941), Factors of Soil Formation, McGraw Hill, New York, USA.

Johnson, A.H. et al (1981), 'Recent changes in patterns of tree growth rate in the New Jersey pinelands: a possible effect of acid rain', J. Envtl. Qual., **10**, 427-30.

Jones, L.H.P. and Harvis, S.C. (1981), 'The fate of heavy metals', in The Chemistry of Soil Processes, eds. Greenland, D.J. and Hayes, M.H.B., Wiley, Chichester, UK, pp. 593-620.

Kendrick, W.B. 91959), 'The time factor in the decomposition of coniferous leaf litter', Can. J. Bot., **37**, 907-12.

Kershaw, K.A. (1973), Quantitative and Dynamic Plant Ecology, Edward Arnold, London.

Kittredge, J. (1976), Forest influences, Reprint, Dover Books.

Kirkby, M.J. (1981), 'The Basis for Soil Profile Modelling in a Geomorphic Context', Working Paper 301, School of Geography, University of Leeds, Leeds, UK.

Kline, J.R. (1973), 'Mathematical simulation of soil-plant relationships and soil genesis', Soil Science, **115**, 240-9.

Koepf, H.H. et al (1976), Bio-dynamic Agriculture, Anthroposophic Press Inc., Spring Valley, New York, USA.

Kononova, M.M. (1966), Soil Organic Matter, 2nd edition, Pergamon Press, Oxford, UK.

Kozlowski, T.T. and Ahlgren, C.E. (1974), Fire and Ecosystems, Academic Press, London.

Larsen, J.A. (1980), The Boreal Ecosystem, Academic Press, London.

Lee, K.E. and Wood, T.G. (1971), Termites and Soils, Academic Press, London.

Levitt, J. (1980), Responses of Plants to Environmental Stresses, 2 vols., Academic Press, 2nd edition, London.

Lewis, W.M. and Grant, M.C. (1979), 'Changes in the output of ions from a watershed as a result of the acidification of precipitation', Ecology, **60**, 1093-7.

Likens, G.E. et al (1970), 'The effect of forest cutting and herbicide treatment on nutrient budgets in the Hubbard Brook watershed ecosystem', Ecol. Monographs, **40**, 23-47.

Likens, G.E. et al (1977), Biogeochemistry of a Forested Ecosystem, Springer Verlag, New York, USA.

Longman, K.A. and Jenik, J. (1974), Tropical Forest and its Environment, Longman, London.

Loughnan, F.C. (1969), Chemical Weathering of the Silicate Minerals, Elsevier, Amsterdam, Netherlands.

Lunt, H.A. (1948), 'Forest Soils of Connecticut', Conn. Agric. Expt. Stn. Bull. 523, Connecticut, USA.

Madge, D.S. (1965), 'Leaf fall and litter disappearance in tropical forest', Pedobiologia, **5**, 273-88.

Mackereth, F.J.H. (1965), 'Chemical investigations of lake sediment and their interpretation', Proc. Roy. Soc. (Lond.) B, **161**, 295-309.

Maltby, E. (1975), 'Numbers of soil micro-organisms as ecological indicators of changes resulting from moorland reclamation on Exmoor, UK', J. Biogeogr., **2**, 117-36.

Manners, G. (1981), 'Our planet's resources', Geogl. J., **147**, 1-22.

McDonald, A.T. and Kay, D. (1982), 'Forest expansion and water resources: the case for the forest industry', Working paper 317, School of Geography, University of Leeds, Leeds, UK.

McFarlane, M.J. (1976), Laterite and Landscape, Academic Press, London.

McKey, D. et al (1978), 'Phenolic content of vegetation in two African rain forests: ecological implications', Science, **202**, 61-4.

Mitchell, J. (1973), 'The bracken problem', in The Organic Resources of Scotland, ed. Tivy, J., Oliver and Boyd, Edinburgh, pp. 98-108.

Moore, P.D. (1975), 'Origin of blanket mires', Nature, **256**: 670-2.

Morgan, R.P.C. (ed.) (1981), Soil Conservation; Problems and Prospects, John Wiley and Sons, Chichester, UK.

Naveh, Z. (1982), 'Landscape ecology as an emerging branch of human ecosystem science', Adv. in Ecol. Res., **12**, 189-237.

Nriagu, J.O. et al (1979), 'Sedimentary record of heavy metal pollution in Lake Erie', Geochim. et Cosmochim. Acta, **43**, 247-58.

Nye, P.H. (1961), 'Organic matter and nutrient cycles under moist tropical forest', Plant and Soil, **13**, 333-46.

Nye, P.H. and Greenland, D.J. (1960), The Soil under Shifting Cultivation, Commonwealth Agricultural Bureaux Tech. Comm. 51, Harpenden, UK.

Odum, E.P. (1969), 'The strategy of ecosystem development', Science, **164**, 262-70.

Odum, E.P. (1971), Fundamentals of Ecology, W.B. Saunders Co, Philadelphia, USA.

Olson, J.S. (1958), 'Rates of succession and soil changes on southern Lake Michigan sand dunes', Botanical Gazette, **119**, 125-70.

Ovington, J.D. (1962), 'Quantitative ecology and the woodland ecosystem concept', Advances in Ecol. Res., **1**, 103-92.

Ovington, J.D. (1965), Woodlands, English Universities Press, London.

Page, G. (1968), 'Some effects of conifer crops on soil properties', Comm. For. Res., **47**(1), 52-62.

Pape, J.C. (1970), 'Plaggen soils in the Netherlands', Geoderma, **4**, 229-55.

Parkinson, D., Gray, T.R.G. and Williams, S.T. (1971), Ecology of Soil Micro-organisms, IBP Handbook No. 19, Blackwell, Oxford, UK.

Philbrick, J. and Philbrick, H. (1971), Gardening for Health and Nutrition, Steiner Publications, New York, USA.

Raison, R.J. (1979), 'Modification of the soil environment by vegetation fires with particular reference to nitrogen transformations: a review', *Plant and Soil*, **51**, 73-108.

Renfrew, C. (1978), 'Trajectory discontinuity and morphogenesis: the implications of catastrophe theory for archaeology', *Amer. Antiq.*, **43**, 203-22.

Richards, B.N. (1974), *Introduction to the Soil Ecosystem*, Longman, London.

Richards, P.W. (1957), *The Tropical Rain Forest*, Cambridge University Press, Cambridge, UK.

Risser, P.G. *et al* (1981), *The True Prairie Ecosystem*, US/IBP Series No. 16, Hutchinson Ross Publ. Co, Stroudsburg, Penn., USA.

Rorison, I.H. (ed.) (1969), *Ecological Aspects of the Mineral Nutrition of Plants*, Blackwell, Oxford, UK.

Rowe, R.K. and Hagel, U. (1973), 'Leaching of plant nutrient ions from burned forest litter', *Australian Forestry*, **37**, 154-63.

Rowell, D.L. (1981), 'Oxidation and reduction', in *Chemistry of Soil Processes*, eds. Greenland, D.J. and Hayes, M.H.B., John Wiley and Sons, Chichester, UK.

Royal Commission (1979), 'Royal commission on environmental pollution', 7th Report, *Agriculture and Pollution*, Cmnd. 7644, HMSO, UK.

Ruhe, R.V. (1975), *Geomorphology: Geomorphic Processes and Surficial Geology*, Houghton Mifflin Co, Chicago, USA.

Russell, E.W. (1973), *Soil Conditions and Plant Growth*, 10th edition, Longmans, London.

Schnitzer, M. (1977), 'Recent findings on the characterisation of humic substances extracted from soils from widely differing climatic zones', in *Soil Organic Matter Studies*, Proc. IAEA/FAO symposium Braunschweig, pp. 117-31.

Schumacher, E.F. (1973), *Small is beautiful*, Sphere Books, London.

Sheals, J.G. (ed.) (1969), *The Soil Ecosystem - Systematic Aspects of the Environment, Organisms and Communities*, A Symposium - The Systematics Assocn., London.

Shimwell, D.W. (1971), *The Description and Classification of Vegetation*, Sidgwick and Jackson, London.

Simmons, I.G. (1979a), 'Physical geography in environmental science', Geography, **64**, 314-23.

Simmons, I.G. (1979b), Biogeography: Natural and Cultural, Edward Arnold, London.

Smalley, I.J. (ed.) (1975), Loess: Lithology and Genesis, Dowden, Stroudsburg.

Smith, R.T. (1975), 'Early agriculture and soil degradation', in The Effect of Man on the Landscape: the Highland Zone, eds. Evans, J.G. et al, Council for British Archaeology Research Report No. 11, pp. 27-37.

Smith, R.T. (1977), 'Bracken in Britain II: Ecological Observations of a Bracken Population over a Six-Year Period', Working paper 190, School of Geography, University of Leeds, Leeds, UK.

Smith, R.T. (1982), 'Towards a More Holistic View of British Vegetation History', Working Paper 328, School of Geography, University of Leeds, Leeds, UK.

Smith, R.T. (1983), 'Aspects of the soil and vegetation history of the Craven District of Yorkshire', in Archaeology in the Pennines, eds. Manby, T. and Turnbull, P., B.A.R., Oxford, UK.

Smith, R.T. and Atkinson, K. (1975), Techniques in Pedology, Elek Science, London.

Smith, W.H. (1981), Air Pollution and Forests, Springer-Verlag, New York, USA.

Soil Association, The (1974), Alternative Agriculture: Organic Farming, Stowmarket, UK.

Sollins, P. and McCorison, F.M. (1981), 'Nitrogen and carbon solution chemistry of an old growth conifer forest watershed before and after cutting', Water Resources Research, **17**, 1409-18.

Specht, R.L. (ed.) (1981), Heathlands and Related Shrublands, Elsevier, Amsterdam, Netherlands (2 vols.).

Spedding, C.R.W. et al (1981), Biological Efficiency in Agriculture, Academic Press, London.

Statham, I. (1977), Earth Surface Sediment Transport, Clarendon Press, Oxford, UK.

Swank, W.T. and Douglass, J.E. (1975), 'Nutrient flux in undisturbed and manipulated forest ecosystems in the southern Appalachian mountains', Publication No. 117 de l'Association Internationale des Sciences Hydrologiques, Symp. Tokyo.

Swank, W.T. and Waide, J.B. (1979), 'Interpretation of nutrient cycling research in a management context', in Forests: Fresh Perspectives from Ecosystem Analysis, Oregon State Univ. Press, pp. 137-58, Oregon, USA.

Szabo, I.M. (1974), Microbial Communities in a Forest - Rendzina Ecosystem, Akademiai Kiado, Budapest, Hungary.

Taylor, J.A. (1973), 'Chronometers and chronicles: a study of Palaeoenvironments in west central Wales', Prog. in Geography, No. 5, 247-334, Edward Arnold, London.

Taylor, J.A. and Smith, R.T. (1980), 'The role of pedogenic factors in the initiation of peat formation and in the classification of mires', Proc. 6th Int. Peat Congr., Duluth, USA, pp. 109-18.

Thom, R. (1975), Structural Stability and Morphogenesis, W.A. Benjamin, Reading, Mass., USA.

Trudgill, S.T. (1977), Soil and Vegetation Systems, Oxford, Clarendon Press, UK.

Trudgill, S.T. (1982), Weathering and Erosion, Butterworths, London.

Turner II, B.L. and Harrison, P.D. (1981), 'Prehistoric raised-field agriculture in the Maya lowlands', Science, **213**, 399-405.

Ugolini, F.C. and Mann, D.H. (1979), 'Biopedological origin of peatlands in southeast Alaska', Nature, **281**: 366-8.

Ulrich, B. and Mayer, R. (1972), 'Systems analysis of mineral cycling in forest ecosystems', in Isotopes and Radiation in Soil-Plant Relationships, including Forestry, Int. Atom. Energy Agency, Vienna, pp. 329-39.

UNESCO/FAO (1978), Tropical Forest Ecosystems, Paris.

USDA (1980), Report and Recommendations on Organic Farming, United States Department of Agriculture, Washington DC, USA.

USDA, Soil Survey Staff (1975), Soil Taxonomy: a Basic System of Soil Classification for Making and Interpreting Soil Surveys, Agriculture Handbook 436, Washington DC, USA.

Vita-Finzi, C. (1969), The Mediterranean Valleys, Cambridge University Press, Cambridge, UK.

Vreeken, W.J. (1975), 'Principal kinds of chronosequences and their significance in soil history', J. Soil Sci., **26**, 378-94.

Waksman, S.A. (1938), Humus, Baillere, Tindall and Cox, London.

Waldron, H.A. (ed.) (1980), Metals in the Environment, Academic Press, London.

Walker, P.H. (1962a), 'Soil layers on hillslopes: a study at Nowra, New South Wales', J. Soil Sci., **13**, 167-77.

Walker, P.H. (1962b), 'Terrace chronology and soil formation on the south coast of New South Wales', J. Soil Sci., **13**, 178-89.

Walsh, P.D. (1980), 'The impacts of catchment afforestation on water supply interests', Aqua, **4**, 82-5.

Watts, D. (1978), 'The new biogeography and its niche in physical geography', Geography, **63**, 324-37.

Wilson, A.G. (1981b), Catastrophe Theory and Bifurcation: Applications to Urban and Regional Analysis, Croom Helm, London.

Woodwell, G.M. and Whittaker, R.H. (1968), 'Primary production and the cation budget of the Brookhaven forest', in Primary Productivity and Mineral Cycling in Natural Ecosystems, Univ. Maine Press, Orono, USA, pp. 151-66.

Woolhouse, H.W. (1981), 'Soil acidity, aluminium toxicity and related problems in the nutrient environment of heathlands', in Goodall, D.W. (ed.) Ecosystems of the World, 9B, Heathlands and Related Shrublands, ch. 21, pp. 215-24, Elsevier, Netherlands.

Yaalon, D.H. (ed.) (1971), Palaeopedology, Israel Univ. Press, Tel Aviv.

Yaalon, D.H. (1982), Aridic Soils and Geomorphic Processes, Israel Univ. Press, Tel Aviv.

Yaalon, D.H. and Yaron, B. (1966), 'Framework for man-made soil changes: an outline of meta-pedogenesis', Soil Science, **102**, 272-7.

Zinke, P.J. (1962), 'The pattern of individual forest trees on soil properties', Ecology, **43**, 130-43.

FURTHER READING

Armson, K.A. (1977), Forest Soils, University of Toronto Press, Toronto, Canada.

Brady, N.C. (1974), The Nature and Properties of Soils, Collier-Macmillan, London (8th edition).

Bridges, E.M. and Davidson, D.A. (eds.) (1982), Principles and Applications of Soil Geography, Longman, London.

Curtis, L.F., Courtney, F.M. and Trudgill, S.T. (1976), Soils of the British Isles, Longman, London.

Duchaufour, P. (1982), Pedology: Pedogenesis and Classification, Allen and Unwin, London (trans. by T.R. Paton).

Etherington, J.R. (1982), Environment and Plant Ecology, John Wiley and Sons, Chichester, UK (2nd edition).

FitzPatrick, E.A. (1983), Soils: their Formation, Classification and Distribution, Longman, London.

Foth, H.D. and Schafer, J.W. (1980), Soil Geography and Land Use, John Wiley and Sons, New York, USA.

Russell, E.W. (1973), Soil Conditions and Plant Growth, Longman, London (10th edition).

Chapter 8

BIOCLIMATES

D. Greenland

INTRODUCTION

It had been traditional in biogeography in the past to regard climates as a major determinant of biological features. Polunin (1960: 10), for example, who refers to 'climate the master', states that 'climate is the most far-reaching of the natural elements controlling plant life ...'. Whether such a deterministic attitude is justified or not, it is certain that the close interrelationship between the biosphere and the atmosphere is of vital importance to life on the planet. Some of the depth and complexity of the interrelationship can be seen by reviewing investigations that have been made of it on the micro-, meso- and macro-, or global, geographical scales. Such a review is the theme of this chapter. Space limitations, however, require only a consideration of plant life, and zoological factors are not treated here, except for some general comments in the concluding section.

The role of the bioclimates in biogeography has been, and is, to explain the distributions and interrelations of, and among, biological phenomena. This role dates from at least the Hellenistic age and the studies of Theophrastus (Glacken, 1967: 68). Since this time, there has been a long and distinguished list of investigators, including such names as Humboldt (1807), de Candolle (1855), Merriam (1889), Köppen (1931), Thornthwaite (1948) and Holdridge (1957 and 1962). These scientists have concentrated, for the most part, on the relationships between macro-scale climate (measured in standard ways) and plant distribution. More recently, attention has been given to the mutual effects existing between meso- and micro-scale climates and plants. The development of surface energy and water budget climatology has greatly aided in this respect (see, for example, Budyko, 1958). Equally of help has been the development of what Holland (1982) has called the 'new' biogeography with its own emphasis on budgeting of energy, nutrients, water and other matter. There is still much to be discovered and one of the major frontiers is an elucidation of the actual mechanisms in which the major climatic elements interact with the plant.

SCOPE AND TRENDS

The scope of the study of bioclimates ranges in space from the micro- to the macro-, or global, scale. It ranges in time from minutes to millenia. Given such a vast array, it is best to start with what might be regarded as central: the growth of a plant and, specifically, the transfer of radiant energy into the production of plant materials by means of photosynthesis.

During the vital photosynthesis process, chlorophyll-containing plants and algae produce primary sugars (carbohydrates) in the presence of light energy (radiation). The reaction may be written as:

$$6CO_2 + 12H_2O \xrightarrow[\text{chlorophyll}]{\text{light (112 Kcal (mole } CO_2)^{-1})} C_6H_{12}O_6 + 6H_2O + 6O_2 \quad . \quad . \quad . \quad . \quad 8.1$$

Note that carbon dioxide is used together with water and the end products are carbohydrates, water and oxygen. The plant materials produced form the raw material for more complex sugars, starches, cellulose and other plant constituents. Thus, photosynthesis is essential for plant growth, but the actual growth processes and synthesis of higher compounds require an input of energy. The oxidation of the carbon compounds produced in photosynthesis is a process called respiration and makes available the energy for the biological and chemical work in the plant. During respiration both carbon dioxide and heat are released. Through photosynthesis, green plants represent the producer stage of the food chain in earth ecosystems. The chemical energy in the form of plant material is passed on to the next stage: the consumer stage or second trophic level, when the plant tissue is consumed by animals. The energy may be passed to higher trophic levels when the animals are consumed by other animals. At each trophic level, respiration occurs, some energy is degraded to the form of heat, and the second law of thermodynamics is obeyed.

The processes of plant photosynthesis are directly and indirectly related to many climatic parameters. Photosynthesis requires carbon dioxide and the concentration of this gas is highly variable in time and space, at least near the earth's surface. The process clearly requires water which will, in most terrestrial cases, be supplied initially by atmospheric precipitation. It equally clearly demands radiant energy, which is also highly variable in time and space on the earth's surface. The plant is usually not very efficient in using solar energy, however. Normally less than 1% of global solar radiation is used in photosynthesis (Sellers, 1965: 102). Generally, the rate of photosynthesis is positively related to light, carbon dioxide and soil moisture availability. Temperature does not strongly affect the photosynthetic reaction as long as extreme temperatures do not prevail. These relationships are not linear. Their complexities are concisely described by Rosenberg (1974: 206-220). Respiration requires water which is essential to many of the chemical processes in the plant and for carrying substances in solution from one part of the plant to another. Temperature, however, is a major control on the rate of respiration since, theoretically, chemical reaction rates are exponentially related to absolute temperature and, in practice, plant developmental and growth rates are linearly related to temperature. The difference between the exponential relation of theory and the linear rate of practice has yet to be explained (Monteith, 1981).

These fundamental relationships between plant growth and climatic parameters operate for the individual plant, groups of plants, and large areas of vegetation. Thus there are different scales of study.

Micro-scale studies cover many aspects of flows of energy and water to and from plants. These studies hinge on the basic energy balance equation:

$$Q^* + H + LE + G = 0 \qquad 8.2$$

where Q* is net radiation, H is sensible heat, LE is latent heat and G is flow of heat into the ground or plant. This describes the law of conservation of energy and the first law of thermodynamics as applied to the earth or plant surface. Such studies would concern themselves with the measurement of these different heat flows under varying conditions (see, for example, Greenland, 1973). This would require detailed studies of incoming and outgoing radiation values, surface and air temperature, atmospheric humidity, wind and turbulent transport, evapotranspiration, and soil temperatures. During the last two decades, investigators in this field have had the benefit of some outstanding texts reviewing the complex subject. These include the works of Gates (1962, 1980), Geiger (1965), Sellers (1965), Rose (1966), Monteith (1973), Rosenberg (1974) and Oke (1978).

Mesoscale investigations usually consider the interaction of bioclimates with groups of plants. Many of these studies are applied in nature. These include the effects of windbreaks and shelterbelts, frost and its control, improving water-use efficiency, and the relations between standard parameters of weather and climate and crop productivity. These kinds of studies have been reviewed by Munn (1970), Maunder (1970), Smith (1975), Rosenberg (1976) and Hobbs (1980). Other mesoscale studies deal with more academic questions, such as plant-climate interactions at the arctic treeline (e.g. Bryson, 1966; Krebs and Barry, 1970; Hare and Ritchie, 1972), or along gradients of environmental stress (Slatyer, 1978).

Global-scale studies may have the longest history in biogeographical literature. At the beginning of the 19th century, Humboldt (1806) established a grouping of vegetational types on a physiognomic basis, together with some effort to express relations between the environment and life-form groups, such as the cactus form and the banana form. Merriam (1895, 1896, 1899) employed the temperature-sum concept to define life zones in North America and believed humidity was secondary to temperature in determining the distribution of vegetation. In the early 20th century, Raunkiaer suggested a plant life-form system consisting of five classes arranged according to increased protection of the buds (Caine, 1950). The widely used climatic classification of Köppen (1931) was derived in part by attempting to relate vegetation boundaries (from the earlier work of de Candolle, 1853) to values of monthly mean temperatures and precipitation totals. One of the guiding principles of Thornthwaite's (1948) climatic classification was also that the plant and the amount of moisture it evapotranspired served to integrate a wide variety of climatic parameters. Budyko (1958) suggested that vegetation zones could be related to the ratio of net radiation receipt to the energy required to evaporate precipitation received: a value he called the Radiational Index of Dryness. A combination of mean annual temperature, precipitation, and the ratio of potential evapotranspiration to precipitation was employed by Holdridge (1962) to establish life zones on an altitudinal basis. Attention continues to be given to this topic, as will be apparent from the later description of the work of Box (1981). Indeed, Hare (1973) has pointed to the study of the interrelationships between the biosphere and the atmosphere as a major direction to be pursued by geographers.

METHODS OF STUDY

Many technological advances have been made in the study of bioclimates in the last two decades, yet there remain significant topics where greater sophistication is needed. One such area is the measurement of primary productivity.

Procedures for measuring primary production on grasslands have been described by Milner and Elfyn Hughes (1968). These include establishing sample plots. Some plots have their vegetation cut at the beginning of the growing season and the plant material is deep frozen or dried within a few hours. The same number of plots are cut at four-weekly intervals throughout the growing season. At each harvest date, the quantity of dead material is determined. Ground vegetation root cores are also taken. The dried material, both roots and shoots, has its calorific value determined using bomb calorimetry. During this process, the temperature of the container surrounding the calorimeter holding the test material is made to rise with that of the calorimeter. This permits direct calculation of the heat generated in combusting the sample material from the temperature rise and the thermal capacity of the calorimeter (Petrusewicz and MacFadyen, 1970: 59). The net primary production is defined as the biomass or total energy content incorporated into a plant community during a specified time interval, less that respired (Milner and Elfyn Hughes, 1968). Its calculation requires estimates of the biomass lost to small herbivore forms during the growing season. This kind of loss is, of course, important in the case of grazed grasslands and here animal exclusion cages have to be employed to determine the biomass lost to grazing. An application of these kinds of procedures was made by Job and Taylor (1978) in their study of upland grazings on Plynlimon in Wales. In contrast to grassland, the determination of primary productivity of forests is more difficult because estimates have to be made of the productivity of all of the forest components, including bud scales, flowers and fruit, leaves and twigs, branches, stems and trunks, roots and litter fall. Many of these components, such as tree roots, cannot be measured practically without expensive destructive sampling. Consequently, many indirect empirical relationships are employed. As an example, the weight, W, of different tree components has been found to be associated with the tree DBH (diameter breast-height) by the relationship:

$$\log W = a + b \log DBH \quad \quad 8.3$$

where a and b are constants for the species (Newbould, 1967: 26). Authors performing or reviewing studies of primary productivity seldom quote accuracy limits for the process as a whole or for the various stages involved in the process.

Compared to measurement of primary productivity, micro-scale bioclimatic studies are usually made in a manner permitting accuracy limits to be known, even if they are not always quoted. It is common for field studies and laboratory investigations to be employed.

Many field studies attempt to evaluate the four terms of the surface energy budget equation (equation 8.2). Net radiation can be measured directly with a net radiometer, of which the most commonly used is the Funk type or variations of it (Funk, 1959). Ground heat flow is directly monitored by heat flow plates or disks commonly only a few centimeters in diameter. If the term is to be evaluated very accurately, the temperature gradient from the surface to below the plate needs to be known so that heat divergence in the vertical plane may be estimated (Wilson and McCaughey, 1971). Sensible heat flux and latent heat flux are called the turbulent heat fluxes because their flows are closely associated with turbulent air motion near the ground. One method of measuring them requires a mast instrumented to record temperature and specific humidity at two heights. Sensible heat transfer can be estimated from an equation of the form:

$$H = \rho C_p K_h \frac{dT}{dz} \quad \quad 8.4$$

where ρ is air density, C_p is the specific heat of air at constant pressure, K_h is an exchange coefficient, and dT/dz is the temperature gradient. The latent heat flux can be estimated from an equation of similar form. In both cases, the main difficulty is the variability of the exchange coefficient. Its values can be estimated from the wind profile data and, if neutral stability (temperature gradient equals the dry adiabatic lapse rate) and equivalence of exchange coefficients are assumed, then the values of LE and H can be found. Since these assumptions do not always hold, many other approaches have been used to make such estimates and these have been summarised by Sellers (1965) and Rosenberg (1974).

Organisation of micro-scale studies along energy budget lines enables concentration to be put on parameters directly affecting plant growth, such as radiation and evapotranspiration.

Consequently, many studies on this scale examine these parameters in great detail. As an example, bioclimatologists are interested in a resolution of net radiation into its components. These are incoming and outgoing shortwave radiation (0.15 μm - 3.0 μm) and longwave radiation (3.0 - 100 μm) (Greenland, 1975). Also of interest is the amount of ultraviolet light arriving at a plant (Caldwell, 1968). Possibly most significant of all is how much radiation is available within wavebands that are photosynthetically active (0.4 - 0.7 μm). Norman et al (1969) have developed an instrument to measure this. Despite the importance of this quality, however, much remains to be learned concerning how much photosynthetically active radiation (PAR) is actually used by different species of plants, and what proportion of total incoming radiation falls in the photosynthetically active waveband. It is common practice to take PAR as 47% of incoming shortwave radiation (e.g. Wassink, 1975), yet Kondratyev (1969) quotes a range of this proportion varying from 25% to 43%. Since photosynthetic and respiratory processes are related to the temperature of the plant, leaf temperatures of plants are often monitored by means of thermocouples (Linacre, 1972) or infrared thermometers (Fuchs and Tanner, 1966).

The existence of so many microclimatic variables in the real world often makes it expedient for biologists to perform laboratory experiments where some of the variables may be held constant. There is a large range in the size of the laboratory environments used and the number of variables that can be controlled in them. At the large end of the scale are the Australian phytotron and the University of Wisconsin biotron, both described by Munn (1970: 80-81) and which occupy entire buildings. At the smaller end of the scale, many organisations have controlled growth chambers ranging from 1 m to 5 m square. An attendant problem with the use of such facilities is relating results of experiments performed in them to real world situations. One interesting example in which this problem has been overcome is the development of a square wave temperature conversion function that provides a good simulation of daily thermograph records (Slatyer, 1978).

Many of the micro-scale bioclimatic studies are performed to increase our understanding of how agricultural production is related to the weather. This problem, however, brings into play meso-scale studies as well. Much of the methodology for forecasting yields a major agricultural crops is statistical in nature. Some relationships have been reviewed by Chirkov (1979: 273-77) who quotes, for example, the relationship of winter wheat yield in Czechoslovakia (y) to the mean October air temperatures (x) as being:

$$y = -29.92 + 8.09\,x - 0.5398\,x^2 \qquad . \quad . \quad . \quad . \qquad 8.5$$

Monteith (1981) has pointed to the limitations of such approaches and notes that the statistical methods are often incorrectly used. Inferences are made that are quite beyond those permissible by the statistical method employed. As in equation 8.5, correlation coefficients and standard estimates of error are frequently omitted despite their importance. Even more important, though, is the fact that the interaction of the physical and physiological mechanisms are ignored. In contrast, Monteith (1981) provides a fine example of how the factors of light and carbon dioxide, precipitation and evaporation, and temperature may be combined in a physiologically meaningful way to make estimates of crop yields for the east Midlands of England. Investigators in the field should be urged to follow this example.

Attempts to relate agricultural yields to climatic and weather variations are in some ways similar to studies which endeavour to relate vegetation type and productivity on a global-scale to spatial climatic variation. Early attempts at pointing to such relationship were qualitative or, at best, semi-qualitative. The work of Köppen, Thornthwaite, Budyko and Holdridge has already been mentioned. An excellent review of these and other important synthesisers has been given to Tuhkanen (1980). One problem has been the lack of a sound method of quantifying vegetation types. A second problem has been the use of climatic parameters that are not directly related to plant growth processes. The plant, for example, is not so much concerned with total precipitation as with how much soil moisture is available to its roots in the growing season. Box (1981) has also alluded to 'problems of scale: micro-climate, heterogeneity, and the availability of data of such smaller scales and for all important macro-scale factors'.

Full quantification has now been achieved, as witnessed by the model of Box (1981) relating physiognomic vegetation types with climate variables. This model treats plant life-forms rather than formation types. It has more vegetation units (41) than most models. Furthermore, it includes more explanatory climatic variables than most other models and uses eight of these. The model consists of tables showing the empirically valuated environmental envelopes with an upper and lower tolerance limit for each life-form with respect to each climatic variable. The climate variables are chosen to express aspects of the annual regimes of temperature, precipitation and potential evapotranspiration. The variables are mean temperature of the warmest and coldest months, annual range of monthly mean temperature, average annual precipitation, a moisture index (the annual precipitation divided by the Thornthwaite estimate of potential evapotranspiration), highest and lowest average monthly precipitation, and the average precipitation of the warmest month.

Other examples of quantification are correlation-regression models between net primary production and actual evapotranspiration (Lieth and Box, 1972), and net and gross primary production with various environmental variables (Lieth and Box, 1977), and terrestrial plant litter production with climatic variables (Meentemeyer et al, 1982). All of these studies are based on the world climatic data base of 1,255 sites and a mapping system assembled by Box (1978). Yet, despite the achievement of quantification, these studies are basically empirical. Box (1981) suggests that subsequent models should seek to relate vegetation to fewer, more integrative variables in order to identify more critical mechanisms and general patterns of environmental limitation. Following Monteith's lead, a move away from empiricism and toward a greater understanding of the actual biospherical pathways leading to the plant environment at this scale would also seem appropriate. Some of what is hidden in the 'integrative variables' can be seen, and an idea of the 'noise' suppressed in the regression methodology can be gained by examining case-studies concerning bioclimates and vegetation productivity.

CASE-STUDIES

Two extreme environments, the alpine tundra and the tropical rainforest, may be selected to demonstrate the difficulties and limitations of specifying bioclimates and determining their relationship to plant productivity. In addition, further insights may be gained by reviewing our knowledge of a biogeographical boundary: the northern treeline, and its interrelationship with the atmosphere.

The alpine tundra ecosystem of Niwot Ridge in the Colorado Rocky Mountains has been a focus of study since 1951 (Halfpenny, 1982). The standard climatic parameters have been recorded of a site called D1 at an altitude of 3,750 m. This site is visited weekly and, despite the extremely severe conditions often prevailing (Greenland, 1977), the site has an almost unbroken climate record. This record shows a mean annual air temperature of 3.8^{o} C and a frost-free period averaging 47 days during the year (Barry, 1973). Mean annual precipitation is 102 cm (40.2 inches), while mean annual wind speed is 10.3 m sec^{-1}. Caldwell (1968) found an increase in ultraviolet B radiation (0.280 µm to 0.315 µm) of 26% between 1,670 m and 4,350 m in elevation on cloudless days. Radiation in these wavelengths can lead to plant damage. However, cloud conditions on the site are such that the higher elevation does not receive a markedly greater amount of global solar radiation over the annual period (Caldwell, 1968, Greenland, 1978).

Billings (1976) described the strategies used by the alpine tundra plants to survive the severe climatic conditions. These include adopting a small and perennial form, having most of its biomass under the ground, adopting prostrate forms in the case of shrubs, having the ability to metabolise rapidly at low temperatures, and the ability to withstand drought.

The bioclimate is further specified by examining the surface energy fluxes. The annual total global solar radiation is 5,525MJ m^{-2} (Barry et al, 1981). Daily values of this parameter on clear days can range from 10.5MJ m^{-2} in the winter to 28MJ m^{-1} in the summer. Values for the surface energy budget during the summer of 1973 were determined by LeDrew (1975). He found average values to be: net radiation 144W m^{-2}, sensible heat 71W m^{-2}, latent heat 55W m^{-2}, and soil heat 18W m^{-2}. These values were shown to be similar to ones recorded in Barrow, Alaska during the summers of 1957 and 1958. Patches of standing water in both these kinds of tundra environment can have a marked effect on the value of sensible heat and latent heat flow. Long-lasting snow patches also affect the energy budget values. The general heterogeneity of the surface thus causes sampling problems.

Despite these factors, it is possible to make some estimates of energy transfers between the atmosphere and the biosphere for this environment. Webber and Ebert May (1977) have estimated the net above-ground productivity to be 100 to 300 g m^{-2} yr^{-1}, given global solar radiation at the site of about 5,525MJ m^{-2} yr^{-1} (Barry et al, 1981) and a calorific value for dry matter of 4K cal gm^{-1} (Caldwell, 1975). This would yield values of photosynthetic efficiency of between 0.03% and 0.09%. This is undoubtedly an underestimate of total efficiency since Webber and Ebert May (1977) reported that much of the biomass is beneath the surface. They estimate aboveground-to-belowground biomass ratios of between 1:3 and 1:25. Thus, significant but unknown amounts of productivity in this biomass occur beneath the ground.

In contrast, the lowest efficiency with respect to new production for forests, quoted by Kira (1975), is 0.50% for young sub-alpine forest, while tropical rainforest has been found to have an efficiency of 0.85%. An exception to this is the rather lower values for three year-old tropical rainforest quoted below. Some of the difficulties in measuring forest productivity and, therefore, photosynthetic efficiency, have already been mentioned. As a result, exact comparison of values estimated by different authors is difficult. Data assembled by Kira (1975), however, indicated that efficiency of gross productivity varied little among forests of different thermal zones, and that forest canopies seem to be able to utilise solar energy equally efficiently under different radiation regimes. The difficulty in measuring productivity in tropical forests is matched by the difficulty in characterising their bioclimate.

Frequently, the bioclimate is not specified further than a general description of the climate and a measurement of solar radiation. This was the case for a field study in Puerto Rico conducted by Jordan (1971). In this study, a pyrheliometer to measure solar radiation had to be mounted on a tower 3.7 m above the forest canopy, which itself could be 60 m high. Annual efficiency at this location varied from 0.15% for one- to three-year-old rainforest to 0.65% for mature rainforest. An idea of the nature of the difficulty of measuring bioclimatic parameters in a tropical rainforest is seen in the investigation of Lemon et al (1970) for a forest in Costa Rica. In this study, two rope-and-pulley systems were rigged to a tall tree to carry sensors and CO_2 samplers. Besides the air samplers, instruments were used to measure net radiation, visible light, water vapour, air temperature and wind speed. It was not possible to obtain data of good enough quality to permit energy budgets to be calculated in the various parts of the forest system which was divided into four layers. However, the CO_2 flux was computed, using wind profiles to estimate the vertical exchange coefficients. One interesting feature encountered was that the wind speed was higher below the canopy than within it: a fact which is in contrast to wind profiles in agricultural crops and complicates the CO_2 profile. Practical difficulties and unique properties of the tropical rainforest therefore make measurements and calculations of its bioclimate difficult. The author is not aware of any full energy budget-productivity study performed in such an environment.

The need for these energy budget-productivity studies, however, is imperative. This is demonstrated by a simulation and field study of bioclimate and production in red mangrove canopies in south Florida (Miller, 1972). In this study, measurements of radiation balance components, air and leaf temperature, humidity and wind were easier to make through a profile that extended only 4 m compared to the profile of over 40 m required for the rainforest. The study argues convincingly that an interpretation of primary production in an area requires detailed knowledge on the radiation fluxes. In particular, air temperature and humidity are critical for comparative production studies since the simulation showed these factors to have the greatest effect on leaf temperature, transpiration and net photosynthesis.

One further case study which clearly demonstrates the two-way interaction between the biosphere and the atmosphere is the examination and attempted explanation of the location of the northern limits of the Boreal forest. Hare and Ritchie (1972) have distinguished four vegetation zones for the northern limits of the Boreal forest in North America. These are: (1) closed forest typified by closed crowns with moist shaded forest floors, (2) open woodland characterised by open forest with lichen floors or a discontinuous shrub layer, (3) forest-tundra consisting of extensive tundra patches with scattered or isolated trees often in prostrate form, and (4) tundra where no trees exist. They describe the Northern Forest line as existing between zones 1 and 2 and the Arctic tree line occurring between zones 3 and 4. Their Figure 4 illustrates a transitional area between the open woodland (zone 2) and the forest tundra (zone 3) in northern Quebec.

Possibly the earliest consideration of this general vegetation boundary was made by Köppen when he was attempting to locate the line between the D and E climates of his classification. He used the location of the 10^{0} C mean temperature of the warmest month of the year as the boundary. Modern studies of the problem may be considered to have started with Hare's (1950) examination of the vegetation boundaries from a water balance point of view. He found a greater correlation between the zonal forest divisions and a thermal parameter (potential evapotranspiration calculated by the Thornthwaite method) than moisture parameters such as precipitation values. There is some correspondence between forest divisions and different potential evapotranspiration values but the divisions are independent of moisture gradients. Before Hare was able to point to the bioclimatic factors that appeared to be more closely related to the vegetation boundaries several interesting synoptic scale studies were performed.

Actually the first of these had taken place as far back as 1928 when Stupart (quoted by Larsen 1974: 345) saw a coincidence between the path of summer cyclonic storms and the position of the forest-tundra ecotone in Canada. Reed (1960) and Bryson (1966) showed an association of the general Boreal forest border in Canada with the summer position of the Arctic Front. The relation of the summer position of the Arctic Front and the Boreal forest boundary is better in some parts of the Northern hemisphere than others as was demonstrated for Canada by Barry (1967) and for Eurasia by Krebs and Barry (1967). At least, for Canada the correspondence was most marked where the forest-tundra boundary is relatively sharp. No investigators have been able to demonstrate a clear physical relationship between the synoptic phenomena and the vegetation boundaries. However, newly available data on the radiation balance of North America lead Hare and Ritchie (1972) to examine the boundary afresh.

Hare and Ritchie noted the strong gradients in albedo, and consequently net and photosynthetically active radiation, existed across the vegetation boundaries of the Boreal forest. These gradients give rise to strong temperature gradients especially in spring when 'only 30 kilometers separate wholly frozen lakes from those only half frozen'. In addition there is a sharp gradient in standing phytomass at the forest-tundra border as one would expect. Interestingly, Warren Wilson (1957) has pointed out that the rate of photosynthesis increases approximately 50% when air temperature rises from 0° C to 1° C whereas at the temperatures common to lower latitude vegetation, photosynthetic rates are much less responsive to temperature change.

In a later study Hare (1973) summarised the atmosphere-biosphere linkages more specifically. First, the vegetation boundaries of the Boreal forest follow specific isolines of annual and growing-season net radiation and the length of season with above freezing temperatures. The vegetation boundaries relate to the structure of the vegetation rather than floristic (actual species) considerations. Second, the strong gradient of standing phytomass values is related to the net radiation field. Third, deeper and more uniform snow cover in the Boreal woodland provides shelter for plants and animals and heavily influences soil temperatures and permafrost distribution. The properties of the snow cover are, in turn, affected by the structure of the vegetation. Fifth, the net radiation values are strongly influenced by surface albedo which is affected by snow cover and again the structure of the vegetation.

The next step in this area of study is to quantify these relationships with actual measurements on the ground. A start has been made on this by the important investigations of Rouse (1977) and his co-workers. They have measured radiation and surface energy budget component values for six different types of surfaces across latitudes 56° to 61° N in Canada for the mid-summer period. They report large differences in albedo and net radiation. Similar studies in the spring season might be very enlightening although rather difficult to perform.

Investigations of the northern boundaries of the Boreal forest demonstrate well the intricate relationships between climate and vegetation while the earlier case-study examples show the very real problems in making measurements of factors specifying bioclimates. This latter topic is examined further in the next section.

STATE OF THE ART AND NEED FOR FUTURE WORK

If the earth is populated by human beings in 200 years' time, the biogeographers among them will undoubtedly classify the present period as one in which their science was primitive. Most of the instrumentation and methods currently used are not very precise. Beyond this fact, there are many other areas that demand more attention. Among these may be considered the physical linkages between the atmosphere and plant and the modelling of such linkages, the need for greater attention to faunal climates, the relationship between plant productivity and social and economic systems, and many aspects of climatic change. Let us look first at accuracy.

Our case-studies demonstrate that accuracy in characterising bioclimates is not high. Most instruments used in measuring bioclimates have known accuracy limits. For instantaneous values, these range from about 0.01% for accurate temperature measurement to about 5% for net radiation measurement. Different time and space scales carry different accuracy ranges. Most studies are complicated, however, by sampling difficulties. Vegetated surfaces and structures are usually so heterogeneous that it is difficult to know how to obtain truly representative samples. Another problem is that many of our bioclimatic parameters are derived and the derivation process often compounds lack of accuracy. On the other side of the coin, the measurement of vegetative productivity is fraught with difficulty. We have seen, for example, how below-ground biomass in the alpine tundra is usually neglected in productivity studies, while in the tropical rainforest, it is most frequently derived empirically. Most productivity studies do not quote the accuracy limits at which they were made. When the results of such studies are extrapolated to one degree of latitude and longitude areas, accuracy problems become even more severe. Clearly, the technology of the science requires much development and the areas where such development might be most needed would be highlighted if accuracy values were attached to bioclimatic studies.

Second, we need much more knowledge on plant-atmosphere interactions on a species-by-species basis. Monteith's (1981) work, quoted earlier, provides some guide lines on how such work might proceed. After much painstaking and laborious work of this kind, there would be two main results. First, we would be able to draw much better and more useful general conclusions on plant-atmosphere interactions. Second, our models would be improved by a substitution of physical relationships for the statistical relationships that are so prevalent in much of the present literature. It may be that later on we would find statistical relationships to be essential, as has been found in other sciences such as nuclear physics, but such a finding should be based on an improvement of our knowledge of the interrelationships. We should not use statistical methods to hide our scientific laziness. Our models also need to be improved by matching their sophistication if they are to be linked together. It is bad practice, for example, to link a complex model for describing radiative flux through a canopy with a simple model for estimating evapotranspiration. Since our present models vary in their sophistication, however, this kind of procedure is sometimes the only one possible.

A third issue is the need for biogeographers to pay more attention to faunal climates. There have certainly been many studies dealing with the relation of the animal kingdom with the atmosphere. On the micro-scale, for example, the energy budget and the ambient microclimatic conditions of the wandering garter snake have been well documented (Scott et al, 1982). A meso-scale example would be the discovery that low-level wind convergence zones represented locations for the effective spraying of locust pests (Rainey, 1973). Typifying the global-scale may be the finding reported by Watt (1973: 134) that species evolve mechanisms for decreased sensitivity to climatic stress in environments where climates fluctuate with wide amplitude. Compared to the literature on vegetation-atmosphere relationships, however, the zooclimatic literature is small. Biogeographers take cognizance of atmospheric factors as, for example, in the work of Simmons (1979) and Putman (in the present volume) but, for the most part, climatologists ignore the field and limit their attention to reviews of the topic such as that of Geiger (1965: 468) and Smith (1975: 104-107).

One of the main reasons for this lack of attention is that zooclimatic studies are often more difficult than those dealing with plants and climate. Energy budget and microclimatic measurements of animals are very complicated as exemplified by the case studies on locusts, lizards and sheep quoted by Monteith (1973: 166). A second problem concerns animal mobility which not only allows animals to use artificial shelter from the elements but also permits migration from one climatic area to another. The fact that response to the atmosphere is only one of many interrelated factors such as territoriality, search for food and predator-prey relationships that might force mobility, is also confusing. A further set of problems arise from the influence of humans on animals. These include the process of domestication and even the provision of shelter for animals, the effects of hunting on animal populations and genetic manipulation have profound implications both in the agricultural and conservational context when attempting to isolate the interrelations between the atmosphere and animal life.

One lesson to be drawn from the zooclimatic literature, however, is that atmospheric events are very important to animals although there is less of a two-way interaction as with vegetation, the only exception to this being the influence of humans on the atmosphere. This atmospheric influence on animal life being the case, there is a pressing need for a more systematic approach to be taken to zooclimates. An important contribution to the field would be a text dealing exclusively with the atmosphere as it affects fauna. Such a work could point to the gaps that have occurred owing to the largely piecemeal and pragmatic, case-by-case approach that has been the situation up to the present.

The consideration of human influence brings up a fourth area of potential future study in the field of bioclimatology. This is the necessity to relate the study of bioclimates to social and economic systems and to human needs. One of the goals of the International Biological Programme was to 'estimate existing and potential plant production in the major climatic regions of the world' (National Academy of Science, 1968), but not too much attention was given to how this information was to be used when it was gathered. Terjung (1976) has outlined a need for climatic studies within a framework of human systems. He defines human-physical process-response systems. There are signs of some work beginning in this direction. Spaeth (1980), for example, has done pioneering work in relating soil moisture and evaporation conditions to agricultural practices in Colorado. Drought-prone areas around the world have been given special attention in considering the interaction between human and physical systems (e.g. Bowden et al, 1981). In addition, some interesting reviews of human-physical system interaction have been forthcoming such as those of Taylor (1970, 1974) relating atmospheric resources and human activity. Bennett and Chorley (1978), however, have discussed the many difficulties of linking human and physical systems, and it will take decades before these difficulties are overcome.

A final point is the need to give more consideration to climatic change in the study of bioclimates. Most bioclimatic studies, of necessity, regard climate as a static variable in the long term since they are usually complex enough without considering such a factor. All of the studies mentioned in this review from the micro-scale process studies to the global interrelation and classification investigations have neglected the long-term climatic variable. Ironically, much evidence for climatic change comes from vegetation-related sources such as palynology, palaeosols and macro-fossils. However, the weak point of these study areas is often the lack of knowledge on 'palaeo-ecology', i.e., what are the actual responses of plants to long-term climatic change? (Burroughs and Greenland, 1979.) This is a difficult question to answer in detail. Nevertheless, it holds great urgency since human activities, such as the removal of tropical rainforests and an increase in atmospheric CO_2, may in themselves trigger climatic change. Furthermore, climatic changes have considerable implications for the agricultural food production of the planet.

Clearly, therefore, there is an immense amount of work to be done. Major advances have been made since the classical studies of the 19th century, but there is a pressing need to improve understanding of plant-atmosphere interrelations on all geographical scales. One of the factors that may have retarded progress in this area is the fact that the study of bioclimates lies at the interface between biologic and atmospheric science. The organisation of research-funding agencies sometimes makes it difficult to allow for this fact to be recognised. Nevertheless, problems of growing world population and the possibility of large-scale modification of environmental systems by human action create a vital urgency for tackling the difficulties that have been outlined in the study of bioclimates.

ACKNOWLEDGEMENT

This chapter was prepared with help in part from the Long Term Ecological Research program of the Institute of Arctic and Alpine Research at the University of Colorado at Boulder. The LTER program is funded by the National Science Foundation. The author is grateful to Dr Tom Veblen of the Department of Geography, University of Colorado, for aid in searching the biogeographical literature, and the present editor for his suggestions on animal climatology.

REFERENCES

Barry, R.G. (1967), 'Seasonal location of the Arctic Front over North America', Geographical Bulletin, **9**(2): 79-95.

Barry, R.G. (1973), 'A climatological transect on the east slope of the Front Range, Colorado', Arctic and Alpine Research', **5**(2): 89-110.

Bennett, R.J. and Chorley, R.J. (1978), Environmental Systems: Philosophy, Analysis and Control, Princeton, J.R.: Princeton University Press, USA.

Billings, W.D. (1974), 'Adaptations and origins of alpine plants', Arctic and Alpine Research, **6**(2): 129-42.

Bowden, M.J., Kates, R.W., Kay, P.A., Riebsame, W.E., Warrick, R.A., Johnson, D.L., Gould, H.A. and Weiner, D. (1981), 'The effect of climate fluctuations on human populations: two hypotheses', in Wigley, T.M.L. et al (eds.) (1981), Climate and History: Studies in Past Climates and their Impact on Man, Cambridge University Press, Cambridge, pp. 479-513.

Box, E.O. (1978), 'Geographic dimensions of terrestrial net and gross primary productivity', Radiation Environment and Biophysics, **15**: 305-22.

Box, E.O. (1981), 'Predicting physiognomic vegetation types with climatic variables', Vegetatio, 45: 127-39.

Bryson, R.A. (1966), 'Air masses, streamlines, and the Boreal Forest', Geographical Bulletin, **8**: 228-69.

Burrows, C.J. and Greenland, D. (1979), 'An analysis of the evidence for climatic change in New Zealand in the last thousand years: evidence from diverse natural phenomena and from instrumental records', Journal of the Royal Society of New Zealand, **9**(3), 321-73.

Cain, S.M. (1950), 'Life forms and phytoclimate', The Botanical Review, **16**(2), 1-32.

Caldwell, M.M. (1968), 'Solar ultraviolet radiation as an ecological factor for alpine plants', Ecological Monographs, **38**, 263-8.

Caldwell, M.M. (1975), 'Primary production of grazing lands', in Cooper, J.P. (ed.) (1975), Photosynthesis and Productivity in Different Environments, IBP 3, 62, Cambridge University Press, Cambridge, UK.

de Candolle, A. (1855), Géographie Botanique ou Exposition des Faits Principaux et des Lois concernant la Distribution Géographique des Plantes de l'Epoque Actuelle, Paris and Geneva, Vol. 1.

Chirkov, V.I. (1979), 'Agrometeorological forecast systems', in Seeman, J., Chirkov, V.I., Lomas, J. and Primault, B. (1979), Agrometeorology, Springer-Verlag, Berlin, New York, USA.

Fuchs, M. and Tanner, C.B. (1966), 'Infrared thermometry of vegetation', Agronomy Journal, **58**, 597-601.

Funk, J.P. (1959), 'Improved polyethylene shielded net radiometer', Journal of Scientific Instruments, **36**, 267-70.

Gates, D.M. (1962), Energy Exchange in the Biosphere, Harper and Row, New York, USA.

Gates, D.M. (1980), Biophysical Ecology, Springer-Verlag, New York, USA.

Geiger, R. (1965), The Climate Near the Ground, English trans., Harvard University Press, Cambridge, Mass., USA.

Glacken, C.J. (1967), Traces on the Rhodean Shore, University of California Press, Berkeley, California, USA.

Greenland, D. (1973), 'An estimate of the heat balance in an alpine valley in the New Zealand Southern Alps', Agricultural Meteorology, **11**, 293-302.

Greenland, D. (1975), 'A study of radiation in the New Zealand Southern Alps', Geografiska Annaler, **57A**, 143-151.

Greenland, D. (1977), 'Living on the 700 mb surface: the Mountain Research Station of the Institute of Arctic and Alpine Research', Weatherwise, **30**(6), 232-8.

Greenland, D. (1978), 'The spatial variation of radiation in the Colorado Front Range', Climatological Bulletin, **24**, 1-16.

Hare, F.K. (1950), 'Climate and zonal divisions of the Boreal Forest formation in Eastern Canada', Geographical Review, **40**, 615-35.

Hare, F.K. (1973), 'Energy based climatology and its frontier with ecology', in Chorley, R.J. (ed.) (1973), Directions in Geography, Methuen, London, pp. 171-192.

Hare, F.K. and Ritchie, J.C. (1972), 'The Boreal bioclimates', Geographical Review, **62**(3), 333-65.

Halfpenny, J.C. (1982), Ecological Studies in the Colorado Alpine: A Festschrift for John W. Marr, Boulder, Colorado, Institute of Arctic and Alpine Research, Occasional Paper No. 37, University of Colorado, USA.

Hobbs, J.E. (1980), Applied Climatology: A Study of Atmospheric Resources, Westview Press, Boulder, Colorado, USA.

Holdridge, L.R. (1975), 'Determination of world plant formations from simple climatic data', Science, **105**, 367-8.

Holdridge, L.R. (1962), Life Zone Ecology, San Jose, Costa Rica Tropical Science Center, San Jose, Costa Rica.

Holland, P. (1982), Review of Vegetation and Productivity, Jones, G.E. (1979), New Zealand Geographer, **38**(1), 34-35.

Humboldt, A. von (1807), Ideen zu einer Physiognomik der Gewachse, quoted by Cain, S.A., 'Life-forms and phytoclimate', The Botanical Review, **16**(2), 1950, 1-32.

Job, D.A. and Taylor, J.A. (1978), 'The production, utilisation and management of upland grazings on Plynlimon, Wales', Journal of Biogeography, **5**, 173-91.

Jordan, C.F. (1971), 'Productivity of a tropical forest and its relation to a world pattern of energy storage', The Journal of Ecology, **59**(1), 127-62.

Kira, T. (1975), 'Primary production of forests', in Cooper, J.P. (ed.) (1975), Photosynthesis and Productivity in Different Environments, IBP 3, Cambridge University Press, Cambridge, UK.

Kondratyev, K. Ya. (1969), Radiation in the Atmosphere, International Geophysics Series, **12**, Academic Press, New York, USA.

Koppen, W. (1931), Grundrisse der Klimakunde, Walter de Gruyter, Berlin, West Germany.

Krebs, J.S. and Barry, R.G. (1970), 'The Arctic Front and the Tundra-Taiga boundary in Eurasia', Geographical Review, **60**(6), 548-54.

Larsen, J.A. (1974), 'Ecology of the northern continental forest border', in Ives, J.D. and Barry, R.G. (eds.) (1974), Arctic and Alpine Environments, Methuen, London, pp. 361-9.

Ledrew, E.F. (1975), 'The energy balance of a mid-latitude alpine site during the growing season, 1973', Arctic and Alpine Research, **7**(4), 301-16.

Lieth, H. and Box, E.O. (1972), 'Evapotranspiration and primary productivity; C.W. Thornthwaite memorial model', Publications in Climatology, **25**(3), Elmer, J.J., Thornthwaite Associates, 36-66.

Lieth, H. and Box, E.O. (1977), 'The gross primary productivity pattern of the land vegetation: a first attempt', Tropical Ecology, **11**, 109-55.

Linacre, T.E. (1972), 'Leaf temperatures, diffusion resistance, and transpiration', Agric. Met., **10**, 365-82.

Maunder, W.J. (1970), The Value of Weather, Methuen, London.

Meentemeyer, V., Box, E.O. and Thompson, R. (1982), 'World patterns and amounts of terrestrial plant litter production', Bioscience, **32**(2).

Merriam, C.H. (1896), 'Laws of temperature control of the geographic distribution of terrestrial animals and plants', Nat. Geog. Mag., **6**, 229-38.

Merriam, C.H. (1895), 'The geographic distribution of animals and plants in North America', in US Department of Agriculture Yearbook (1896), 203-14.

Merriam, C.H. (1899), 'Life zones and crop zones', Bull. of Biol. Surv., **10**, 9-79.

Milner, C. and Elfyn Hughes, R. (1968), Methods for the Measurement of the Primary Production of Grassland, IBP Handbook No. 6, Blackwell Scientific Publications, Oxford, UK.

Monteith, J.L. (1973), Principles of Environmental Physics, Edward Arnold, London.

Monteith, J.L. (1981), 'Climatic variations and the growth of crops', Quart. J. Roy. Met. Soc., **107**(454), 769-74.

Munn, R.E. (1970), Biometeorological Methods, Academic Press, New York, USA.

National Academy of Science (1968), Man's Survival in a Changing World, National Academy of Science-National Research Council, Division of Biology, IBP Office, Washington DC, USA.

Newbould, P.J. (1967), Methods for Estimating the Primary Production of Forests, IBP Handbook No. 2, Blackwell Scientific Publications, Oxford, UK.

Norman, J.M., Tanner, C.B. and Thurtell, G.W. (1969), 'Photosynthetic light sensor for measurements in plant canopies', Agron. J., **61**, 840-3.

Oke, T. (1968), Boundary Layer Climates, Methuen, London.

Petrusewicz, K. and Macfayden, A. (1970), Productivity of Terrestrial Animals: Principles and Methods, IBP Handbook No. 13, F.A. Davis Company, Philadelphia, USA.

Polunin, N. (1960), Introduction to Plant Geography, Longmans, London.

Rainey, R.C. (1973), 'Airborne pests and the atmospheric environment', Weather, **28**, 224-39.

Reed, R.J. (1960), 'Principal frontal zones of the Northern Hemisphere in winter and summer', Bull. Amer. Met. Soc., **41**(11), 591-8.

Rose, C.W. (1966), Agricultural Physics, Pergamon Press, Oxford, UK.

Rosenberg, N.J. (1974), Microclimate: the Biological Environment, John Wiley and Sons, New York, USA.

Rouse, W.R., Mills, P.F. and Stewart, R.B. (1977), 'Evaporation in high latitudes', Wat. Res. Res., **13**(6), 909-14.

Scott, J.R., Tracy, R. and Pettus, D. (1982), 'A biophysical analysis of daily and seasonal utilisation of climate space by a montane snake', Ecol., **63**(2), 482-3.

Sellers, W.D. (1965), Physical Climatology, University of Chicago Press, Chicago, USA.

Simmons, I.G. (1979), Biogeography: natural and cultural, Edward Arnold, London.

Slatyer, R.O. (1978), 'Altitudinal variation in the photosynthetic characteristics of Snow Gum, Eucalyptus pauciflora Sieb. ex Spreng VII', Austral. J. Bot., **26**, 111-21.

Smith, K. (1975), Principles of Applied Climatology, John Wiley and Sons, Chichester, UK.

Spaeth, H.J. (1980), Die Agro-olologische Trockengrenze in den Zentralen Great Plains von Nord-Amerika, Franz Steiner Verlag GmbH, Wiesbaden, West Germany.

Taylor, J.A. (ed.) (1970), Weather Economics, Pergamon Press, Oxford, UK.

Taylor, J.A. (ed.) (1974), Climatic Resources and Economic Activity, David and Charles, Newton Abbott, UK.

Terjung, W.H. (1976), 'Climatology for geographers', Annals of the Association of American Geographers, **66**(2), 199-222.

Thornthwaite, C.W. (1948), 'Approach towards a rational classification of climate', Geog. Rev., **38**, 55-96.

Tuhkanen, S. (1980), 'Climatic parameters and indices in plant geography', Acta Phytogeographica Suecica, **67**, Svenska Vaxtgeografiska Sallskapet, Uppsala, Sweden.

Warren Wilson, J. (1957), 'Observations on the temperatures of arctic plants and their environment', J. of Ecol., **45**, 499-531.

Wassink, E.C. (1975), 'Photosynthesis and productivity in different environments - Conclusions', in Cooper, J.P. (ed.) (1975), Photosynthesis and Productivity in Different Environments, IBP 3, Cambridge University Press, Cambridge, UK, pp. 675-87.

Watt, K.E.F. (1973), Principles of Environmental Science, McGraw-Hill, New York, USA.

Wilson, R.G. and McCaughey, J.H. (1971), 'Soil heat flux divergence in developing corn crops', Clim. Bull., **9**, 9-16.

FURTHER READING

Gates, D.M. (1980), Biophysical Ecology, Springer-Verlag, New York, USA.

Monteith, J.L. (1973), Principles of Environmental Physics, Edward Arnold, London.

Monteith, J.L. (1981), 'Climatic variations and the growth of crops', Quart. J. Roy. Met. Soc., **107**(454), 769-74.

Rosenberg, N.J. (1974), Microclimate: the Biological Environment, John Wiley and Sons, New York, USA.

Tuhkanen, S. (1980), 'Climatic parameters and indices in plant geography', Acta Phytogeographica Suecica, **67**, Svenska Vaxtgeografiska Sallskapet, Upsalla, Sweden.

Chapter 9

THE MAN/LAND PARADOX

J. A. Taylor

THE MAN-LAND RELATIONSHIP

Biogeography, as a sub-field of geography *as a whole*, has inherited a commitment to study and re-appraise the traditional man-land relationship. Glacken (1967) has provided detailed perspective on the history of ideas surrounding this relationship but concedes that, although it is axiomatic that man is an integral part of nature, final arguments gain cogency only when human cultures are set off from the rest of natural phenomena (Preface, p. x). Chorley (1973) also argues that 'social man is being more and more set apart from the physical and biological environment'. Simmons (1970), taking land-use ecology as a theme, concludes that biogeography can be differentiated from ecology only when ecological systems are looked at from a human standpoint. These essentially one-sided views of the man-land relationship simply fail to focus on its *reciprocity* as well as its inevitable universality. The need to assess human performance in ecological terms provides the counterpoint for a final resolution of both sides of the man/land equation.

Human activity must be gauged against the state of the environmental condition before, during and after its impact. Equally, the nature of environmental conditioning or change, operating independently of man's activities, should be evaluated not only *per se* but also in relation to its reciprocal impact on those activities, as affecting, for example, their location or functioning or efficiency. In this way, the stubborn intellectual unconformity at the man/land interface can be bridged or at least finessed. Thereafter, the old bogey of crude, one-way, environmental determinism (as witnessed in such concepts as 'spring-line villages' (Eyre, 1964) although threatening a revival in the contemporary archaeological philosophy of *cultural materialism* (Conrad, 1981), can be finally laid. At the same time, the counterpoint, anthropocentric possibilism, now operating on floating isotropic planes (as implied in Openshaw's study (1982) of the geography of reactor siting policies in the UK) should be recast on the more resilient and realistic platform of environmental resource-usage. Again, the fundamentally false view that man, because of his superiority over other species, is therefore different from other animals and has become increasingly technologically independent of nature, may be corrected in the scientific and philosophical awareness that the survival of man and his cultures is synonymous with the survival of nature, whether it is perceived in terms of location, space or resource.

The risk that this very broad remit (deriving very much from Western philosophy and its educational traditions) (Black, 1970) might diffuse into the worst kind of omniology can be successfully avoided by focussing biogeographical inquiry right at the heart of the man/land relationship, however subtle it may become for sophisticated, urbanised and industrialised cultures or situations. In this context, biogeography is not entitled to any uniquely integrative role as between physical and human geography: but integrative it most certainly is and, more properly, its mainstream objective should be the reciprocal measurement of man/land linkages, involving such concepts as land-use optimisation, agricultural efficiency, long-term natural resource evaluation, and the marrying of ecological, economic and bureaucratic principles in resource management and planning strategies. Is marriage possible between such diverse partners of utterly different origins, contrasted objectives (either prescribed by nature from within or determined by human decision-making from without), speaking such different languages, and evincing such incongruous sets of phenomena? Chorley (1973), even within the flexible scope of his beloved systems approach, thinks not, mainly because natural ecosystems are conceived in terms of homeostacy and abundant negative feedback loops whilst social systems possess powerful built-in positive feedback mechanisms, including a time-dependence which often dominates them. The reconciliation of the short-term and long-term stabilities of both natural and man-made systems is confounded by the accelerating modification and destruction of this planet's store of natural ecosystems by man's increasingly powerful appetites and aspirations. The so-called natural world is but a shadow of its former self. Every nook and cranny of the biosphere is accessible, directly or indirectly, to human-derived activity, pollution and artificialisation. Thus, increasingly, the confrontation is now between man and man-adapted nature. The transformation of natural ecosystems, especially via domestication, is of the very essence of cultural prehistory and history expressed in ecological terms (Ucko and Dimbleby, 1969). The imprint of successive transformations on the landscape, especially via deforestation, agriculture and settlement, reinforced by the impact of natural processes (e.g. slope development, river channel change, climatic variability, soil change, floral and faunal change), together bestow the cumulative legacy of environmental and cultural history in the modern landscape (Evans et al (eds.), 1975; Limbrey et al, 1978; Taylor (ed.), 1980).

The apparent imminent standardisation of natural ecosystems should invoke the direst warnings from the traditional conservationists who bemoan the continual extinction of species, the reduction of genetic diversity and the irreversible decline in finite organic resources (Eyre, 1978). However forceful these conventional arguments may be, the increasing velocity and frequency of man-induced ecological changes are a reality that must be accepted. Equally real now is the recent accumulation of evidence that self-generated environmental changes are also just as short-term, episodic and variable as social changes. The nineteenth-century image of natural ecosystems attaining identity by aggression and monopoly over space and persistence over time is due for revision in the light of recent repetitive confirmation of the fact that environmental changes, both climatic and biological, do occur, did occur (and presumably will occur) on short, medium and long time-scales over short, medium and long distances (Taylor, 1975; Taylor (ed.), 1980; Kershaw, 1978; Woillard, 1078; Flenley, 1979a, 1979b; Street, 1981). This dynamic reappraisal implies that natural ecosystems have also been at the mercy of short-term environmental change, variability and hazards since time immemorial and, initially, without the interference of man. It follows that the traditional concept of climatic climax vegetation, with its associated faunal hierarchies and soil maturation, must be less tenable and less applicable in the real world. The prospect of polycyclic ecosystems in combination with an accelerating complexity of resource management is truly daunting but can any inherent ecological constants be extracted from such a large host of variables?

CONCEPTS OF PRODUCTIVITY

Inherent Productivity

Productivity is a relative rather than an absolute concept but, assuming constancy of climate, flora, fauna and human usage, any location or area of the earth's (or the sea) surface may be regarded as possessing a particular, and normally quasi-constant, production potential, no matter whether the system is natural or man-made. Under such hypothetically controlled conditions, are there any inherent qualities of location or area which are, in any sense, immutable?

Firstly, the three coordinates of latitude, longitude and altitude provide a fixed geometrical location for any point on the earth's surface. Ignoring the variable of the soil/vegetation/land-use layer and local microclimatic deviants (such as aspect, slope) and mesoclimatic deviants (such as distance from the sea, cloudiness, etc.), it is possible to estimate the net radiant-energy available throughout the year, based on calculations of latitudinal and seasonal variations in the angle of the sun striking a uniform plane tangential to the curvature of the earth at the given location.

Net Radiation Inputs: the Primary Variable in Growth Potential

Following Collingbourne (1976), the rate at which radiant energy is received at a given planar unit area or point is referred to as the irradiance. Global solar radiation (G) is the solar irradiance of an assumed plane symmetrically tangential to the receiving hemisphere. Solar radiation usually includes two components: first, direct insolation (I) from the sun, and second, diffuse radiation (D) which, under overcast conditions, constitutes effectively 100% of the net energy receipt at the earth's surface. However, direct solar radiation (I) is the solar irradiance received on a surface perpendicular to the sun's direction and is expressed in terms of the angle of incidence ('h') of the sun's beam as follows:

$$G - D = I \sin h \qquad (1)$$

Solar radiation receipts on a horizontal surface outside the earth's atmosphere (I_o) vary on two time-scales. First, the earth's rotation imposes a diurnal variation on the sun's elevation from a maximum at local apparent noon to zero at local sunset. The maximum intensity occurs at the maximum solar elevation (h_o) at midday and is given by:

$$G = I_o \sin h_o \qquad (2)$$

However, a second time-scale involved is the motion of the earth along its orbit round the sun. Thus, h_o varies from a maximum in midsummer to a minimum in mid-winter. If O is the sun's elevation and L is the latitude, then:

$$h_o = (90 - L) + \sigma \qquad (3)$$

σ varies from $+23^o27'$ to $-23^o27'$, and the earth's slightly elliptical orbit introduces a variation of ±3.3 per cent in the mean value of I_o, with a minimum in early July and a maximum in early January. A diagram by Petherbridge (1969) displays sterographic sunpaths for latitude 51^oN for a selection of dates throughout the year. The total radiation received on a horizontal surface in a day outside the atmosphere for all latitudes is shown in Figure 9.1 (Petherbridge, 1969) in megajoules per square metre (MJ m^{-2}). One Joule (J) is defined as an energy rate of 1 watt lasting for 1 second. The unit of irradiance is joules per square (J m^{-2}) which is so small that one megajoule per square metre (MJ m^{-2}) is normally used. The variation of direct radiation on cloudless days on north, east and south slopes at 50^oN has been summarised in Geiger's (1965) diagram (which uses cal cm^2 h^{-1}, see Table 9.1) (Figure 9.2) and discussed by Jones (1976). The symmetry of these diurnal curves for the north and south slopes is in contrast to the asymmetry of the east (and west, not shown) slopes. Holland and Steyn (1975) have shown that, in the mid-latitudes, the aspect variable in topoclimates imposes a greater differential on radiation receipts and, consequently, on growing season, than in the higher and lower latitudes. Hare and Ritchie (1972) have mapped the radiation climates for Canada (Figure 9.3) in kilolangleys (Kly) (see Table 9.1).

Figure 9.1: Daily totals (from January to December) of global solar radiation received outside the atmosphere ($MJ\ m^{-2}$) (after Collingbourne, 1976)

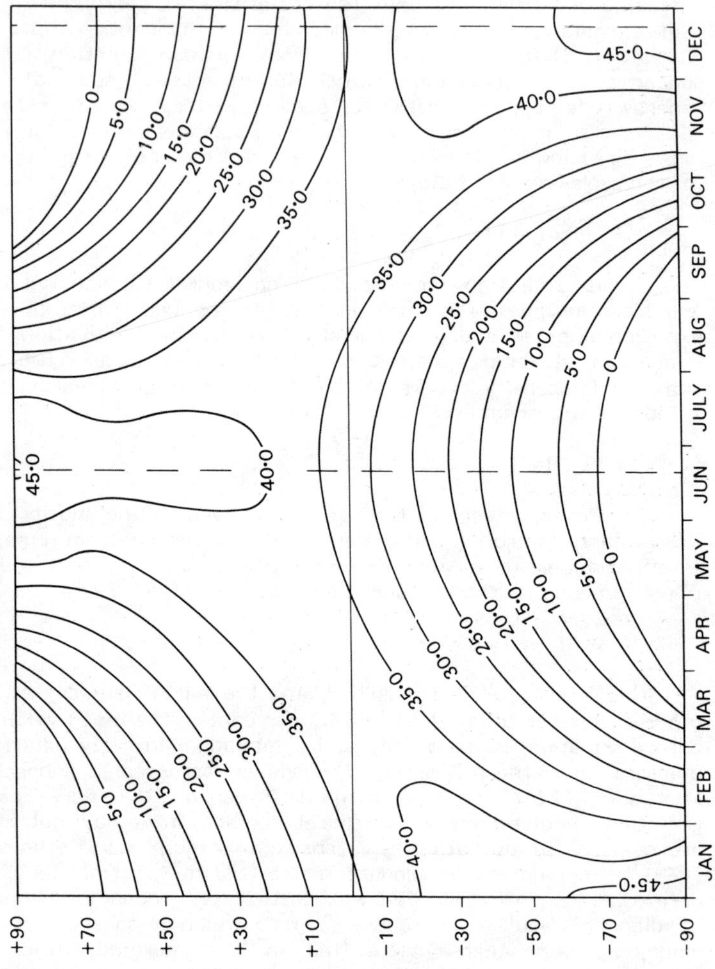

Figure 9.2: Direct radiation (cal cm^{-2} h^{-1}) on cloudless days on north, east and south slopes at *circa* 50°N (after Geiger, 1965 using the data of W. Kaemfert)

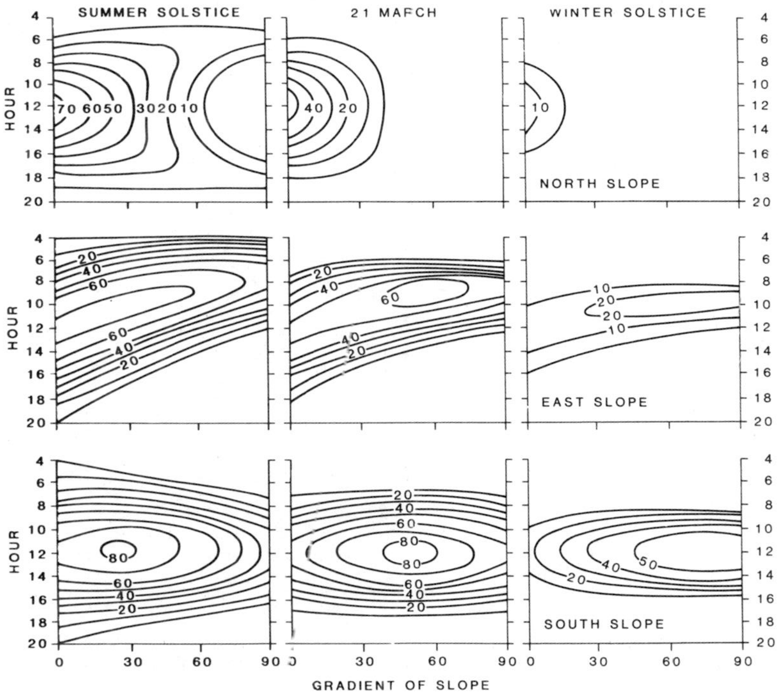

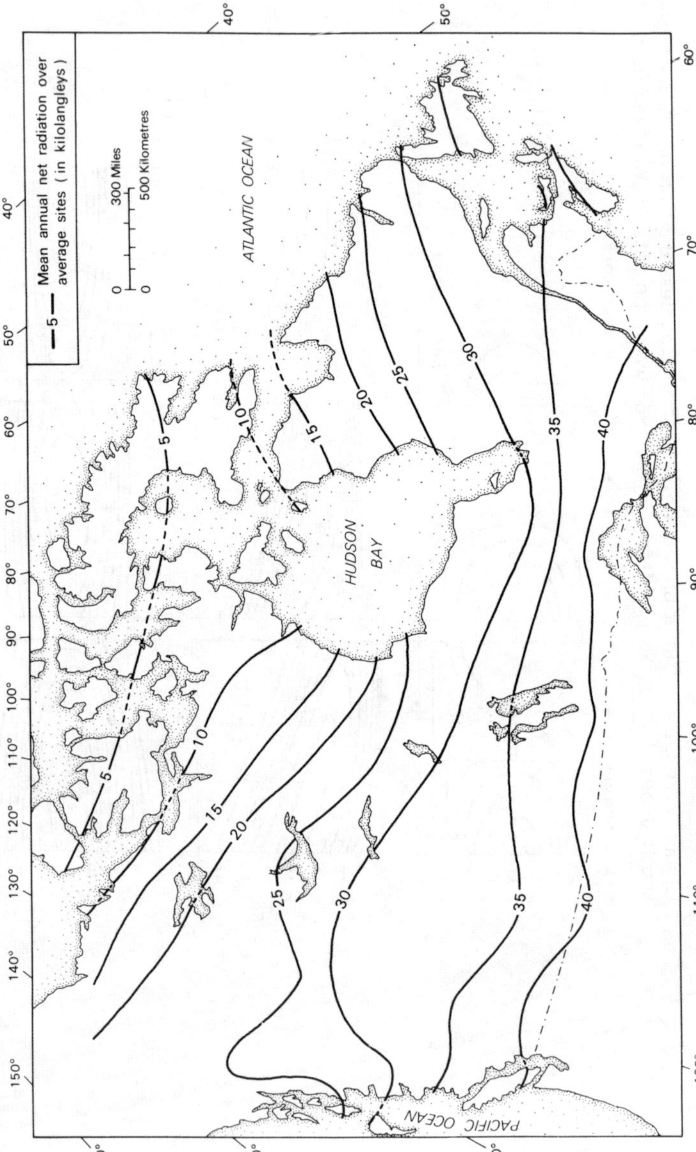

Figure 9.3: Canada: mean annual net radiation over average sites (in kilolangleys) (after Hare and Ritchie, 1972)

Table 9.1: Conversion rates for units of energy (after Collingbourne, 1976)

1 MJ m^{-2}	= 278 W m^{-2}	= 23.9 cal cm^{-2}	= 23.9 ly
1 W m^{-2}	= 0.0036 MJ m^{-2}	= 0.0860 cal cm^{-2}	= 0.0860 ly
1 cal cm^{-2}	= 0.0419 MJ m^{-2}	= 11.6 W m^{-2}	= 1 ly
1 ly	= 1 cal cm^{-2}	= 0.0419 MJ m^{-2}	= 11.6 W m^{-2}

Notes:

1. The SI unit for irradiance is 'watts per square metre' (W m^{-2}) and is convenient for general usage. The solar constant (it varies by less than 1: Coulson, 1975), i.e. the mean value of direct solar radiation received outside the earth's atmosphere, is estimated at 1,353 W m^{-2}.

2. The unit calorie per square centimetre (cal cm^{-2}) has traditionally been used and is equivalent to the Langley (Ly) used in the USA. It is now recommended that the Joule (an energy rate of 1 watt lasting for 1 second) and the corresponding unit of irradiance, i.e. Joules per square metre (J m^{-2}) and, in particular, the larger unit of megajoules per square metre (MJ m^{-2}), be universally adopted.

These measures of the net radiation received at given locations or areas of the earth's surface are primary measures of the potential energy climate and of the inherent site productivity, assuming no macro-climate change and accepting any inherent variability of the ambient in situ meso-and micro-climates and the biophysical properties of the site.

Biological Productivity

A second method of estimating inherent productivity is through the response of vegetation to the available physical energy input. The rate at which growth processes occur in an organism or ecosystem is termed the biological productivity. It is usually expressed in quantity of dry matter produced over unit area and unit time, e.g. Kg/ha/yr. Gross Primary Productivity (GPP or, simply, P) is the gross production of plants as represented in the total energy assimilated by the organism in unit time. That proportion of the assimilated energy converted by the plants into heat or mechanical energy or used in life processes is respiration (R). The equation provides a simple derivation of Net Primary Productivity (NPP):

$$\text{NPP (or N)} = \text{GPP (or P)} - \text{R} \quad . \quad . \quad . \quad . \quad . \quad (4)$$

Although up to about 9% (60×10^{22} joules) of the total radiation received (520×10^{22} joules) in the outer atmosphere is potentially available for photosynthetic conversion in the biosphere, only 1% or less (commonly about $\frac{2}{3}$, or less, of 1%) is normally used by the vegetation and agricultural systems of the world (Holliday, 1966; Monteith, 1965, 1966; Penman, 1968; Taylor, 1974a). This indicates the conservational nature of energy use in the biosphere and, at the same time, the inefficiency of our atmosphere as a heat engine (Oort, 1970). Table 9.2 (after Packham and Harding, 1982) summarises other methods of measuring growth rates using leaf growth indices as well as increment in biomass.

Table 9.2: Definitions used in growth analysis (after Packham and Harding, 1982)

Relative growth rate (RGR) is the rate at which a plant increases its dry weight per unit dry weight. RGR can, at a particular instant, be resolved into three components:

1. Unit leaf rate (ULR) = $dW/dt \times 1/L_A$
2. Leaf weight ratio (LWR) = L_w/W
3. Specific leaf area (SLA) = L_A/L_w

where L_A = total leaf area
L_w = total leaf dry weight
W = total plant dry weight
dW/dt = rate of dry weight increase of the whole plant

RGR = ULR x LWR x SLA

Leaf area ratio (LAR) is a morphological index of plant form, the leaf area per unit dry weight of the whole plant.

4. Leaf area ratio (LAR) = L_A/W = LWR x SLA

In contrast, unit leaf rate (ULR), the rate of increase in dry weight of the whole plant per unit leaf area, is a physiological index closely connected with photosynthetic activity. A high LAR together with low ULR is characteristic of heavily shaded woodland herbs in temperate forests.

Relative leaf growth rate (RLGR) is analogous to RGR and is the rate of increase in leaf area per unit leaf area.

5. Leaf area index (LAI) = L_A/ground area occupied by plant

Stomatal index = Number of stomata per unit area (number of stomata per unit area + Number of epidermal cells per unit area)

In simple terms, the Leaf Area Index (LAI) is the total area of leaves growing above a given area of ground divided by the area of the ground itself. Net radiation inputs in combination with biological productivity measures enable a biophysical estimation to be made of the inherent productivity of sites or areas on the earth's surface. What of the contribution of the surface and subsurface terrestrial component, expressed in the site topography and soil condition, as derived in part from its parent material, the geological context, the ground hydrology and the complex heritage of environmental and land-use history, e.g. effects of glaciation, slope processes, water movements, cycles of previous cultivation or grazings or natural vegetation effects, etc.?

Land Productivity

Any location or area of land possesses a precise topography and a soil resource bestowing a series of fixed, semi-fixed and unfixed properties. These range from stable, immutable or resistant properties to instable, easily changed or exchangeable properties. The former, at their most extreme, are difficult and expensive to modify and may constrain soil potential and restrict the range of land use despite successions of advancing cultural technologies. The latter are more adaptable to change of use and may enable soil survival of a diverse land-use history and management. However, such soils must be comprehensively buffered by such assets as loamy texture, stable structure, high organic content and nutrient status, etc., to sustain intense or diverse land-use programmes.

A given location or area of land is thus endowed with a series of more-or-less fixed resources which not only bestow a level or range of use-potential but also condition, both positively and negatively, the operation of the energy exchanges and growth processes in the ecosystems functioning on the land surface. The three basic coordinates of latitude, longitude and altitude together with site topography provide the stage on which the drama of climate is enacted. The macroclimate is modified by local exposure or shelter and by local smoothness or irregularity of relief. The solid geology will present rock materials of specific structure, geo-chemical composition and weatherability which, in turn, will bequeath specific characteristics to overlying regolith, soil parent materials and, eventually but less consistently, to the soils themselves. Care must be taken here not to resuscitate the early twentieth-century notion that soils are fundamentally a direct expression of their geological substrates. The effects of slope processes, laterial colluviation and the lateral transfer and mixing of surface deposits, e.g. as a result of glaciation, river action, land slips, vegetation or land-use changes, etc., mean that many apparently in situ parent materials have been imported and are liable to continual modification, especially on slopes. This demonstrates the increasing susceptibility of soil and subsoil to change, thus representing the semi-fixed or unfixed attributes of a given site. Nonetheless, the stoniness, the trace element content, the characteristic mineral range and, thereby, any aspect of the inherent fertility potential of a soil will be related, directly or indirectly, to its parentage. Again, in combination with relief, the site hydrology and the soil profile drainage will be characteristically derived in relation of the intrinsic texture and structure of the soil parent materials. For example, Silurian and Ordovician grits and shales weather down initially to silts, although eventually to clays. Thus, soils from such rocks frequently contain a characteristic silt fraction regardless of their other textural properties; they range from coarse, gritty silt loams to heavy, impeded silty gley soils. Soils derived from chalk and limestone contain high levels of calcium carbonate, $CaCO_3$, and are initially buffered against leaching but, in the higher limestone uplands of Britain, the very wet climate may induce a slow decalcification of a shallow surface veneer of the rock, allowing a succession from calcicole to calcifuge mosses to be followed by Calluna and, locally, even shallow peat formation, a process which is accelerated should a thin drift cover be present on the limestone, as at Fountains Fell on Malham Moor in North Yorkshire. This illustrates how ecosystem processes may counter and eventually reverse the initial impact of parent material. Peat formation is normally associated with siliceous mineral substrates (Taylor and Smith, 1980).

In summary, despite the varied legacy of vegetation and land-use history, implying a variety of appraisals of the same land quality, a nucleus of inherent site properties, biological as well as physical, exists in variably fixed, semi-fixed and unfixed forms, providing a constant baseline from which the alternating productivities of different land-use systems can be assessed, assuming no major environmental changes have taken place to modify the inherent capacity of a given location. The resistance and longevity of such capacities are an index of the stability, flexibility, accessibility and improvability of the site from the point of view of land usage, both ancient and modern. Let us take case-studies for the past and the present.

MAN-ADAPTED ECOSYSTEMS

The Renaissance of Cultural Biogeography

A central sub-field of biogeography which is currently experiencing a renaissance is cultural biogeography which focusses essentially on man's changing relationships with nature over time. Harris (1969 and 1981) has reviewed this development, quoting the initial imaginative ideas of Rousseau (1755) (*vide* Masters, R.D. and Masters, J.R., 1964) and Humboldt (1807) on the origins of agriculture and the domestication of plants and animals, the speculations of Darwin (1868), the penetrations of Hahn (1896), the evolutionary and diffusionist models of Childs (1936 and 1942), the seminal writings of Sauer (1952) and the scientific approaches of Zeuner (1946) and Dimbleby (1962). The recent emergence of theoretical and environmental archaeology (e.g. Clarke (ed.), 1972; Evans, 1975; Taylor (ed.), 1980) has provided more diverse reconstructions of the logical transitions that must have taken place from (a) natural ecosystems, which merely accommodated the pristine forms of land use (e.g. collecting, hunting for food), to (b) variously adapted natural ecosystems, manipulated and then changed by man and, finally, to (c) man-created ecosystems, involving domesticated plants and animals, i.e. the ancestral forms of modern agricultural systems. To begin to understand the ecological basis of the first primitive forms of land use, let us follow the guidelines provided by Harris (1969, 1973 and 1977).

The Manipulation and Modification of Natural Ecosystems by Man

Harris (1969) makes the broad initial distinction between 'generalised' and 'specialised' ecosystems, both natural and agricultural. Generalised ones contain large numbers of species of plants and animals but small numbers of individuals of each species; they thus have a high diversity index, i.e. the ratio of numbers of species to number of individuals. Specialised ecosystems have, in contrast, a low diversity index, with small numbers of species but large numbers of individuals. Generalised ecosystems normally have high net primary productivity, abundant ecological niches and elaborate food webs, all lending to stability or homeostasis. Specialised ecosystems, on the other hand, are less productive, less versatile, less stable and more vulnerable to change from within as well as from the external pressures exerted by other ecosystems or environmental change.

Harris quotes the tropical rain forests as the best example of a highly generalised and very productive ecosystem with a maximum diversity index but even their stability has recently been brought into question with the discovery of short-term periodicities in development (Flenley, 1979a and 1979b). Such periodicities could be due to short-term climatic variability or could be interpreted as ecological 'noise' within the long-term and intensive, internal competitiveness of the forests which nonetheless are externally vulnerable to rapid decline should climate change or deforestation take place. The tundra formation, in contrast, illustrates the comprehensive fragility of the specialised ecosystem with a very low ratio of species to individuals and an increasing vulnerability to mid-latitude pollution sources (Hare, 1970; Hare and Ritchie, 1972). The montane vegetation of the British uplands may be similarly categorised (Job and Taylor, 1978).

Newbould (1971) warned of the dangers of oversimplistic comparisons between natural and man-made ecosystems. Available data are generally imprecise and vary with the investigational techniques used: they should go beyond simple estimates of production biomass. It was hoped that the IBP (International Biological Programme) would go some way towards rectifying this situation but results published so far have been disappointing. There was over-optimism concerning the prescribed modelling approach and inevitable frustration in the acquisition of an adequate field data bank (Persson (ed.) 1980). Nevertheless, Rodin and Basilevich's pioneer paper (1968), Westlake's estimate (1963) of

Table 9.3(a): Production estimates for selected ecosystems (after Westlake, 1963)

No.	Location	Dominant Species	Annual production or annual mean productivity (mt/ha^{-1} year^{-1})
1.	Java	Sugar cane	94
2.	Hawaii	Sugar cane	83
3.	Marshall Islands	Green algae	49
4.	Holland	Maize and rye	39
5.	Georgia	Spartina alterniflora	37
6.	England	Grand Fir	35
7.	New Zealand	Lolium spp. (good temperate grassland)	32
8.	Scotland	Bracken	26
9.	Minnesota	Typha reedswamp	25
10.	Holland	Sugar beet	18
11.	England	Alder, 0-22 yr	16
12.	England	Scots Pine	16
13.	England	Birch, 0-24 yr	9
14.	Sweden	diatoms (blue green and flagellate algae	c.2

production for a wide variety of ecosystems and Cannell's (1982) collation provide valuable global scale and perspective on biomass production estimates (Table 9.3 (a) and (b)).

The First Manipulations of Natural Ecosystems

Harris (1969) contemplates three possible ways in which natural ecosystems underwent successive conversions to agricultural systems. Firstly, general natural ecosystems would have been converted into specialised artificial ecosystems, as is very apparent in modern times. Selective domestication of plants and animals reduces species diversity very markedly and creates monocultural areas or fields. Then species adapt to the disturbed environments, emerging as plant or animal 'weeds' which add to the specialised nature of the agricultural ecosystem at the expense of earlier dominant and diverse groups of species which become suppressed. Net primary productivity is usually reduced except in the most intensive forms of agriculture. Ovington et al (1963) discovered that Hawaiian sugar cane production, at its most intensive, could equal that of the tropical rain forests and, again, that heavily fertilised maize growing

Table 9.3(b): World Ecosystems: net primary production estimates (adapted from Rodin and Basilevich, 1968)

	Ecosystem		Net primary production (mt/ha^{-1} year^{-1}
1.	Arctic tundra		1.0
2.	Dwarf shrub tundra		2.5
3.	Boreal Forests	(North taiga	4.5
4.		(Middle taiga	7.0
5.		(South taiga	8.5
6.	Oak forests		9.0
7.	Beech forests		13.0
8.	Steppes	(temperate, dry	11.2
9.		(dry	4.2
10.		(Dwarf, semi-shrub	1.2
11.	Deserts	(Dwarf, emphemera semi-shrub	9.5
12.		(Sub-tropical	2.5
13.	Savannahs, dry		7.3
14.	Savannahs		12.0
15.	Sub-tropical forests		24.5
16.	Tropical rain forests		32.5
17.	Sphagnum bogs, afforested		3.4
18.	Mangroves		9.3

in Minnesota could match the productivity of adjacent protected oakwood (cf. Table 9.3(b)). Secondly, but much less likely and more rarely, a few specialised natural ecosystems could have been converted into more generalised agricultural systems. The insertion of mixed farming, involving animals, crops and weeds, into the mid-latitude prairies of North America and the pampas of Argentina created a greater species and land-use diversity than what preceded it. Again, irrigation agriculture has variegated the desert margins but these are exceptions to the extensive cereal-based monocultures which have created soil problems in parts of the drier prairies. Thirdly, natural ecosystems could have undergone a much gentler process of amendment by selective, small-scale gardening or controlled hunting and collecting, designed not to disturb the overall structure and stability of the parent ecosystem which is simulated in pockets of cultivation or browse and not in any sense replaced. The best, and presumably the earliest, forms of agricultural land use probably adopted this ecologically sympathetic strategy, using garden-style cultivation and supervised grazing/browsing. They, in particular the garden-cultivation, have been discovered today in remote forest areas in Brazil, north of the Amazon, and in parts of New Guinea. They may be regarded as ancestral to the widespread forms of shifting agriculture (swidden) and sedentary agriculture

Figure 9.4: Neolithic clearance phases in Northern Ireland (after Pilcher *et al*, 1971)

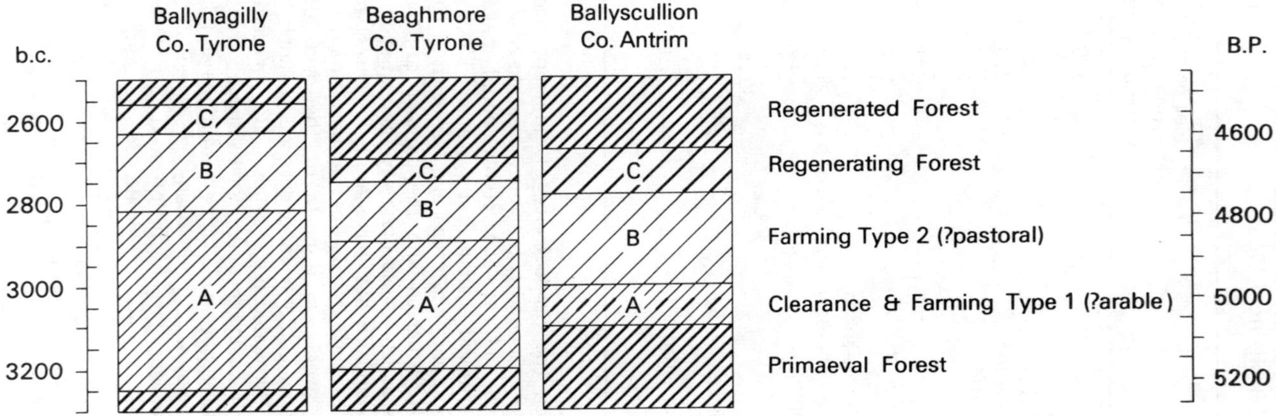

The diagram shows the durations of the phases within the elm-decline clearances at three sites in the north-east of Ireland; the boundaries of the stages are dated from deposition rate graphs and must not be taken as fixed points. Within the limits of the method, the beginnings of the clearance phases could have been contemporaneous.

which now occur in tropical environments. Moreover, they provide the much needed longer time-scales that are required for the evolution of the domestication process to be fulfilled.

Harris (1969) proposes that generalised natural ecosystems were very likely to have been the scene of the pristine manipulations described above. They, together with their climates, offered more options for experiments in the adaptation of plants and animals, especially in the forest habitats of the low and middle latitudes. The specialised tundra and taiga environments induced a highly seasonal and very mobile commitment to hunting or fishing strategies which allowed little or no opportunity for agricultural experimentation, notwithstanding the short growing season of these high latitudes. Little wonder these areas have yielded, as yet, so little archaeological evidence of early occupations. On the other hand, the more generalised, pre-agricultural, hunter-fisher-collector groups in the warmer latitudes had more diverse food sources and more varied and more mobile lifestyles. The time, opportunity and potential for a graduation to garden-based cultivation and semi-domesticated browsing existed. The favoured locations would be forest-edge, forest clearing, river-bank, for example, where habitat boundaries converge. Accessibility to a range of territories is implied, i.e. aquatic-terrestrial, woodland-grassland, upland-lowland: in other words, what may be termed 'cross-road' habitats, positioned in ecotones at the margins of major ecosystems. Such sites allow for alternatives in food supply, hunting manoeuvre, protection strategies. Risks are reduced; survival chances are enhanced.

Jacobi (1980a and b) describes the availability of such versatile habitats in the climatically 'early' littorals in south-west Dyfed where colonisations by Upper Palaeolithic and early Mesolithic man were congregated. Again, Pilcher _et al_ (1971), as amplified by Smith (1975), postulates a three-fold pattern of local forest clearances of the _landnam_ type during the earliest Neolithic settlement phases in Northern Ireland (Figure 9.4). Stage A presents a mosaic of woodland with cultivated clearings which could conceivably have been of the garden type. Stage B, when plantain pollen replaces cereal pollen in the fossil record, is a pastoral stage with more extensive clearances of former arable land, but by Stage C, woodland has regenerated extensively, the entire cycle taking between 100 and 400 years. Clearly, forest edge and clearings would have been optimal primary sites for these pioneer domestications, involving both plants and animals.

To what extent do such site appraisals recur along the succession of advancing cultures in a given area and to what extent are they consistent? The answers to these questions lie at the heart of the historical relationship between man, his inherited environment and the environment he bequeaths to the next generation.

THE ECOLOGICAL FACTOR IN LAND USE HISTORY

The Search for Continuity in Land Evaluation

It is often assumed that improved technology has emancipated man from the constraints and controls imposed by his natural environment. It is argued that increasing regional specialisation in land use and the growth of interregional and international exchange of specialised products and services provide buffers against shortages and surpluses and moderate the impacts of climatic variability and environmental hazards on productivity. Even allowing for the inherent variability of the politico-economic factors that exert primary controls on the management of land use systems at all scales, it emerges that vulnerabilities not only persist but may well be more pernicious. Specialisation, although progressive, creates new sensitivities to change: this is a law of nature and it could well apply to human society, although such a controversial issue invokes deep philosophical as well as historical and ecological debate. The environmental resource interacts with the culture variable through the technology of the contemporary land-use. Assuming an inherent constancy of environmental resource which is mostly immune to major environmental change, can any continuity of resource appraisal be detected over prehistoric and historic time? Let us take a series of case-studies.

Prehistoric and Early Historic Land Options

Taylor (1980) states that 'the man-environment equation involves relative rather than absolute terms, and variables rather than constants'. Assuming that environment and technology may, on their own terms, be ascendant or descendant, four different combinations may theoretically occur as follows: (A) ameliorating environment; (B) deteriorating environment; (C) advancing technology; and (D) declining technology. Depending on the varying coincidences of (A) and/or (B) with (C) and/or (D) in both time and space, different states of conservation or decline of both environment and culture may be generated. For example, the optimum conservational sequence would involve a prolongation of (A) accompanied by either (D) or an early version of (C). One of the nearest approximations to this was the Mesolithic culture as it developed in Britain between *circa* 10,000 and 6,000 BP. In contrast, a combination of (B) and (C) would induce maximum exploitation of dwindling environmental resources unless (C) had the capacity for alternative technology based on unused resources. The succession of culture stages from the Neolithic, through the Bronze Age and Iron Age between 6,000 and approximately 2,000 BP follows this particular model, as do many stages of the historic period.

The historic record is conventionally rich in evidence of man's continual revaluation of site or resource. Yet an underlying thread of continuity of evaluation can be detected where site or resource is of the highest inherent quality or where the range of environmental options is especially narrow or especially broad. The intermediate case, as usual, displays less sharp and less predictable relationships. Vita-Finzi (1969) quotes spectacular examples of the dependence of irrigation-based cultures on the geomorphology and hydrology of the flood plain and its terraces regardless of the ambient climate. The initial prerequisite for flood-water farming is a wide, undissected, alluvial valley floor over which flood waters can be diverted as required. Subsequent trenching by river erosion eventually restricts the area available for flood agriculture and its population capacity until a new alluvial plain becomes wide enough for a second cycle of water diversions to occur. Vita-Finzi quotes two examples, one in the Tesuque Valley of New Mexico, settled between 2,230 and 700 years BP, and another in the upper Wadi Hasa in Jordan, settled between 10,000 and 2,000 years BP. The small Welsh colony in the lower Chubut Valley of Patagonia in the 1870s also exploited similar hydrogeomorphological opportunities in a climate otherwise too dry for the regular cultivation of crops (Bowen, 1966).

This is very good illustration of an admittedly specialised environmental resource, a flood plain, at a particular stage of its development, which has been comparably evaluated and exploited de novo by different cultures in different regions at different times. Yet we must be careful at all stages to differentiate 'environmental inducement' from 'human motivation' (Taylor, 1965).

Ellison and Harris (1972) and Ball (1964) have demonstrated how inherent land quality may be assessed with some consistency by widely divergent cultures. Ellison and Harris (1972) critically qualify the fashionable theme of 'cultural continuity', beloved of many archaeologists and some historical geographers. They develop locational models for prehistoric and early historic settlements and successfully apply them to the notion of proportional availability of land of different inherent quality within prescribed 'catchments', incorporating Chisholm's (1962) conclusion that the average distance from settlement to cultivated land ranges from 1 to 3 or 4 km, regardless of culture type and technology. Beyond the perimeter of postulated circular catchments then, they adopted a 2 km radius as typical of a compact medieval parish in southern England and therefore applicable, in their view, to pre-medieval settlements in the two counties they studied, i.e. Wiltshire and south central Sussex. Figure 9.5 summarises their findings for Wiltshire. They concede that their sources of evidence are, to say the least, diverse. They show, however, that accessibility to the full range of land quality is a recurrent requirement: as also embodied in the concept of the self-sufficient medieval parish. The proportion of High-Quality-Arable is relatively constant throughout the period taken, and the proportion of Medium-to-Good-Quality-Arable used is reasonably consistent from Iron Age to Romano-British and Saxon times; but it was much less preferred by the Romans who, along with the Saxons, elected, and were more technically able to adapt, the valley and vale landscapes when later deforestations of the thicker oakwoods and the first clearings of the heavy clay vales took place, especially in Saxon times. For south central Sussex, Figure 9.6 reveals an increasing

Figure 9.5: Land use combinations from the Iron Age to the Saxon periods in Wiltshire (after Ellison and Harriss, 1972)

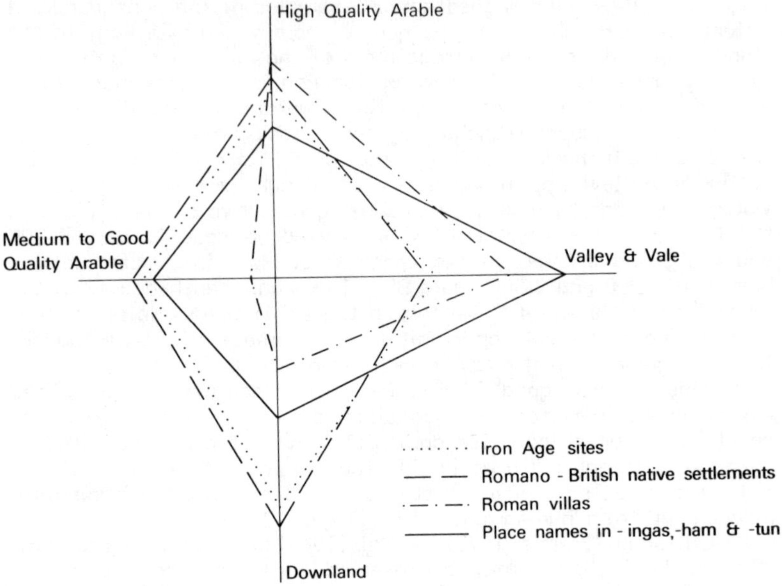

The four axes show the percentage of the total number of sites having some area of each category of land use within their catchments. Note that the 'High Quality Arable' axis maintains its proportion within the combination throughout the time period presented and so does the axis for 'Medium to Good Quality Arable', except during Roman times when villas were most consistently associated with the 'High Quality Arable' axis. The 'Valley and Vale' axis shows increasing representation over time as the improved technologies of Roman and Saxon times were able to deforest and adapt the lowlands for settlement and land use.

intake of the High-Quality-Arable land from the Iron Age onwards and especially for the siting of Roman villas (not the Roman native settlements) and the Early Saxon settlements. It also emerges, significantly, that distance of settlements from the intersection points in soil boundaries (these are analogous 'ecotone' or 'cross road' sites, see p. 269) exerts a regular control on the degree of concentration of settlements, especially the 'native' ones with relatively inferior technology and resources and therefore the greater

Figure 9.6: Percentage of 'Higher Quality Arable' land for selected site catchment areas in south-central Sussex from the Bronze Age to Early Saxon times (after Ellison and Harriss, 1972)

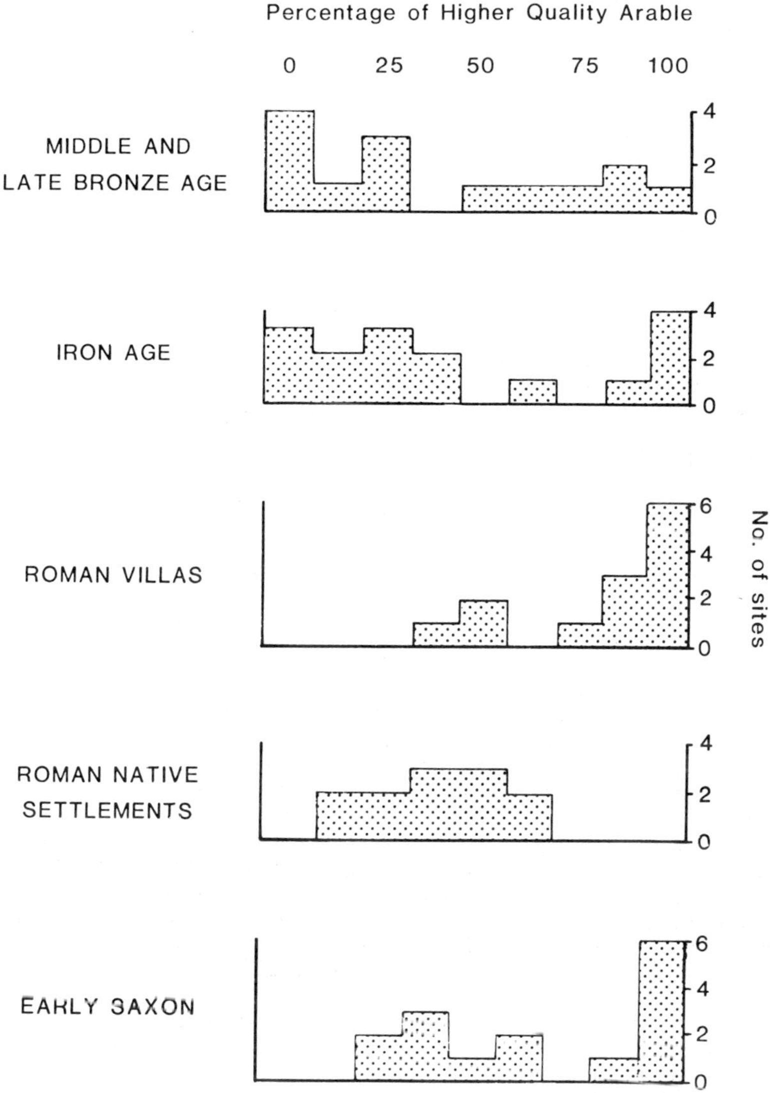

Note the increasing preference over time for the best arable land, except for Roman native settlements which were relegated to inferior sites.

Table 9.4: Soils and settlement in Medieval Anglesey (after Ball, 1964 and Jones, 1966)

Soil Series	Medieval Settlements			Clarke Profile Index	Lands of the Bishop of Bangor	
	Number	Acres/ Settlement	Acres /Unit		Average Holding Bovates	Carucates
Flint*	15	200	79	471	1.16	-
Penrhyn	20	273	111	468	-	-
Gaerwen	83	307	143	424	1.11	0.077
Castleton	5	270	104	418	1.50	-
Arvon	18	506	228	389	-	0.051
Pentraeth*	16	698	272	384	2.20	-
Maelog	1	500	167	351	-	-
Hendre*	1	620	207	320	-	-
Colyn	12	305	131	240	-	0.228
Rocky Gaerwen	15	412	167	205**	2.08	-

* Indicates relatively high base status
** Estimate only because of variations in soil depth

vulnerability. This reinforces earlier arguments propounded when discussing options open to Mesolithic and earlier settlers (see p. 270 *et seq*.).

Ball uses a simple but workable formula devised by Clarke (1951) to evaluate soil texture and soil drainage (based on depth of gleyed horizon) for selected soil series in Anglesey (Roberts, 1958). He then ranks the soils in terms of the Clarke Profile Index, the name given to the formula, and compares the order of quality with a measure of the intensity of medieval settlement devised by Jones for the same area (1955; 1966). This measure assumes that the mean acreage required per unit of medieval settlement is inversely correlated with the quality of the soil. This relationship is partly borne out in Table 9.4 which corroborates the view that our medieval forebears empirically adjusted their choice of location of settlements and associated land use to their perpetual need to satisfy local food requirements as reliably as possible. Clearly, their perception of inherent soil quality in this instance anticipated the professional classification of soil properties in the twentieth century.

THE ECOLOGICAL FACTOR IN MODERN LAND USE

Modern Manifestations

Modern land use systems may be construed as creations of profit-motivated management but it is often in the most intensive or specialised systems that environmental relationships over space and time, and functionally, can be detected, focussed and calibrated. The best examples often combine sharp differentiations in relief, soil or climate around a tightly defined core or zone of environmentally 'optimal' conditions for a specialised form of agricultural production serving a highly competitive market. Fringing 'suboptimal' zones (following McCarty and Lindburgh's (1966) model and as further discussed by Grigg (1982)) are narrow, with increased liability to environmental hazards and loss of productivity. A dramatic illustration of the point occurred during the great British drought of 1976 when the other major area to produce satisfactory maincrop potatoes was the mossland of south-west Lancashire (Taylor, 1952) where the abundant ground-water stores within the reclaimed basin peats help to avoid the moisture stress induced by the prolonged drought on adjacent sandy soils where potato yields were far less satisfactory. The south-west Lancashire plain also happened to escape the extreme impact of the drought which centres on south-east England and coastal parts of adjacent Europe on the south side of the English Channel (Doornkamp *et al* (eds.), 1980). Unprecedented price-levels for potatoes were attained in 1976 and, in the seasons immediately following, persisted, later to decline as farmers in suboptimal and even new areas flooded the market in the much less droughty seasons that followed. Ironically, the excessively wet and late spring of 1983 delayed access and planting of all crops, including potatoes, on the Lancashire mosslands so severely that major financial losses were incurred on many of the same mossland farms that reported record profits in 1976. Potato prices soared again in late 1983 to high levels in the UK.

It should be asserted, however, that this is a most exceptional case. For the majority of agricultural systems on the majority of soils for the majority of seasons, the relationships between environment, production and marketing are less direct and more complex and often buried deep within a host of management variables. But invisibility does not mean inoperability of the ecological variable in land-use, which itself is never 'switched-off' except under conditions of controlled environments, like factory farms or growth chambers as used in plant breeding experiments, where it may be enhanced, substituted or rendered variation-free and hazard-free for experimental design purposes.

Land-Use Patterns on the Lancashire Plain of North-West England

The Lancashire plain, in particular that part lying in south-west Lancashire (now vastly eaten into by the Merseyside and Greater Manchester conurbations) and the Fylde (behind Blackpool), is richly patterned in diverse glacial and post-glacial deposits, including clays, sands and peats, which have bestowed an unusually sharp differentiation into soil types over short distances. Although these three deposits, clay, sand and peat, are laterally associated both stratigraphically and as mixes, they do impose over substantial areas a faithful and exceptional relationship between parent material and soil attributes.

In 1948, Taylor (1952) mapped the field distributions of crops in representative areas of the plain and identified a restricted cropping range on peat, dominated by main crop potatoes, oats and hay poor in clover. In contrast, the sands supported a greater variety of crops and longer, more flexible rotations, including early vegetables, wheat, peas, brassicas and hay rich in clover. The clays were stubbornly in permanent grass except where local outcrops on individual arable farms were obliged to carry their share of crops rotating round the farm. The heavy clays were more easily adapted to non-arable, grass-based dairy farming especially in eastern Lancashire where they outcrop extensively; in the Flyde in particular a major secondary enterprise was poultry-keeping. Both the industrial towns of Lancashire and the resort towns provided handy markets for milk, eggs, etc. The sands were eminently suited to intensive cash crop production, essentially large-scale market gardening: being light textured, not late, freely drained and amenable to regular, heavy doses of fertiliser. The peats, however, although eminently arable, suffered from a late frosty start to the growing season, which rules out early crops, and involved high costs of initial reclamation and subsequent management. Inherent soil acidity was expensive to correct, and comprehensive programmes of mineral fertiliser are regularly required to maintain crop yields. Inevitably, a restricted cropping regime became characteristic of mossland farms unless they were very highly maintained or specialised in suitable horticultural crops such as celery, lettuce, rhubarb and, more recently, onions and carrots.

Expressed in management terms, the farmers of south-west Lancashire, over a period embracing the last hundred years or so, have empirically adapted their techniques of husbandry and cropping selections to the advantages and restrictions of whatever proportion of soil types were available on the farm. These policies became consistent on most farms of similar soil resource and, having stood the test of time and the continuous challenge of competitive marketing, at first local but now well beyond the county limits, impose crop distributional patterns which, on a field and map basis, presented self-evident and predictable spatial associations between drift parent material, soil type and crop association (Taylor, 1949, 1952).

In 1960-61, as an integral part of the Second Land Utilisation Survey of Britain, questionnaires were obtained for a sample of 337 farms, selected on the basis of non-random national grid coordinates, for the Lancashire plain (Aitchison, 1964).

Table 9.5: Chi-square statistics for selected relationships between soil type and crop, livestock and enterprise combinations for samples of Lancashire farms (after Aitchison, 1964)

(a) No. of crops in combination	Generalised Soil Type* Sands	Peats	Clays	Totals
0-2	5	17	75	97
3	15	15	15	45
4	25	8	4	37
5	19	4	2	25
Totals	64	44	96	204
$\chi^2 = 104.23$: degrees of freedom = 6 : very significant at 0.001 level				
(b) No. of livestock combinations	Sands	Peats	Clays	Totals
0	20	7	3	30
1	32	19	38	89
2	17	7	40	64
3-5	6	2	20	28
Totals	75	35	101	211
$\chi^2 = 31.57$: degrees of freedom = 6 : very significant at 0.001 level				

(c) First dominant enterprise	Sands	Peats	Clays	Totals
Dairy or Beef	4	5	52	61
Poultry	5	4	36	45
Roots	27	21	6	54
Vegetables or Market Gardening	37	5	5	47
Totals	73	35	99	207

χ^2 = 127.58 : degrees of freedom = 6 : very significant at 0.001 level

(d) No. of farm enterprises	Sands	Peats	Clays	Totals
1	7	5	27	39
2	34	15	51	100
3	27	14	20	61
4	6	2	4	12
Totals	74	36	102	212

χ^2 = 14.7 : degrees of freedom = 6 : very significant at 0.05 level

The original sample was for 337. Farm units from the Lancashire plain where soils are predominantly on boulder clay, fossil blown sands (the Shirdley Hill Sands) or peats (both 'low moor' and 'raised bog'). Only those farms exclusively on one of these three parent materials were included in the subsamples used in the four tabulations above. Subsequent analysis of the farm data and the construction of a farm-enterprise regional map (Taylor, 1967), for south-west Lancashire alone, confirmed and quantified the earlier spatial associations identified on a field basis by Taylor (Figure 9.7) (Aitchison, 1968).

Table 9.5 shows an additional statistical analysis of data for selected subsamples of farms and referring to the Lancashire plain as a whole. Chi-square tests on four relationships between soil type and farm data show that they are all very significant. The frequency distribution of crops in combination per soil type (Table 9.5(a)) peaks at 4 crops on the versatile sands but at 0-2 on the more restrictive peat soils. The clays also peak low at 0-2 and the overall pattern of Figure 9.7 is confirmed. In Table 9.5(b), livestock combinations are low, scoring only 1 for the dominantly arable farms on the sands and peats whereas the clays, with their commitment to permanent grass and dairy cattle and, locally, beef cattle and poultry, record higher

Figure 9.7: Land use combinations on a selection of south-west Lancashire farms (after Aitchison, 1964)

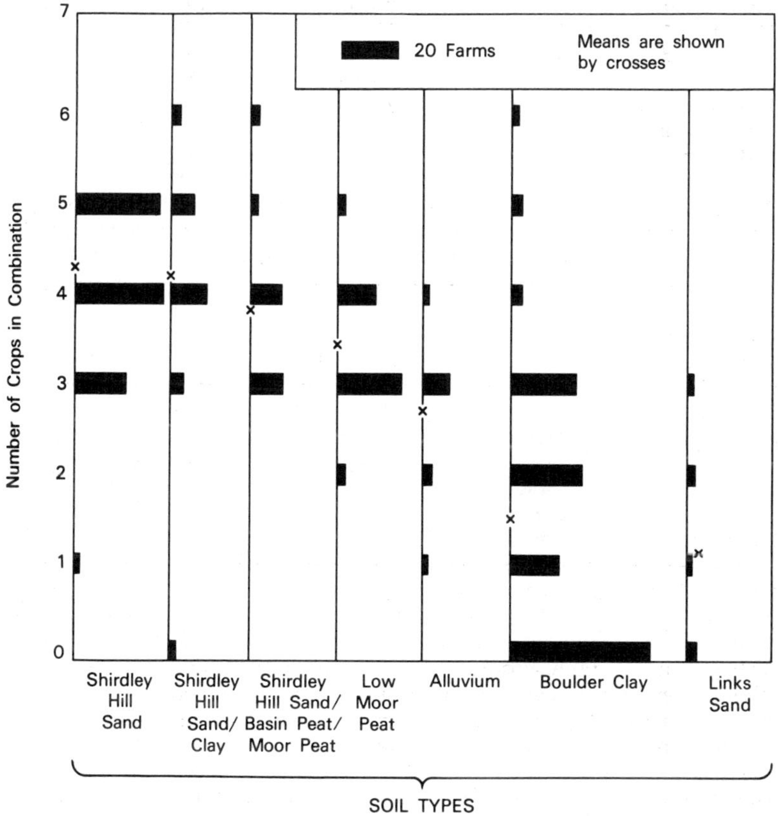

combinations at the 2 and 3-5 levels. The relationship of soil type to first-dominant farm enterprise (Table 9.5(c)) is the most significant of the tour. The sands record the highest values for vegetables-or-market-gardening with roots a good second. Roots dominate the rankings on the sands but dairy-or-beef and poultry dominate on the clays. Finally, Table 9.5(d) shows that the rather more complex parameter of number of farm enterprises, regardless of arable or pastoral commitment, is still significantly related to soil type. Both sands and peats score highest at 2 and 3 enterprise combinations but the clays peak at 2 and are quite well represented at both the 1 and 3 combination levels.

Thus, it may be concluded for the Lancashire plain that, within the context of intensive commercial farm systems, continuously geared to the objective of maximising profits, the availability of specific and dominant soil resources currently predisposes the initial choice between crops and grass and conditions the ultimate choice of specific crop and stock combinations within the context of market, technical and personal priorities. In areas of less clearly differentiated soil types and less specialised agriculture, the relationships with soil resource availability are less clear, more random and may be variably negative as well as positive. But the environmental factors do continue to register, however invisibly, in the variability of productivity of farm enterprise combinations and in the making of farm-management decisions.

Soil Variables and Profitability on Londonderry Farms, Northern Ireland, UK

One of the most successful studies of the linkages between soil parameters and agricultural profitability is that of Cruickshank and Armstrong (1971) who studied a sample of 70 farms and associated soils in County Londonderry, Northern Ireland. On the negative side, one of the conclusions drawn was that soil series maps were of limited use when applied to agricultural land classification and, moreover, that the statistically demonstrated relationship between soil variables and agricultural profitability did not correspond with the boundaries of the soil series units. The spatial expression of relationships between ecological factors and land-use is always difficult to refine, measure and map. The problem is the unconformity between the fragmented, institutional network (field, farm, tenure, etc.) with its pockets of decisions and policies, and the continuity, universality and independence of the ecological framework (relief, climate, soil, etc.) which finds spatial expression in naturally homogeneous units (valley bottom, lower slope, hill top, etc.) that rarely find complete identity with institutional units. Clearly, scale problems, degrees of generalisation, reconciling point information (soil pit, soil sample, climatological station, stream flow gauge, etc.) with area information (the farm unit, the field unit, the soil series, etc.): all help to frustrate the search for relationships between information on static points or areas in space and data referring to functions, economic or technical, in space. For instance, part of field 'w', growing crop 'x' on soil series 'y' on farm 'z' has to be reconciled with economic data referring to either the one farm (z), as a whole, or the total area in crop 'x', which may also occupy land on a second, quite different soil series. The difficulties are truly formidable but Cruickshank and Armstrong were able to establish a number of relationships between selected soil properties and gross margins for selected crops and for one selected farm unit.

A stratified random sample of ten comparable farms on each of seven soil series was selected for the Co. Londonderry study. Table 9.6(a) shows average gross margins from sampled farms, grouped by soil series, for oats, barley, potatoes, livestock and for one specific farm of 40.5 ha. The analysis of variance reveals that there are significant differences among the soil series for all enterprises with the sole exception of oats, which may be due to its tolerance of a wide range of soil conditions. Five stepwise multiple regressions were carried out on the gross margins from these enterprises against the values of the eight properties of subsoil samples taken from each farm (Table 9.6(b)). The sub-soil was wisely selected as more reliable source of representative samples than the soil itself which has clearly suffered greater interference and change. These sub-soil samples explained between 35 per cent and 43 per cent of the variations in gross margins, the highest value being attained for the individual farm unit. Again, all eight properties were limiting on one or more of the enterprises. However, analyses of variance showed that only livestock production was significantly related to differences in soil series in regard to gross margins on each enterprise. But for livestock, the soil series present a quite different ranking from those shown in Table 9.6(a). It is this inconsistency which casts serious doubts on the use of soil series divisions and distributions for agricultural classification purposes. Nonetheless, specific relationships were proven between selected soil characteristics and variation in the gross margins and profitability of selected enterprises. It is in this specific direction that soil productivity studies should be developed.

The Location of Early Potato Fields in Western Pembrokeshire, Dyfed, Wales, UK

Early potato growing in western Pembrokeshire (south-western Dyfed) is so intensely specialised for such a highly competitive market that the 'inherent' earliness (through sea-proximity, slope, aspect and soil) and 'operational' earliness (the actual growing season and husbandry practices adopted) are intimately associated with variations in productivity and profits (Tyrrell, 1970; Thomas, 1972). A residual measure of the performance of any one early growing season, both environmentally and managerially, is the date of first lifting of the early potato crop which varies from late May, through June and into early July. The greatest profits accrue from a combination of early harvest and heavy yield. Figure 9.8 shows the extent to which lifting dates and yields vary, with soil type available on the farm and distance of the farm from the sea, for a random sample of 33 farms located within the coastal zone of south-west Pembrokeshire. In addition, different symbols on the graph identify (a) earlier farms on sandy loams and/or within two miles of the coast and (b) 'later' farms on heavier soils and/or more than two miles from the coast. Of the two regression lines plotted, the steeper one isolates the majority of the 'earliest' farms which are almost exclusively on sandy loams. The less steep line segregates the majority of the 'latest' farms which are mostly on heavier soils and more than two miles from the coast. In this category, lateness is somewhat compensated by the heavy yields. The promotion or

Table 9.6(a): Average gross margins and selected soil parameters for a selection of Londonderry (NI) farms (after Cruickshank and Armstrong, 1971): average gross margins from sampled farms, grouped by soil series

Soil series	Oats £/ha	Barley £/ha	Potatoes £/ha	Livestock £/ha	Farm of 40.4 ha £ (100 acres)
Marine alluvium gley	None	61.2	None	36.2	1870.8
Mixed till gley	32.5	None	184.5	42.7	1714.4
Basalt till gley	None	44.0	176.3	58.7	2211.5
Schist till mineral gley	36.0	52.3	226.0	26.5	1935.4
Schist till peaty gley	31.2	None	260.5	31.7	1604.5
Carboniferous sandstone till gley	37.0	54.2	230.0	49.0	2124.7
Sand and gravel podzols	44.0	54.0	181.0	55.2	2433.2
Inter-series variance estimate, and degrees of freedom (in parentheses)	21(4)	44(4)	928(5)	129(6)	489(6)
Inter-series variance estimate, and degrees of freedom (in parentheses)	23(25)	11(21)	33(23)	29(43)	180(44)
Variance ratio F	1.0	4.0	28.0	4.4	2.8
Per cent confidence level at which means may be considered significantly different	<95.0	97.5	99.9	99.0	97.5

Table 9.6(b): Average gross margins and selected soil parameters for a selection of Londonderry (NI) farms (after Cruickshank and Armstrong, 1971): average gross margins calculated from multiple regression equations

Soil series	Oats £/ha	Barley £/ha	Potatoes £/ha	Livestock £/ha	Farm of 40.4 ha £ (100 acres)
Marine alluvium gley	None	61.8	None	48.7	2330.4
Mixed till gley	22.7	None	186.5	36.2	1543.0
Basalt till gley	None	41.2	176.0	57.5	2260.4
Schist till mineral gley	35.7	54.2	163.5	50.2	1994.0
Schist till peaty gley	34.0	None	237.0	38.0	1914.0
Carboniferous sandstone till gley	33.0	53.2	219.0	45.0	1873.6
Sand and gravel podzols	40.7	54.0	196.2	51.3	2318.2
Inter-series variance estimate, and degrees of freedom (in parentheses)	151.1(4)	36.0(4)	624.0(5)	83.0(6)	350.0(6)
Inter-series variance estimate, and degrees of freedom (in parentheses)	8.5(25)	16.5(21)	739.0(23)	16.0(43)	990.0(44)
Variance ratio F	1.8	2.1	<1.0	5.2	< 1.0
Per cent confidence level at which means may be considered significantly different	<95.0	<95.0	<95.0	99.0	< 95.0

Figure 9.8: West Pembrokeshire, Dyfed, Wales: lifting dates and yields of the earliest field per farm (after Tyrrell, 1970)

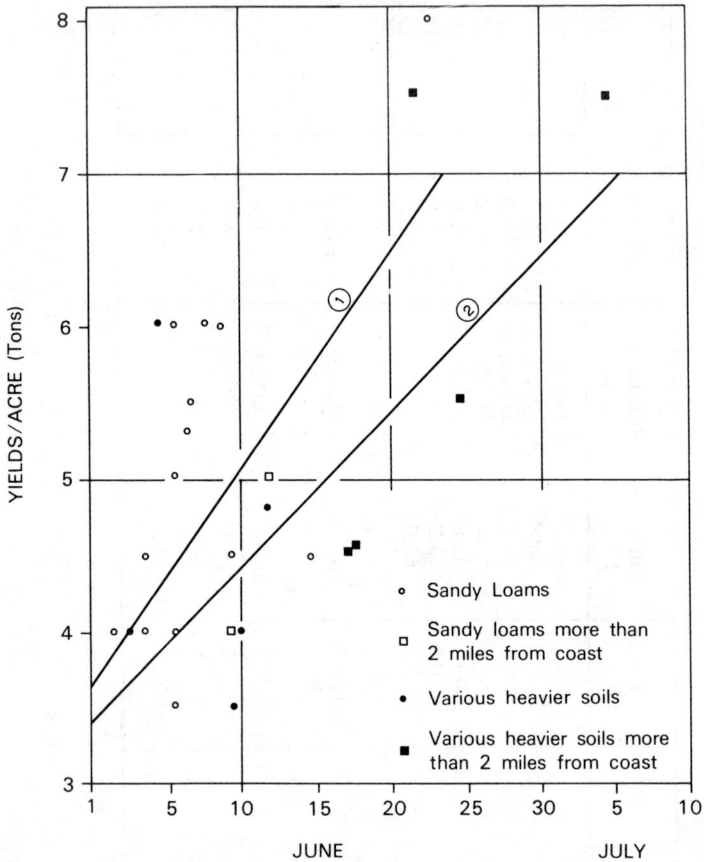

(1) SITES WITHIN 2 MILES OF THE COAST $y = 0.12x + 3.91$
(2) SITES MORE THAN 2 MILES INLAND $y = 0.134x + 2.824$
where y is the yield (in tons per acre) for the earliest field of each farm in the sample and x is a measure of time from the beginning of June, when June 1 is day 1 etc.

The point of origin of the two regression lines would occur in May if the time axis were extended. This would have little significance, however, as tubers in May in both location groups were very small and of no commercial value. Thereafter, the difference in bulking rates increases to the advantage of the coastal, sandy-loam sites. The regression lines have been plotted as far as the date of the latest 'first lifting' but not beyond, since the lines will not necessarily behave in the same way beyond this point.

demotion of earliness by soil texture, land aspect and distance from the sea produces hierarchies of gradients on different scales according to the combination of advantages or disadvantages per farm and per season. At maximum a range of four weeks in the time lifting dates may occur in this comparatively small area, and profits on early farms may be twice and sometimes three times those on late farms on the average. Thus, a full array of environmental and site factors controls variations in profitability. The scale of operations of the ecological factors is closely mirrored in the differential profitability of individual farms in individual seasons. The calibration of environmental processes, e.g. rate of soil warming, incidence of water stress, is directly paralleled by the price mechanisms as operated by economic and institutional forces, ranging from farm level to that of the Potato Marketing Board and beyond (Tyrrell, 1970; Thomas, 1972).

THE MANAGEMENT VARIABLE IN ECOLOGICAL CONTEXT

The Management Variable

The extent to which ecological resources are efficiently converted into production depends primarily and extensively on management, its objectives, motivations and equipment. The relationship is well shown when ecological resources are of intermediate quality, without extreme advantages or disadvantages of soil, slope, climate or location, or when such moderate ecological resources are not liable to much variation per hectare of farm unit. Daw (1964) makes the point very nicely for a sample of some 32 farms on the relatively low-quality soils on sandstones in 'sand land' area north of Nottingham (Table 9.7). The relative consistency of terrain and soils allowed Daw to assume a quasi-constant initial land resource per acre against which he could assess the variable efficiency of management. In Figure 9.9, gross margins per acre at 1964 prices ranged from £33 profit per acre at best to an actual running loss of £4 per acre at worst. Daw assembled data on the age, intelligence, etc. of farmers and extent to which they used public and/or private information in running their farms. The correlations revealed are positive and consistent, the more intelligent and more information-using farmers attaining the highest profits (Figure 9.10).

Duckham (1967) takes the argument further by showing how the management variable is related to a variable ecological resource. Table 9.7 (initially based on work by Bone and Tayler, 1963) presents averaged data from long-term records for three different stocking rates, A_1, A_2 and A_3, on Reading University's Sonning Farm in Berkshire, for total profit expectation in £/acre over a ten-year period, incorporating the probable combination of three wet (W_1), four medium (W_2) and three dry (W_3) seasons.

The most intensive and most skill-demanding strategy, A_1, incurs the highest variable costs for the heaviest fertiliser programmes and laid-on irrigation equipment, etc. In wet years, (W_1), over-capitalisation, in the form of unused irrigation, and

Figure 9.9: Financial results arrayed in profit groups for a selection of farms near Nottingham (after Daw, 1964)

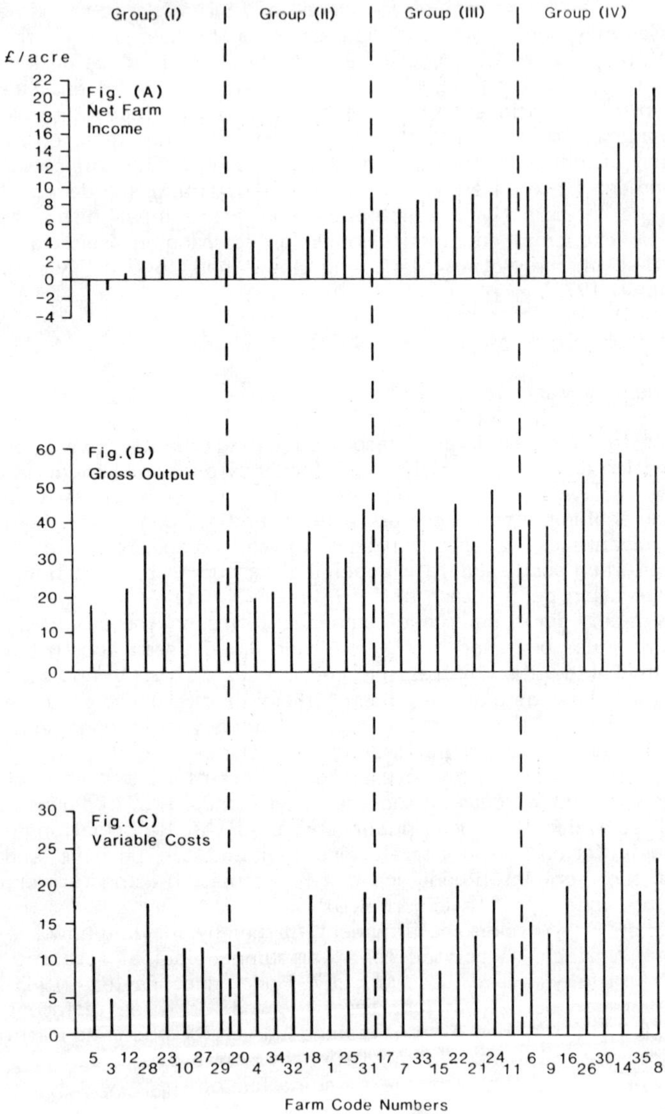

(A) Net Farm Income; (B) Gross Output; (C) Variable Costs; (D) Gross Margin; (E) Fixed Costs; (F) Total Costs; (G) Management and Investment Income

Figure 9.9 cont'd.

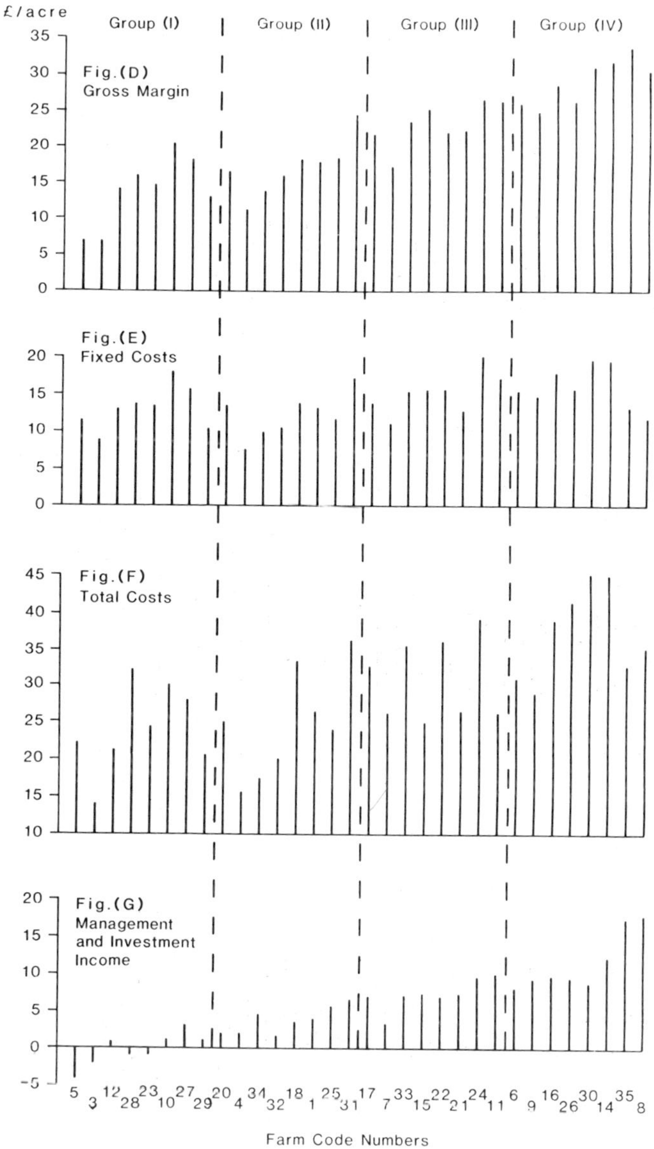

Table 9.7: Stocking rates, management variables, weather variables and profit expectation on Reading University Farm, Berkshire, England (after Bone and Tayler, 1963 and Duckham, 1967)

	W_1 Wet years	W_2 Med. years	W_3 Dry years	Mean annual profit expectation over 10 years
Frequency of W_1, W_2, W_3 in 10 years	3	4	3	
Alternatives:				
A_1 Very High Stocking (2 cows/acre) high nitrogen plus irrigation	£6 (600)	£14 (650)	£12 (650)	£11 (CV about 30%)
A_2 High Stocking (1½ cows/acre) medium nitrogen	£24 (600)	£4 (400)	-£16 loss (200)	£4 (CV about 400%)
A_3 Medium Stocking (1 cow per acre) low nitrogen	£8 (400)	£8 (400)	£8 (400)	£8 (CV Nil)

CV = Coefficient of Variation

poaching damage detract from profitability but, in medium (W_2) and dry (W_3) years, this system attains consistently high productivity. Overall, for the ten-year period, however, there is substantial variability (30%) of profits as between wet and dry years. In contrast, the least intensive and least skill-demanding strategy, A_3, remains consistently profitable, if at a modest level, regardless of the variations in summer rainfall, and has the lowest and the most consistent variable costs, with no overstocking or over-capitalisation problems. Strategy A_2, the intermediate stocking rate, combines the best and the worst of both worlds. In wet years, with no irrigation investment or costs, a very high profitability is obtained: four times that of A_1. In dry years, however, very substantial losses are suffered and in medium years profits are very low. Mean annual profit expectation with the intermediate system overall shows the greatest sensitivity (400%) to variations in summer rainfall and the lowest average net annual rate of profitability over the ten-year period.

The principle emerging from Duckham's revealing analysis is

Figure 9.10: Correlations between farm management characteristics and gross margins for a sample of farms near Nottingham (after Daw, 1964)

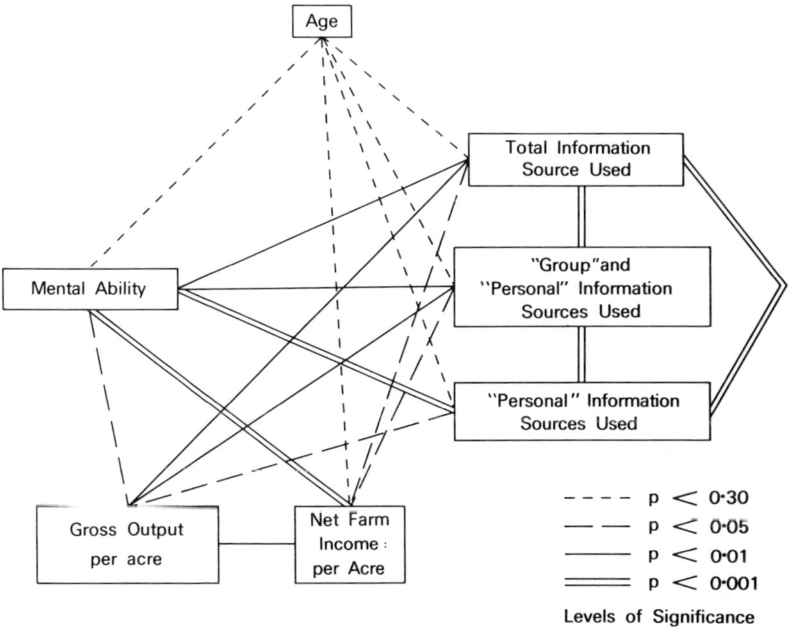

The relationships with age are in a negative direction. All other relationships are positive.

that the assessment of the probable variability of an ecological resource over a prescribed time-period is very relevant to the level and intensity of management strategy to be adopted on the individual farm in relation to capital and technical resources available and levels and variability of profits to be expected or tolerated. Strategy A_1 runs the highest risks to win the highest overall profits but the operational risks vary each year with the weather and so do the profits. The long-term risks relating to sward and soil deterioration and to animal health risks must also be taken into consideration. Strategy 3 involves minimal risks but steady, reliable and predictable, if modest, profits. This conjures up the image of the older, traditional, conservational farmer whereas strategy A_1 could imply the younger, innovatory farmer, able and willing to attract the necessary bank loans to introduce a high-risk/high-profits system, not least perhaps because of restrictions in grazing land available. Strategy A_2 implies a lack of financial or technical resources or only a half-hearted motivation to intensify the system, resulting in extreme vulnerability to the weather and depending heavily on dry years to compensate for wet ones. The overriding conclusion is that land-use systems are best adapted to either maximum or minimal intensification levels of management; intermediate levels would appear to be ill-advised. Aside from variations in growing season, the same point applies to seasonal or punctual accessibility to specific land, soil or grazing resources, e.g. at lambing time, for out-wintering or few autumn housing or spring dehousing of stock, for the preparation of land for ploughing, sowing, etc. Duckham's data reveal the subtle, criss-crossed interplay between environmental resource availability (or its predicted availability) and use-strategy adjustments to maintain profitability. This principle lies at the very heart of land-use ecology and of the traditional balancing of the man-land equation.

The Interlocking of Environmental and Management Variables

Lest the thrust of this essay be lost in a sequence of local and specialised case-studies, a final disentanglement of the environmental and management factors, which together impinge on variations in production, marketing and price of coffee on the world's markets, will be attempted with particular reference to the changing circumstances of Brazilian production (Taylor, 1974b).

Figures 9.11(a) and 9.11(b) reveal successive shifts in the location of coffee production from the middle Paraiba valley behind Rio de Janeiro in the mid-nineteenth century to São Paulo State from about 1880 to 1950 and, thereafter, an increasing preference for the more extensive terra roxa soils on the volcanic 'trap' outcrop of northern Parana where, however, the greater risk of July frosts has become a major factor in causing variations in production with repercussions for price-levels. Figure 9.11(c) demonstrates the association of frost years with sharp price rises in the years following. The proximity of the first occurrence in Brazil of coffee leaf-rust disease in 1970 to the July frosts of 1971 reduced the country's store of beans to only 4 or 5 years' supply. As 4 years' stock is regarded as the minimum to maintain a competitive position in the world's markets, a good harvest was required in 1972 which, unfortunately, was another severe frost year. World coffee prices soared to unprecedented levels and Brazil was obliged to reappraise her own internal economic priorities towards greater diversification of export commodities and greater independence of the impact of fluctuations of world coffee prices. Angolan beans were imported to produce 'Brazilian' Nescafe for the European market, and northern Parana, by the early 1980s, had recorded considerable rural depopulation and also out-migration to renewed older coffee areas in Sao Paulo State and to new fazendas in Minas Gerais. The International Coffee Organisation, which has headquarters in London,

Figure 9.11(a): The changing locations and levels of coffee production in south-east Brazil as related to selected environmental

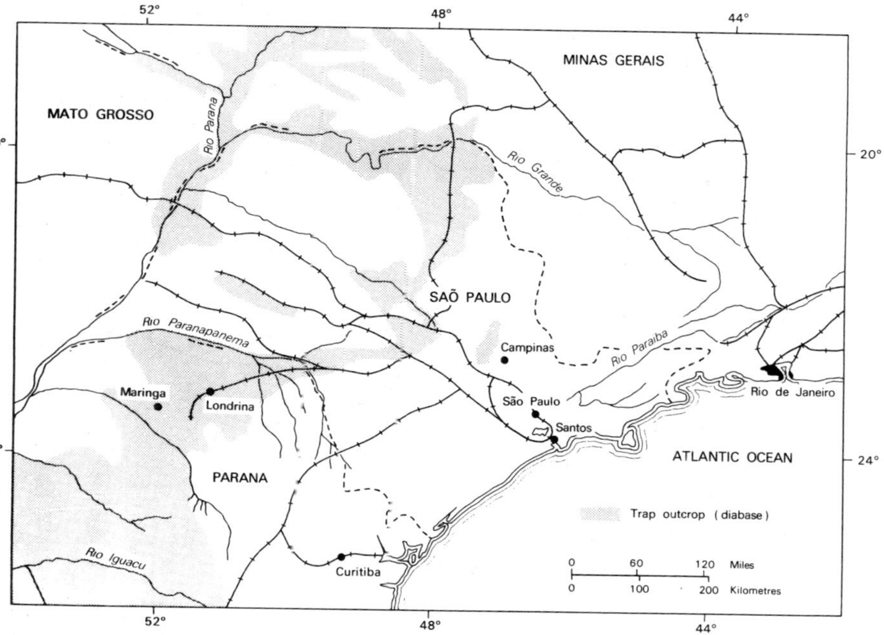

(a) Distribution of the volcanic 'trap outcrop' (diabase) forming the parent material of the <u>terra roxa</u> soils.

Figure 9.11(b)

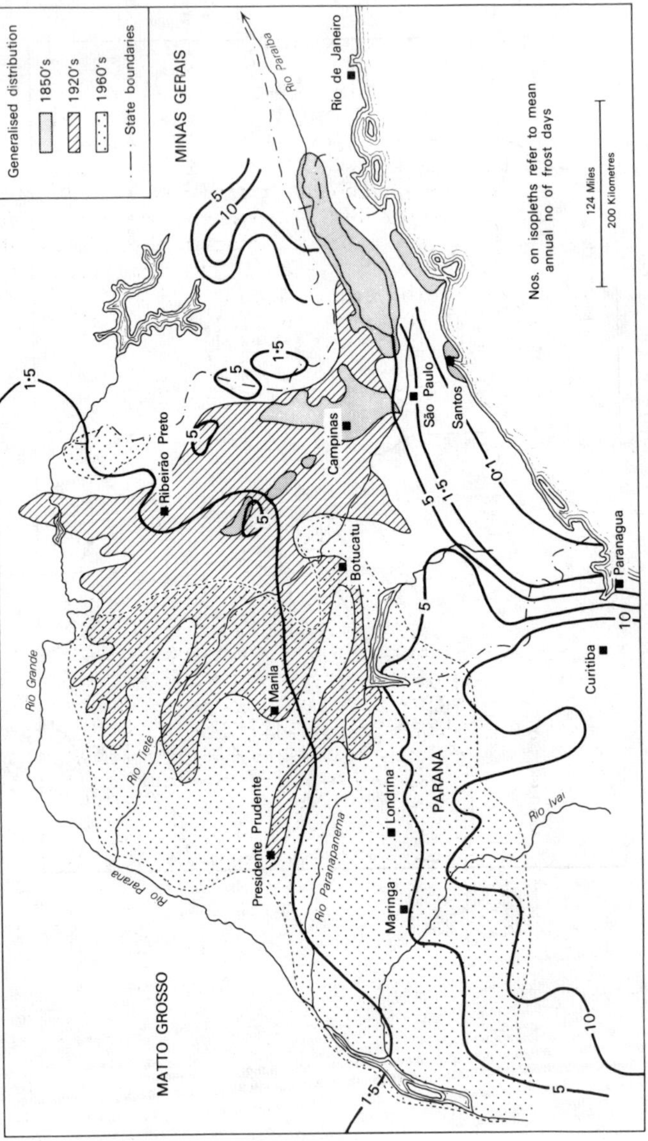

(b) Locations of coffee production in (i) the 1850s, (ii) the 1920s and (iii) the 1960s. The isopleths refer to the mean annual number of frost days.

Figure 9.11(c)

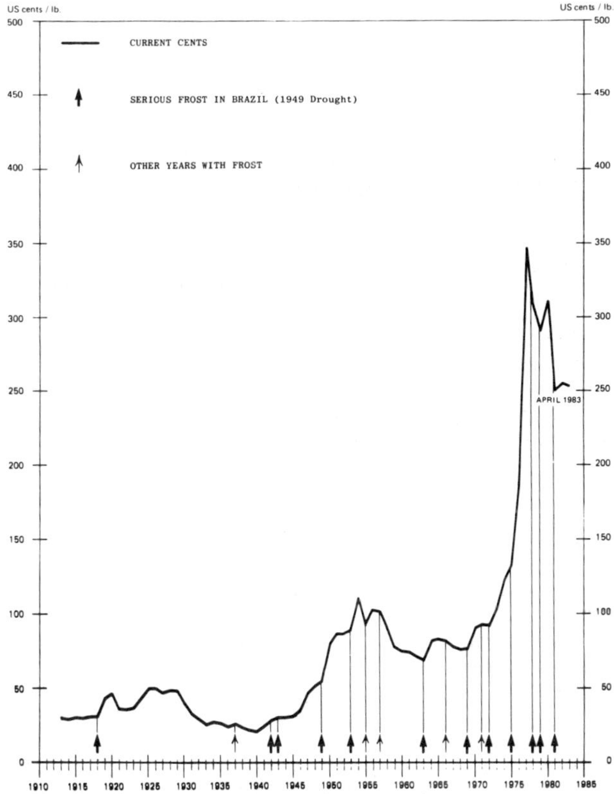

(c) Annual averages from 1913 to 1983 for retail prices of coffee in the USA in current and constant terms. Years of 'serious frost' are shown by the black arrows, except that 1949 refers to the drought hazard of that year. (Graph compiled and reproduced by kind permission of the International Coffee Institute, London.)

Note, following Taylor (1974), that less severe frosts, shown by the thin arrows, also occurred in 1937, 1955, 1957, 1966 and 1971. The dramatic mid-70s quadrupling of retail coffee prices is well shown. It should be emphasised that the apparent increase in the frequency of severe frosts since the early 40s, and especially since the late 60s, has coincided with an increasing switch, especially since the early 60s, of coffee production from São Paulo state into northern Parana where frost frequencies are normally greater, independently of any climatic trend (Figure 9.11b).

has the annual problem of planning production quotas and marketing projections, a problem which is almost permanently insoluble when confronted with the uncertainties about the likelihood of frosts recurring in northern Parana and the maximum price levels the market will tolerate. Chaos, Latin American and international style, is thus further confounded by local environmental vicissitudes, which have global impacts on commodity prices and marketing efficiency, not to mention the fact that the Brazilian economy in mid-1983 became so weak as to cause a major international monetary crisis. The ecological base continues to assert itself throughout the production and marketing systems.

Under the more primitive subsistence cultivation systems of the tropics, the interplay between environment and land-use management is equally sharply displayed. The scale and rate of land rotation is related to the productivity of land in cultivation, the recoverability of land in fallow, and the population numbers that can be sustained per unit land area. Allan (1965) has discussed land potential models as applied to shift agriculture in tropical Africa. Gourou (1962), as reported in Henshall (1967), has proposed the formula:

$$P = A \times \frac{C}{B} \quad . \quad . \quad . \quad . \quad . \quad . \quad . \quad . \quad . \quad (5)$$

where P is the population capacity, A is the amount of cultivable land expressed as a proportion of all the land available, B is the number in years occupied by the full rotation, i.e. cultivation period plus fallow period, and C is the sustainable population number per unit land area cleared each year. The process of colonisation has brought improved medical services, increased life expectancy and increased population pressure on the same land resources. Again, intensification of land use has resulted from the introduction of commercial, as distinct from subsistence, crops and, on occasions, from the local introduction of fertilisers. The consequent increased turn-over in land rotation has abbreviated fallow periods and induced degradation of the forest and soil resources. The situation can be avoided by the use of the 'corridor system' (Dumont, 1957) as operated in Zaire, when it was the Belgian Congo. Adjacent substantial strips of cultivation are alternated with surviving forest belts. Commercial and subsistence crops are intermixed and an annual record of land use is kept. An ordered, monitored, controlled system is established, combining the best of the traditional and modern methods of land-use. Food production is stabilised and predictable so population levels can be maintained. This is a good example of optimisation of land-use resources and systems and of the reconciliation of ecological and socio-economic priorities.

CONCLUSION

Geographical situations or patterns result from the operation of either, or the interaction of both, of two variables, i.e. the ecological/environmental variable or the human/technology variable, e.g. Hare _et al_ (1977). Each of these variables is subject to its own inherent variability but on different timescales, especially in the context of man-controlled, land-use systems which involve an accumulated legacy of man/land interactions. However, the slowness, spontaneity and normally prolonged timescales of environmental changes allow the bestowal to given locations of a quasi-permanent biophysical stock of resources, productivity and potential, which constitute a constant ecological base-line against which the pressure or efficiency of the land-use succession can be assessed.

Individual cultures, both primitive and advanced, will evaluate the same ecological base on their own cultural terms to the extent, perhaps, of negating its existence. The power of advancing technology may modify or even replace the ecological base but it is the vulnerability or resilience of the latter to land-use pressure which is part and parcel of that base, thus providing a functional measure of its quality. Again, it is the technical, economic and social costs of the conversion of ecological resources which is the mirror image of their inherent capacities.

The search for continuity in land evaluation is best pursued in identifying relationships between land potential and land-use with reference to successions of natural ecosystems, man-adapted ecosystems and man-made ecosystems. Cultural history may be summarised as a perpetual spatial search, first for adaptation and stabilisation and, finally, for maximisation and, ideally, optimisation of land-use resources for man's needs. Trudgill and Briggs (1977-81) have presented a comprehensive and critical review of the soil factor in land potential, and in the principles and objectives involved in land capability surveys, as operated in both the developed and the developing world. Their major conclusions are sympathetic to the arguments evolved in this essay. These are, basically, that the more permanent 'inherent' properties of the soil environment should form the basis of land assessment in the long-term because ecological resource is thereby better used and conserved. Short-term objectives, such as increased food production or immediate yield predictions, place such long-term policies at risk, as management priorities change so rapidly in response to economic, social or political pressures. Equally, policies to maintain environmental quality survive better when based on long-term ecological principles. The review also points to the effectiveness of the use of functional soil or site properties in assessing use-potential. For example, accessibility, erodability, poachability, stability are measurable properties which lie at the heart of sensible soil use and efficient husbandry. On the other hand, the concept of improvability linkages the soil capacity with soil performance. Physical land classification schemes are incomplete until they are expressed per unit area of output, which is usually expressed in gross margins of the land-use enterprises they are supporting, e.g. the expression of the gross margins of farm enterprises a, b and c per acre of the land types p, q and r involved for farm x or parish y or region z (MAFF, 1966). These combinations enable the overall appraisal of field or farm productivities and, subsequently, their degree of improvability can be derived (Taylor, 1974(c) and 1978). Expressed simply, the economic law of diminishing returns operates in relation to soils of varying quality. Good soils, per unit investment over time, will respond greater and longer than poor soils where the response per the same unit of investment is smaller and short-lived (Taylor, 1978).

Johnston (1983) concludes that resource studies have so far failed to achieve the much needed re-integration of physical and human geography. It is in fact the armies of pure physical geographers and pure human geographers who, since the mid-1950s, have sacrificed integrative and regional studies on the altar of ever-increasing specialisation, even to the extent of aping adjacent physical and social scientists and almost breaking off diplomatic relations between physical and human geographers. Johnston condemns the highly selective set of resource studies he has consulted as lacking in the full consideration of both the natural and social processes operating behind the interface of resource analysis and use, thereby creating the need for the 'unifiers', 'polymaths' or 'superscientists' which he later concludes are roles for which a geographical training would not be, in any automatic sense, exclusively appropriate. For the geographer to be both natural and social scientist is 'both utopian and arrogant', he suggests! He is nearer the point, however, when he mocks Park's (1980) honest approach, in his book entitled Ecology and Environmental Management, which is 'not solely or even primarily physical or human but founded in one and written for the other'. The problem is twofold: one of academic depopulation within geography from its traditional middle to its physical and human sectors and, second, of the powerful forces of inertia in our education system which traditionally separates the arts from the sciences.

Biogeographical approaches to the integrative study of ecological and economic systems require joint first-degree training in ecology and economics: strange bed-fellows at any university. Thus, the laws and processes of the natural environment could be understood alongside the laws and processes of socio-economic systems. It is essential that the techniques of analysing both ecological and economic data are comparable and, as appropriate, identical. Conversion to energy equivalent, as demonstrated by Nilsson (1976) for the Swedish Forestry industry and its environmental base, is a valuable first step. Haggett and Chorley (eds., 1969) have discussed models which may be adopted for study in both physical and human geography and, ipso facto, for land-use analysis. Isard (1969) has studied specific linkages between ecological and economic systems (Figure 9.12). Woldenburg (1968), although concerned with hydrology rather than land-use, has demonstrated that both stream order and market areas can be treated as nested hierarchies, Christaller-style. Peters (1970) has reviewed the main economic methods of land-use analysis which may be related to land potential. These include the well-known technique of cost-benefit analysis, which Rubra (1970) has usefully oriented to three different levels of reference: first, the entrepreneur himself, the land user and the location of his operation; second, the

Figure 9.12: Modelling relationships between ecologic and economic systems (after Isard, 1969)

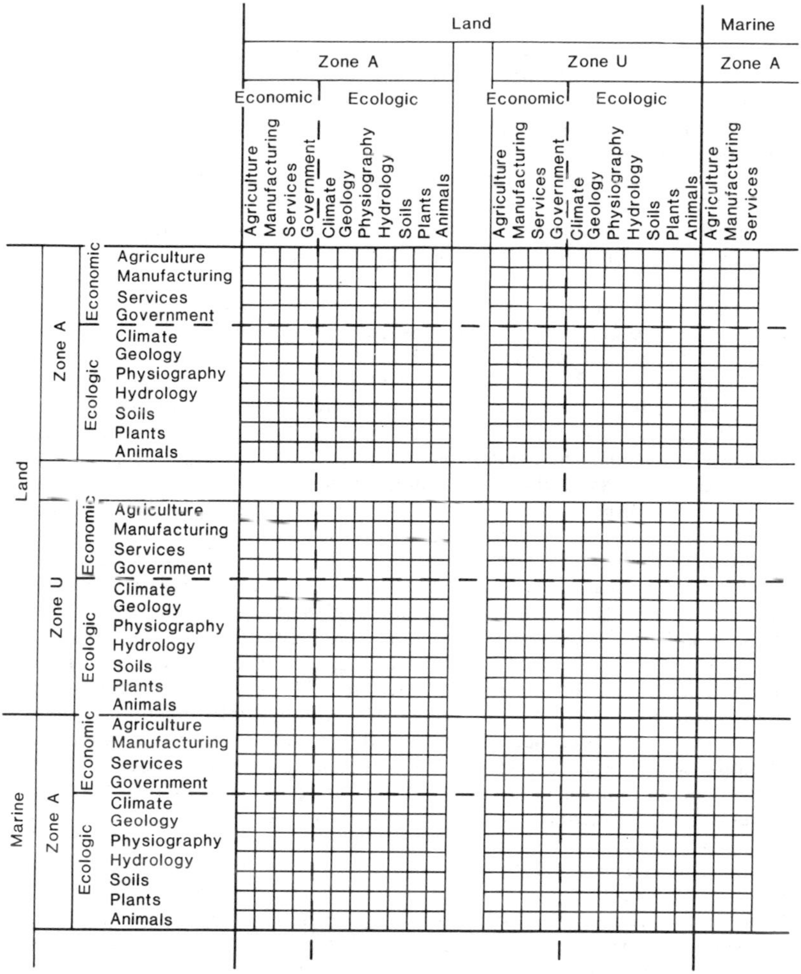

The columns in the table represent sectors (examples of economic activities and ecological processes), and the rows represent selected commodities (both man-made and natural) associated with these sectors, as outputs and resources. The interprocess coefficients of the ecologic systems (bottom-right) parallel the interindustry coefficients of the economic systems (top-left). The coefficients for the inputs and outputs of economic systems (bottom-left) complement those for ecologic systems (top right).

population or group of land users, as a whole, and the general or regional location of their operations; and third, the national community, as a whole, as affected by market forces and price-level mechanisms, including supra-national agencies, as appropriate. Through this hierarchy, costs and benefits may well be unequally distributed and often man-controlled at different levels. Although Rubra's discussion is confined to weather impacts on agriculture, the principles involved (e.g. the acceptance of perceived risks and good and bad years or the introduction of physical or monetary protection to reduce losses in bad years) apply to land-use studies in general, as already shown by Duckham (Table 9.7). For the analysis of soil/land-use interrelationships the case-study by Cruickshank and Armstrong (1971) of Londonderry farms, discussed earlier, clearly points to the way forward.

Land-use studies, like resource analysis, also require a wide range of expertise to identify the reciprocal relationships between the ecological base and the ways it has been, is and should be adapted for land-use purposes. It is essential in the scientific long-term to reconcile the pragmatic objectives of much ecological land evaluation work, which may be designed to serve land users rather than land-use research, with the theoretical and philosophical need to go beyond the secondary matching of primary information inputs, derived from environment, on the one hand, and management, on the other. Perhaps the emergent science of information technology will facilitate the processing of combinations of environmental and management data. This would strengthen even more the fundamental need to extend the data-base (using point-sample, ground survey and remote sensing and more representative samples of farms, forest holdings, etc. (Taylor, 1974d; Darch, Chapter 10 herein)) on both sides of the equation. The latter, thereby, would become more soluble and more applicable to sound long-term, ecologically-based land-use planning, to which biogeographers, within teams of environmental and social scientists, can offer a continuing professional and communicational role in both pure and applied research.

ACKNOWLEDGEMENTS

The Editor would like to thank the following for their generous assistance in the preparation of this Chapter: Richard Smith, for useful comments on an early draft; Carol Parry and Linda James for typing the manuscripts; Michael Gelli Jones and Huw Hughes for drawing the illustrations and David Griffiths and Anthea Cull for photographing them.

REFERENCES

Aitchison, J.W. (1964), 'Farm units and productivity in west central Lancashire', unpublished MA thesis, UCW, Geography Department, Aberystwyth, Wales, UK.

Aitchison, J.W. (1968), 'The land factor and agricultural production in west central Lancashire', in Report No. 5, Agric. Geog. I.G.U. Symposium, Dept. of Geog., Univ. of Liverpool, Liverpool, UK, pp. 3-13.

Allan, W. (1965), The African Husbandman, Oliver and Boyd, London.

Ball, D.F. (1964), 'Soil classification, land-use and productivity', in Ball, D.F. (ed.), Land Classification in Relation to Productivity, Welsh Soils Discussion Group Report No. 5, pp. 1-15.

Black, J. (1970), The Dominion of Man: the Search for Ecological Responsibility, Edinburgh University Press, Edinburgh, Scotland and New York, USA.

Bone, J.S. and Tayler, R.S. (1963), 'The effect of irrigation and stocking rate on the output from a sward', J. Brit. Grassl. Soc., **18**, 190-96 and 295-9.

Bowen, E.G. (1966), 'The Welsh colony in Patagonia, 1865-1855: a study in historical geography', Geog. J., 132, (1), 16-31.

Cannell, M.G.R. (1982), World Forest Biomass and Primary Production, Acad. Press, London.

Childs, G. (1942), What Happened in History?, Penguin Books, Harmondsworth, UK.

Childs, G. (1963), Man Makes Himself, Watts, London.

Chisholm, M. (1962), Rural Settlement and Land-use: an Essay in Location, Hutchison, London.

Chorley, R.J. (1973), 'Geography as human ecology', in Chorley, R.J. (ed.), Directions in Geography, Methuen, London.

Clarke, D.L. (ed.) (1972), Models in Archaeology, Methuen, London.

Clarke, G.R. (1951), 'Evaluation of soils', J. Soil Sci., **2**, 1, 50-60.

Collingbourne, R.H. (1976), 'Radiation and sunshine', in Chandler, T.J. and Gregory, S. (eds.), The Climate of the British Isles, Longmans, London **4**, pp. 74-95.

Conrad, G.W. (1981), 'Cultural materialism, split inheritance and the expansion of ancient Peruvian empires', American Antiquity, 46 (1): 3-26.

Coulson, K.L. (1975), Solar and Terrestrial Radiation, Academic Press, New York, USA.

Cruickshank, J.G. and Armstrong, W.J. (1971), 'Soil and agricultural land classification in County Londonderry', Trans. Inst. Brit. Geogrs. **53**, 79-94.

Darwin, C. (1868), The Variation of Animals and Plants under Domestication, 2 vols., John Murray, London.

Daw, M.E. (1964), Benefits from Planning, Dept. of Agric. Economics, Univ. of Nottingham, Farm Reports: No. 155, Nottingham, UK.

Dimbleby, G.W. (1962), The Development of the British Heathlands and their Soils, Oxford Forestry Memoirs No. 23, Clarendon Press, Oxford, UK.

Doornkamp, J.C., Gregory, K.J. and Burn, A.S. (1980), Atlas of Drought in Britain, Inst. Brit. Geogrs., London.

Duckham, A.N. (1967), 'Weather and farm management decisions', in Taylor, J.A. (ed.), Weather and Agriculture, Pergamon Press, Oxford, pp. 69-80.

Dumont, R. (1957), Types of Rural Economy: Studies in World Agriculture, Methuen, London.

Ellison, A. and Harriss, J. (1972), 'Settlement and land use in the pre-history and early history of southern England: a study based on locational models', in Clarke, D.L., Models in Archaeology, Methuen, London, **24**, pp. 911-62.

Evans, J.G. (1975), The Environment of Early Man in the British isles, Elek, London.

Evans, J.G., Limbrey, S. and Cleere, H. (eds.) (1975), Impact of Man on the Landscape: the Highland Zone, Research Report No. 11, Council for British Archaeology, London.

Eyre, S.R. (1964), 'Determinism and the ecological approach to geography', Geography, **49**, 369-76.

Eyre, S.R. (1978), The Real Wealth of Nations, Edward Arnold, London.

Flenley, J. (1979a), 'The late Quaternary vegetation history of the equational mountains', Prog. in Phys. Geog., **3**, 488-509.

Flenley, J. (1979b), The Equatorial Rain Forest: a Geological History, Butterworth, London.

Geiger, R. (1965), The Climate near the Ground (trans. from 4th German edition of 1961 by Scripta Technica, Cambridge, Mass., USA), Harvard University Press.

Glacken, C.J. (1967), Traces on the Rhodian Shore, University of California Press, USA.

Gourou, P. (1962), 'Agriculture in the African Tropics: observations of a geographer', paper read at Univ. of Oxford and referred to in Henshall (1967).

Gourou, P. (1966), The Tropical World: its Social and Economic Conditions and its Future Status, 4th edition, Longmans, London.

Grigg, D. (1982), The Dynamics of Agricultural Change, Hutchinson, London.

Haggett, P. and Chorley, R.J. (1969), Network Analysis in Geography, Edward Arnold, London.

Hahn, F. (1896), Die Haustiere und ihre Beziehungen zur Wirtschaft des Menschen, 2 vols., J. Englehorn, Stuttgart.

Hare, F.K. (1970), 'The tundra climate', Trans. Roy. Soc. Canada, Series IV, vol. VIII, 393-9.

Hare, F.K. and Ritchie, J.C. (1972), 'The Boreal bioclimates', Geog. Rev., **62**, 333-65.

Hare, F.K., Kates, R.W. and Warren, A. (1977), 'The making of deserts: climate, ecology and society', Econ. Geog., **53**, 332-46.

Harris, D.R. (1969), 'Agricultural systems, ecosystems and the origins of agriculture', in Ucko, P.J. and Dimbleby, G.W. (eds.), The Domestication and Exploitation of Plants and Animals, Duckworth, P. 40, London, pp. 4-15.

Harris, D.R. (1973), 'The prehistory of tropical agriculture: an ethno-ecological model', in Renfrew, C. (ed.), The Explanation of Culture Change: Models in Prehistory, Duckworth, London, pp. 391-417.

Harris, D.R. (1977), 'Settling down: an evolutionary model for the transformation of mobile bands into sedentary communities', in Friedman, J. and Rowlands, M.J. (eds.), The Evolution of Social Systems, Duckworth, London, pp. 401-7.

Harris, D.R. (1981), 'Breaking ground: agricultural origins and archaeological explanations', Bull. No. 18, Institute of Archaeology, University of London, pp. 1-20.

Henshall, J.D. (1967), 'Models in Agricultural Activity', in Chorley, R.J. and Haggett, P. (eds.), Models in Geography, Methuen, pp. 425-58.

Holland, P.G. and Steyn, D.G. (1975), 'Vegetational responses to latitudinal variations in slope, angle and aspect', J. Biogeogr., 2, 179-83.

Holliday, R. (1966), 'Solar energy consumption in relation to crop yield', Agric. Progress, **XVI**, 24-34.

Humboldt, A. von (1807), Essai sur la Géographie des Plantes (facsimile published by the Society for the Bibliography of Natural History, 1959), Levrault, Schoell & Compagnie, Paris.

Isard, W. (1969), 'Some notes on the linkage of ecologic and economic systems', Papers and Proc. of the Reg. Sc. Ass., **22**, 85-96.

Jacobi, R.M. (1980a), 'The Upper Palaeolithic in Britain with special reference to Wales', in Taylor, J.A. (ed.), Culture and Environment in Prehistoric Wales, British Arch. Reports, Oxford, pp. 15-99.

Jacobi, R.M. (1980b), 'The early Holocene settlement of Wales', in Taylor, J.A. (ed.), Culture and Environment in Prehistoric Wales, British Arch. Reports, Oxford, pp. 131-206.

Job, D.A. and Taylor, J.A. (1978), 'The production, utilisation and management of upland grazings on Plynlimon, Wales', J. Biog., **5**, 173-91.

Johnston, R.J. (1983), 'Resource analysis, resource management and the integration of physical and human geography', Prog. in Phys. Geog., **7**, No. 1, 127-46.

Jones, G.R.J. (1955), 'The distribution of medieval settlements in Anglesey', Trans. Ang. Ant. Soc., pp. 27-96.

Jones, G.R.J. (1966), 'Rural Settlement in Anglesey', in Eyre, S.R. and Jones, G.R.J. (eds.), Geography as Human Ecology: Methodology by Example, Edward Arnold, London, pp. 199-230.

Jones, M.E. (1976), 'Topographic climates: soils, slopes and vegetation', in Chandler, T.J. and Gregory, S. (eds.), The Climate of the British Isles, Longman, London, pp. 288-306.

Kershaw, A.P. (1978), 'Record of last interglacial-glacial cycle from north-eastern Queensland', Nature, **272**, 159-61.

Limbrey, S. and Evans, J.G. (eds.) (1978), The Effect of Man on the Landscape: the Lowland Zone, Res. Report No. 21, Council for British Archaeology, London.

MAFF (1966), Agricultural Land Classification, Tech. Rep. No. 11, MAFF, London.

Masters, R.D. and Masters, J.R. (eds.) (1964), Jean-Jacques Rousseau. The First and Second Discourses (English trans.), St Martin's Press, New York, USA.

McCarty, H.H. and Lindberg, J.B. (1966), A Preface to Economic Geography, Prentice Hall, NJ, USA.

Monteith, J.L. (1965), 'Radiation and crops', Exper. Ag. Rev., **1**, 241-51.

Monteith, J.L. (1966), 'The photosynthesis and transpiration of crops', Experim. Agric. Rev., **2**, 1-14.

Newbould, P.J. (1971), 'Comparative production of ecosystems', in Wareing, P.F. and Cooper, J.P. (eds.), Potential Crop Production, Heinemann, London, **15**, 228-38.

Nilsson, P.O. (1976), 'The energy balance in Swedish Forestry', in Tamm, C.O. (ed.), Man and the Boreal Forest, Ecol. Bull. NFR 21, Stockholm, pp. 95-101.

Oort, A.H. (1970), 'The energy cycle of the earth', Sci. Amer., **223** (3), 54-63.

Openshaw, S. (1982), 'The geography of reactor siting policies in the UK', Trans. Inst. Brit. Geogrs., **7**(2): 150-162.

Ovington, J.D. (1964), 'Prairie, savanna and oakwood ecosystems at Cedar Creek', in Crisp, D.J. (ed.), Grazing in Terrestrial and Marine Environments, Oxford, UK, pp. 43-53.

Ovington, J.D., Heitkamp, D. and Lawrence, D.B. (1963), 'Plant biomass and productivity of prairie, savanna, oakwood and maize field ecosystems in central Minnesota', Ecology, **44**, 52-63.

Packham, J.R. and Harding, D.J.L. (1982), Ecology of Woodland Processes, Edward Arnold, London.

Park, C.C. (1980), Ecology and environmental management, Dawson, Folkestone, UK.

Penman, H.L. (1968), 'The earth's potential', Sc. J., **4**, No. 5, 43-7.

Persson, T. (ed.) (1980), Structure and Function of Northern Coniferous Forests - an Ecosystem Study, Ecol. Bulls. No. 32, Swed. Nat. Sc. Res. Council, Sweden.

Peters, G.H. (1970), 'Land use studies: a review of the literature with special reference to the applications of cost-benefit analysis', J. Agric. Econ., **XXI**, 171-214.

Petherbridge, P. (1969), Sunpath diagrams and overlap for solar heat gain calculations, HMSO, London.

Pilcher, J.R. et al (1971), 'Land clearance in the Irish Neolithic: new evidence and interpretation', Science, **172**, pp. 560-62.

Roberts, E. (1958), The County of Anglesey; Soils and Agriculture, Soil Survey Report, HMSO, London.

Rodin, L.E. and Basilevich, N.I. (1968), 'World distribution of plant biomass', in Eckhardt, F.E. (ed.), Functioning of Terrestrial Ecosystems at the Primary Production Level, UNESCO, Paris, pp. 45-52.

Rousseau, J.J. (1755), Discours sur l'Origine et les Fondements de l'Inegalité parmi les Hommes, Marc Michel Rey, Amsterdam.

Rubra, N. (1970), 'Economic Postscript', in Taylor, J.A. (ed.), Weather Economics, Pergamon, Oxford, pp. 107-19.

Sauer, C.O. (1952), Agricultural Origins and Dispersals, American Geog. Soc., New York, USA.

Simmons, I.G. (1970), 'Land use ecology as a theme in biogeography', Canadian Geographer, XIV, 4, 309-22.

Smith, A.G. (1975), 'Neolithic and Bronze Age landscape changes in Northern Ireland', in Evans, J.G. et al (eds.), The Effect of Man on the Landscape: the Highland Zone, Research Report No. 11, CBA, pp. 64-74.

Street, F.A. (1981), 'Tropical palaeoenvironments', Prog. in Phys. Geog., **5**, No. 2, Edward Arnold, London, pp. 157-85.

Tamm, C.O. (ed.) (1976), Man and the Boreal Forest, Ecol. Bulletin/NER 21, Stockholm, Sweden.

Taylor, J.A. (1949), Mossland Farming in South-West Lancashire, unpublished MA thesis, University of Liverpool, Liverpool, UK.

Taylor, J.A. (1952), 'The relationship of crop distributions to the drift pattern in south-west Lancashire', Trans. IBG., **18**, 77-91.

Taylor, J.A. (1965), 'Implications and Conclusions', in Taylor, J.A. (ed.), Climatic Change with special reference to Wales and its Agriculture, Memo. No. 8, Geog. Dept., UCW, Aberystwyth, Wales, UK, pp. 93-104.

Taylor, J.A. (1967), 'The Shirdley Hill Sands: a study in changing geographical values', in Steel, R.W. and Lawton, R. (eds.), Liverpool Essays in Geography: a Jubilee Collection, Longmans, London, pp. 441-59.

Taylor, J.A. (1974(a)), 'The atmosphere as a resource', in Taylor, J.A. (ed.), Climatic Resources and Economic Activity, David and Charles, Newton Abbott, England, pp. 21-45.

Taylor, J.A. (1974(b)), 'Current problems in Brazilian coffee production', Geogr. Tijdsch., VIII, No. 1, 40-46.

Taylor, J.A. (1974(c)), 'Marginal physical environments', in Jenkins, D.A. (ed.), Marginal Land: Competition or Integration?, Potassium Institute, Bangor, Wales, UK, pp. 10-29.

Taylor, J.A. (1974(d)), 'The ecological basis of resource management', Area, **6**, (2), 101-6.

Taylor, J.A. (1975), 'The role of climatic factors in environmental and cultural changes in prehistoric times', in Evans, J.G. et al (eds.), The Effect of Man on the Landscape: the Highland Zone, CBA Res. Rept. No. 11, pp. 6-19.

Taylor, J.A. (1978), 'The British upland environment and its management', Geography, **63**, (4), 338-53.

Taylor, J.A. (1980), 'Man-environment relationships', in Taylor, J.A. (ed.), Culture and Environment in Prehistoric Wales, British Arch. Reports, Oxford, UK, pp. 311-36.

Taylor, J.A. (ed.) (1980), Culture and Environment in Prehistoric Wales, British Archaeological Reports, Oxford, UK.

Taylor, J.A. and Smith, R.T. (1980), 'The role of pedogenic factors in the initiation of peat formation and the classification of mires', Proc. Sixth Int. Peat Congress, Duluth, USA, pp. 109-18.

Thomas, W.L. (1972), 'The value and relevance of weather study and weather forecasting in the profitable production of early potatoes', in Taylor, J.A. (ed.), Weather Forecasting for Agriculture and Industry, David and Charles, Newton Abbott, England, pp. 86-98.

Trudgill, S.T. and Briggs, D.J. (1977-81), 'Soil and land potential', Prog. in Phys. Geog., **1**, 319-32; **2**, 321-32; **3**, 283-99; **4**, 282-95; **5**, 274-91.

Tyrrell, J.G. (1970), 'A note on the areal patterns in the value of early potato production in south-west Wales, 1967', in Taylor, J.A. (ed.), Weather Economics, Pergamon Press, Oxford, pp. 51-65.

Ucko, P.J. and Dimbleby, G.W. (eds.) (1969), The Domestication and Exploitation of Plants and Animals, Duckworth, London.

Vita-Finzi, C. (1969), 'Geological opportunism', in Ucko, P.J. and Dimbleby, G.W. (eds.), The Domestication and Exploitation of Plants and Animals, Duckworth, London, 31-34.

Westlake, D.F. (1963), 'Comparisons of plant productivity', Biol. Rev., **38**, 385-425.

Woillard, G.M. (1978), 'Grande Pile peat bog: a continuous pollen record for the last 140,000 years', Quat. Res., **9**, 1-21.

Woldenburg, M.J. (1968), 'Energy flow and spatial order: mixed hexagonal hierarchies of central places', Geog. Rev., **58**, 552-74.

Zeuner, A.F.E. (1946), Dating the Past: an Introduction to Geochronology, Methuen, London.

FURTHER READING

Anderson, J.M. (1981), Ecology for Environmental Sciences: Biosphere, Ecosystems and Man, Edward Arnold, London.

Ball, D.F., Radford, G.L. and Williams, W.M. (1983), A Land Characteristic Data Bank for Great Britain, Bangor Occ. Paper No. 13, ITE, Bangor, Wales, UK.

Bayliss-Smith, T.P. (1982), The Ecology of Agricultural Systems, Pp. 112, CUP, Cambridge, UK.

Beckett, R.H.T., Webster, R., McNeil, G.M. and Mitchell, C.V. (1972), 'Terrain evaluation by means of a data bank', Geog. J., **138**, 430-56.

Black, J. (1970), The Dominion of Man: the Search for Ecological Responsibility, Edinburgh University Press, Edinburgh, Scotland and New York, USA.

Bridges, E.M. and Davidson, D.A. (1982), Principles and Applications of Soil Geography, Longmans, London.

Childs, G. (1963), Man Makes Himself, Watts, London.

Chorley, R.J. (1973), 'Geography as human ecology', in Chorley, R.J. (ed.), Directions in Geography, Methuen, London.

Collingbourne, R.H. (1976), 'Radiation and sunshine', in Chandler, T.J. and Gregory, S. (eds.), The Climate of the British Isles, London, **4**, pp. 74-95.

Cruickshank, J.G. and Armstrong, W.J. (1971), 'Soil and agricultural land classification in County Londonderry', Trans. Inst. Brit. Geogrs., No. 53, 79-94.

Davidson, D.A. (1980), Soils and Land Use Planning, Longmans, London.

Dent, D. and Young, A. (1981), Soil Survey and Land Evaluation, Allen and Unwin, London.

Duckham, A.N. and Masefield, G.B. (1970), Farming Systems of the World, Chatto and Windus, London.

Edwards, G. and Wibberley, G.F. (1971), A Land Budget for Britain for the Year 2000, Univ. of London, Wye College Publication, London.

Isard, W. (1969), 'Some notes on the linkage of ecologic and economic systems', Papers and Proc. of the Reg. Sci. Ass., **22**, 85-96.

Isard, W. et al (1972), Ecologic-Economic Analysis for Regional Development, The Free Press, New York, USA.

McCrae, S.G. and Burnham, C.P. (1982), Land Evaluation, OUP, Oxford, England.

Ovington, J.D., Heitkamp, D. and Lawrence, D.B. (1963), 'Plant biomass and productivity of prairie, savanna, oakwood and maize field ecosystems in central Minnesota', Ecology, **44**, 52-63.

Park, C.C. (1979), Ecology and Environmental Management - a Geographical Perspective, Dawson, Folkestone, England.

Parry, M.L. (1978), Climatic Change, Agriculture and Settlement, Dawson, Folkestone, England.

Peters, G.H. (1970), 'Land use studies: a review of the literature with special reference to the applications of cost-benefit analysis', J. Agric. Econ., **XXI**, 171-214.

Rodin, L.E. and Basilevich, N.I. (1968), 'World distribution of plant biomass', in Eckhardt, F.E. (ed.), Functioning of Terrestrial Ecosystems at Primary Production Level, UNESCO, Paris, pp. 45-52.

Rubra, N. (1970), 'Economic Postscript', in Taylor, J.A. (ed.), Weather Economics, Pergamon, Oxford, pp. 107-19.

Sauer, C.O. (1952), Agricultural Originals and Dispersals, American Geog. Soc., New York, USA.

Selman, P.H. (1981), Ecology and Planning: an Introductory Study, George Godwin Ltd., London.

Simmons, I.G. (1979), Biogeography: Natural and Cultural, Edward Arnold, London.

Simmons, I.G. (1981), The Ecology of Natural Resources, Edward Arnold, London.

Taylor, J.A. (1978), 'The British upland environment and its management', Geography, **63**, (4), 338-53.

Taylor, J.A. (1979), Recreation Weather and Climate: a State of the Art Review, SC/SSRC Panel, London.

Taylor, J.A. (ed.) (1980), Culture and Environment in Prehistoric Wales, British Archaeological Reports, Oxford, UK.

Trudgill, S.T. and Briggs, D.J. (1977-81), 'Soil and land potential', Prog. in Phys. Geog., **1**, 319-32; **2**, 321-32; **3**, 283-99; **4**, 282-95; **5**, 274-91.

Wareing, P.F. and Cooper, J.P. (eds.) (1971), Potential Crop Production, Heinemann, London.

Westlake, D.F. (1963), 'Comparisons of plant productivity', Biol. Rev., **38**, 385-425.

Chapter 10

REMOTE SENSING IN BIOGEOGRAPHY

J. P. Darch

INTRODUCTION

Remote sensing is the process whereby information is acquired about the earth or other planets by non-contact methods using electromagnetic energy or seismic, gravitational, vibrational or magnetic forces. More specifically, remote sensing is now regarded as the sensing of electromagnetic energy, particularly visible and infrared light, thermal energy and the microwave or radar regions of the spectrum. For the remote sensing of vegetation the most useful parts of the spectrum are the visible and near infrared: from 0.4 µm to 2.5 µm, but in this spectral region cloud cover can be a major problem so, in areas of the world prone to cloudiness, radar imagery is sometimes favoured.

THE SPECTRAL PROPERTIES OF VEGETATION

A typical plant reflectance spectrum is shown in Figure 10.1. The various peaks and troughs are associated with areas of high reflectance and low reflectance, respectively. The incident radiant energy falling on vegetation can be subdivided into energy reflected back into the atmosphere, energy which is absorbed by the plant and energy that is transmitted through the leaves.

In the visible part of the spectrum (0.4 to 0.7 µm) leaf reflectance is comparatively low, with a peak in the green region, at about 0.5 µm. This green reflectance peak gives vegetation its green colour. Most of the incident visible light falling on vegetation which is not reflected is absorbed by pigments contained in the plant mesophyll layers. Of particular importance are chlorophyll a and b which absorb blue and red light, carotenoids which absorb blue light and xanthophylls which absorb blue and green light. In contrast, in the near infrared part of the spectrum, between 1.1 µm and 2.5 µm, reflectance is high and there is a sharp break between the low reflectance in the visible and the high reflectance in the near infrared at approximately 0.7 µm. This has been called 'the red edge' (Horler et al, 1980, 1983). The red edge is not prone to shifts in its position due to background noise, such as variations in the amount of soil or lignified tissue viewed by the sensor (Dockray, 1981). In the infrared the high reflectance from plants is interrupted by narrow bands where water absorbs the energy; these are centred at 0.85 µm, 0.95 µm, 1.10 µm, 1.40 µm, 1.90 µm and 2.40 µm. Beyond 2.5 µm green plants act virtually as black bodies, i.e. they absorb all of the energy incident on them (Howard, 1970).

The remotely sensed information that can be obtained about vegetation differs according to the part of the spectrum sensed. Between 0.4 and 0.75 µm plant pigments are dominant; between 0.75 and 1.35 µm leaf structural properties control reflectance, and between 1.35 and 2.50 µm leaf water content is most important. Therefore, if one wishes to study moisture stress in vegetation, one would select to study the part of the spectrum most sensitive to plant water content, i.e. the 1.35 to 2.50 µm region. Alternatively, if one were studying leaf structural damage, say as a result of disease, the 0.75 to 1.35 µm region would be of most help.

THE SENSOR SYSTEM

Introduction

The electromagnetic spectrum consists of a variety of energy forms which can be remotely sensed but, because the forms of energy are so diverse, no one remote sensing instrument can sense the whole spectrum. Therefore, the part of the spectrum to be used has to be defined before the sensor system can be chosen. The choice is dependent on a number of parameters which include:

(i) the size of the area to be remotely sensed;

(ii) the minimum ground resolution or scale of the imagery that is acceptable;

(iii) whether cloud cover is likely to be frequent;

(iv) whether repetitive remote sensing imagery is required for environmental change monitoring;

(v) whether the data is required in image or spectral form;

(vi) how much time is available for data collection;

(vii) the costs and money available.

The choice of remotely sensed data currently available is large. One of the major distinctions that can be made is between imaging and non-imaging sensors. All imaging systems, whether they be ground, air or space platforms, provide data in a pictorial fashion, although often the original data are in a digital format. In contrast, non-imaging systems record remotely sensed data as spectra, i.e. as percentage reflectance plotted against wavelength. The latter are very useful for studying the reflectance of small areas in great detail, for example, to distinguish moisture stress in a field of wheat, whereas imaging systems are more applicable for small scale-large area mapping requirements.

Non-Imaging System

One of the most common sets of spectral remote sensing instruments are the radiometers. There are two basic types: (i) spectrophotometers and (ii) spectroradiometers. The former are laboratory bench type instruments which contain their own light source and measure hemispherical reflectance, that is, the reflectance has no angular dependence. These instruments have an extremely small field of view, only being able to measure the reflectance of one square centimeter of a leaf at a time. Therefore, it is often difficult to extrapolate spectrophotometer results to a naturally growing vegetation canopy, even if a large number of leaf samples are taken. This is because vegetation canopy reflectance does not consist of an individual leaf's reflectance; it also includes the reflectance of dead tissue, bark, flowers, fruits, soil, rock and so on. However, spectrophotometers have a role in the remote sensing of natural vegetation. They are often used in controlled laboratory studies, aimed at elucidating the complex plant environment relationships found in the field, such as the differentiation of the effects of moisture stress, salinity stress or heavy metal stress on the basis of plant reflectance, so that the results can be applied to field studies (Gates, 1970; Myers, 1970; Gausman, 1974; Press, 1974; Gausman et al, 1978; Chang and Collins, 1980; Horler et al, 1980). In contrast, spectroradiometers are designed as field instruments which can be handheld, tripod, mast, platform, helicopter or aircraft mounted, according to the field of view required. Spectroradiometers use the sun as their light source, so the reflectance is angularly dependent on both the sun and the sensor angles; such reflectance is called bidirectional reflectance. Examples of particular spectroradiometers and their usefulness in vegetation studies are given by Jackson et al, 1980; Milton, 1980; Tucker et al, 1980; Budd and Milton, 1982. Spectroradiometers cover various sections of the spectrum and they are capable of producing a wide variety of detailed spectral information about a target, both in terms of spectral and spatial resolution. All of the instruments have some form of calibration so that results taken at different times under contrasting lighting conditions are comparable. For example, in the Milton radiometer, calibration is achieved by alternatively allowing the sensor to view the target and a grey card and then ratioing the target and grey card readings. Another example, the Barringer Refspec, uses an integrating sphere to ratio the incoming radiation with the reflected radiation.

Imaging Systems

Imaging systems for remote sensing can be divided into two groups: (i) the photographic systems which use light-sensitive film to record the image, and (ii) non-photographic imaging systems.

Photographic imaging systems. Photographic systems are restricted to the visible and near infrared parts of the spectrum (0.4 to 0.9 µm), that is, they measure only reflected light energy. Photographic imaging systems were the first widely used forms of remote sensing, and the identification and interpretation of aerial photographs for vegetation studies are well documented (American Society of Photogrammetry, 1983). Photographic remote sensing uses various film and filter combinations. The main choices of film are between black and white panchromatic, true colour, false colour infrared and black and white infrared. In addition, the primary colours which comprise visible light can be sensed separately instead of as an entity, and this is done using colour filters to exclude the unwanted light rays. This is called multispectral photography. For vegetation studies, the green, red and reflective infrared wavelengths yield the most information, so filters are used to highlight these light rays, either together or independently, and to exclude blue light. It has been found, for many biogeographical studies, that false colour infrared film gives the best discrimination of vegetation because vegetation has a higher reflectance in the reflective infrared (0.7 to 0.9 µm) than in the visible wavebands. However, the human eye is not sensitive to reflective infrared light so it is only with the aid of specially sensitive film that we can study infrared reflectance. Because the reflective infrared reflectance of vegetation is actually higher than green reflectance, healthy vegetation is red on false colour infrared film. In this part of the spectrum, the reflectance of vegetation is particularly sensitive to changes in plant water content and cellular structure.

Both colour and false colour infrared film have three light-sensitive emulsion layers. The difference between them is that true colour film emulsions are sensitized to blue, green and red light (0.4 to 0.7 µm), while colour infrared film is sensitive to green, red and infrared light (0.5 to 0.9 µm) (Table 10.1). Therefore, on a colour infrared film the blue light is excluded, using a yellow filter (usually a Wratten 12) and the blue emulsion is made sensitive to green light, the green emulsion to red light and the red emulsion to the reflective infrared. No film is sensitive beyond 0.9 µm. One of the main discriminatory features of vegetation on colour infrared film is between healthy chlorophyll-containing plants and leafless, diseased or dead non-chlorophyll-containing plants, as it is the chlorophyll pigments which create the high infrared reflectance of vegetation. For instance, a winter colour infrared photograph of a mixed deciduous-coniferous forest will give an accurate record of the distribution of the leafy, chlorophyll-containing coniferous trees and the leafless deciduous trees.

Table 10.1: Sensitivity of colour and false colour infrared film emulsions

Wavelength (μm)	Light Colour	True Colour Film Emulsions	False Colour Infrared Film Emulsions
0.4 to 0.5	blue	blue	excluded
0.5 to 0.6	green	green	blue
0.6 to 0.7	red	red	green
0.7 to 0.9	reflective infrared	excluded	red

Air photography, taken from either an oblique or a vertical prospective, offers a flexible remote sensing system in terms of wavelengths that can be sensed, scales at which targets can be photographed and times of the day or the year at which photographs are required to be taken. One problem encountered, though, is that none of the reflectance data recorded on film are calibrated to a standard grey scale, which means that quantification of reflectance between scenes, or of one scene photographed on consecutive dates, can be achieved only subjectively.

Non-photographic imaging system. Non-photographic remote sensing systems use electronic energy detectors which have a far greater spectral range than film. Of the non-photographic imaging systems available, three are of particular interest to vegetation studies: (i) the visible and near infrared sensors of the Landsat, Tiros-N and the NOAA satellites and the planned French satellite, SPOT, (ii) airborne multispectral scanners and (iii) satellite or airborne radar systems.

A summary of the satellite-borne visible and near infrared imaging systems are given in Table 10.2. From this table it is evident that each satellite has been designed for different purposes. The number and width of the wavebands, the resolution of each satellite and the speed at which repetitive coverage can be obtained vary with each system. The present generation of satellite imaging systems has a range of sensitivites from 0.3 to 14.0 μm, although few instruments cover the whole of this range. Scene reflectance data are recorded for each multispectral band, and each data set is internally calibrated to a standardised grey scale to allow scene comparisons to be made objectively by statistical analysis. The reflectance values can be reconstructed into images after they are transmitted back to earth from the satellite.

The position and the spectral range of the Landsat 1, 2 and 3 multispectral scanner bands were somewhat arbitrarily chosen for general earth resources studies but it has been found that bands 5 and 7 are the best for vegetation studies. The impetus for the development of the Landsat 4 thematic mapper, launched in July,

Table 10.2: Satellite Imaging Systems

Satellite	Wavebands (μm)	MSS Band Numbers	Resolution	Repetitive Coverage
Landsat 1, 2 and 3 multispectral scanner (MSS)	0.5-0.6	4	79m	18 days
	0.6-0.7	5	79m	18 days
	0.7-0.8	6	79m	18 days
	0.8-1.1	7	79m	18 days
Landsat 3 Return Beam Vidicon (RBV)	0.475-0.575	1	40m	18 days
	0.580-0.680	2	40m	18 days
	0.690-0.830	3	40m	18 days
Landsat 4 MSS	0.5-0.6	1	80m	16 days
	0.6-0.7	2	80m	16 days
	0.7-0.8	3	80m	16 days
	0.8-1.1	4	80m	16 days
Landsat 4 Thematic Mapper (TM)	0.45-0.52	1	30m	16 days
	0.52-0.60	2	30m	16 days
	0.63-0.69	3	30m	16 days
	0.76-0.90	4	30m	16 days
	1.55-1.75	5	30m	16 days
	10.40-12.50	6	120m	16 days
	2.08-2.35	7	30m	16 days
NOAA-6 and 7 Advanced Very High Resolution Radiometer (AVHRR)	0.58-0.68	1	1.1 km	1 day
	0.725-1.10	2	1.1 km	1 day
	3.55-3.93	3	1.1 km	1 day
	10.5-11.5	4	1.1 km	1 day
	11.5-12.5	5	1.1 km	1 day
TIROS-N AVHRR	0.55-0.90	1	1.1 km	1 day
	0.725-1.10	2	1.1 km	1 day
	3.55-3.93	3	1.1 km	1 day
	10.5-11.5	4	1.1 km	1 day
SPOT - Colour mode (to be launched in 1984)	0.50-0.59	1	20m	26 days
	0.61-0.68	2	20m	26 days
	0.79-0.89	3	20m	26 days
SPOT - black and white mode	0.51-0.73		10m	26 days

1982, came from the inadequacies of the position and resolution of the earlier MSS bands, their spatial resolution being too coarse for many purposes. The wavebands on both Landsat 4 and the planned SPOT satellite have been specially chosen for vegetation studies, and the resolution of these sensors is more appropriate to ecological studies than the earlier systems (Anon, 1982; Traizet, 1981). On the NOAA 6 and 7 satellites, bands 6 and 7 of the AVHRR are the best for vegetation mapping (Townshend and Tucker, 1981), and the equivalents on the TIROS-N satellite, bands 1 and 2, are also useful. However, only small scale studies can be conducted as the maximum ground resolution is 1.1 km.

Airborne multispectral scanners operate as line-scanning systems with a rotating mirror which moves the field of view along a scan line perpendicular to the direction of flight. The forward motion of the scan lines is caused by the flight of the aircraft. The system is similar to a satellite multispectral scanner, but the number of MSS bands is generally larger, maybe 10 or 12, the spectral range of each band is somewhat narrower and the range of the spectrum covered is usually wider. Like the satellite MSS data, airborne multispectral scanner data are digitally recorded and calibrated. Its advantages, compared with satellite MSS imagery, are that the flight times, dates and scale of the imagery can be chosen by the individual user.

An increasing amount of valuable biogeographical information is being acquired by air- and space-borne radar sensors. The aircraft-borne systems are called side-looking airborne radar (SLAR) and the space-borne systems, such as the Seasat 1 satellite of the space shuttle, are called synthetic aperture radar (SAR). Both are active remote sensing systems which propagate a radar signal and measure its return. They operate in the microwave region of the spectrum using wavelengths within the 1 mm to 1 m range. Radar imagery has certain advantages over visible and infrared imagery, as it can penetrate cloud, haze and rain. In addition, the system is non-light dependent, so it can operate during the day or night. The images produced by radar remote sensing often look very similar to black and white aerial photographs but the properties of the surface measured are totally different, so care in interpretation is needed. Radar is sensitive to surface roughness and the electrical properties of the surface, the latter being controlled by its dielectric constant or its electrical conductivity. A surface's water content directly affects its electrical conductivity, wet surfaces having a much higher radar return than dry surfaces. Thus, the presence of moisture in soil and vegetation can significantly increase radar reflectivity, i.e. the more moisture, the lighter the tone of the vegetation on the resulting image. In fact, plants are particularly good radar reflectors because of their high moisture content.

REMOTE SENSING OF VEGETATION: A CONSIDERATION OF THE APPROPRIATE REMOTE SENSING SYSTEMS FOR MAPPING VEGETATION AT SELECTED SCALES

Any of the remote sensing systems that have already been discussed can be successfully used for the remote sensing of vegetation. The choice of sensor is great and the selection of the best type of imagery is dependent to a great extent on the scale at which vegetation maps are required. Table 10.2 summarises the resolution of the main satellite systems, the range being from the small scale resolution of 1.1 km of the AVHRR on the NOAA and TIROS-N satellites, the slightly finer 80 m resolution of the Landsat MSS, the 30 m resolution of Landsat 4 thematic mapper and the finer resolutions of airborne imagery. The range of scales at which remotely-sensed data exists encourages a regional approach to the remote sensing of vegetation to be taken, mapping at the formation or association level, but it is equally feasible to map at the plant society level by choosing a sensor that yields images at the appropriate scale.

At the plant formation level, one is concerned with physiognomically distinct, natural vegetation categories which occupy a habitat of constant general characteristics, that is, vegetation which is structurally distinct, such as tropical forest, deciduous forest or grassland. This type of mapping is most appropriately conducted at scales greater than 1:50,000. The two most useful remote sensing systems for use at this scale are the Landsat MSS and the AVHRR. For some vegetation formations, a ground resolution of 1.1 km is too coarse but in many parts of the world, where there has been minimum interference with the vegetation, homogenous plant formations of such dimensions are common, for instance in semi-arid or tropical forest regions. In these instances, mapping of the formation boundaries can be accomplished quite adequately at small scales. In fact, if information on vegetation boundaries only is required small-scale/large-area imagery such as the AVHRR format is more economical than the larger scale imagery. More expensive Landsat imagery will often not yield substantially more information in proportion to its price. In parts of the world where vegetation formations are less extensive, the use of larger scale imagery will be necessary to accomplish the same job. Here, Landsat MSS imagery would be appropriate. At the scale of Landsat MSS imagery, it is possible to map very small vegetation formations, often those that have been substantially modified or dissected by man.

In a case-study in North Wales, where much of the natural vegetation has been modified or replaced by forestry plantations or pasture, an accurate map of the vegetation was constructed using Landsat MSS imagery acquired in spring (Darch, 1982). Five natural or semi-natural formations were identified on the imagery and ascertained by field checking; these were coniferous forest, acid grassland, upland heath, rough grazing and permanent pasture. However, during the study it became evident that Landsat scenes, taken at other times of the year, did not allow the discrimination of all of the known vegetation formations. It was concluded that the degree of discrimination was governed to a large extent by vegetation phenology.

The North Wales study was conducted in an area 15 x 13 km, so it was unlikely that environmental conditions, particularly weather conditions, would influence the interpretability of a scene, but comparisons of the grey scale ranges for the same area in the same month in wet and dry years showed that, under wetter conditions, scene reflectance and the range of grey scale values were lower than in a dry year (Table 10.3).

Table 10.3: Changes in reflectance with local weather conditions

MSS Band	April 1975	April 1978
4	18-51	11-91
grey scale range	33	80
5	13-81	1-105
grey scale range	68	104
6	14-123	0-105
grey scale range	109	151
7	4-130	1-190
grey scale range	126	189
Total rainfall for the month preceding imaging	176.4 cm	71.3 cm

This could be extrapolated to indicate that the classification of vegetation on a whole Landsat scene, which covers 185 x 185 km, or an AVHRR scene measuring 3000 x 3000 km could also be affected by the moisture status of the vegetation. Nevertheless, Townshend and Tucker (1980) drew favourable comparisons for the use of AVHRR for mapping vegetation both in southern Italy and in the Imperial Valley, California. Quantitative comparisons between AVHRR and Landsat MSS imagery also revealed strong correlations between the vegetation communities classifiable on both image types. This suggests that, for the mapping of larger vegetation formations, AVHRR imagery provides a satisfactory remotely sensed data source, where really small formations demand Landsat MSS.

Mapping of plant associations, which are the floristic groups which comprise formations, is best done at larger scales, preferably between 1:10,000 and 1:50,000. Within an association, all of the plants will be of a similar structure so the aim of remote sensing at the association level is to identify the floristic groups that characterise each association. This may best be done by distinguishing the dominant species of the association, for instance oak woodland, Calluna heath. At this scale the tone and the texture of the image are important criteria. The land area occupied by associations varies enormously, so the choice of imagery must to some extent be determined by the unique qualities of the area. Strahler (1981) found that natural vegetation associations in the Klamath National Forest, northern California, could be effectively classified by combining Landsat MSS imagery and automated digital analysis of tone, texture and terrain formation. Both the tone and texture data were derived from geometrically corrected images and the digital terrain model was constructed using elevations, slope angles and aspects extracted from US National Cartographic Information Center digital terrain tapes. All of the data were combined to produce a fourfold plant association classification, on which Douglas Fir, Ponderosa Pine, Red Fir and mixed coniferous associations were identified. However, because the spatial variability of vegetation associations is diverse, problems of mapping at a maximum ground resolution of 79 m arose, as the reflectance values of a pixel are an average reflectance of a segment of ground measuring 79 x 56 m. This averaging of reflectance creates 'mixed pixels' on the boundaries of associations which contain an average reflectance derived from more than one association. Therefore, if the vegetation associations are small or dissected, problems result, indicating that a larger scale of imagery should have been used. Another consideration is that a small association will be represented by only a few pixels and these will not comprise a representative sample. Curran (1979), for instance, found that a 13 hectare site at Shapwick Heath, Somerset, was represented by only 37 pixels, many of which consisted of a mixed reflectance from contrasting associations. Any attempt to classify the associations using Landsat MSS data would have been fruitless.

Few results from the Landsat thematic mapper have yet been published, but this imagery will allow the classification of areas of 2 to 4 hectares, a more appropriate scale at which to map vegetation associations like those at Shapwick Heath in Somerset. A combination of increased ground and spectral resolution, with MSS bands placed in regions of the spectrum that are sensitive to vegetation biomass and health, suggests a vastly improved system of vegetation association mapping from space (Tucker, 1978). The potential of thematic mapper bands 1 to 4 for mapping saltmarsh associations were evaluated by Budd and Milton (1982). They found that, by approximating the wavebands with a portable Milton spectroradiometer, the vegetation associations were distinguishable but the radiometer was not operating at the equivalent resolution to the thematic mapper.

At the scales needed to map vegetation associations, satellite imagery has therefore to some extent superseded air photography, but photography is still useful for studying small associations of less than a hectare, as found by Cole et al (1974).

Plant society and species mapping usually takes place at scales larger than 1:10,000, and studies at this scale still rely heavily on airborne imagery. However, even when using very large-scale images, field studies are required to name individual species. At this scale, remote sensing is best regarded as a confirmatory aid to fieldwork, thus enhancing its efficiency. At very large scales, vegetation societies can be mapped by measuring their spectra with an airborne spectroradiometer. Such studies are particularly useful for distinguishing diseased from healthy plants, or biogeochemical anomalies (Collins, 1976; Collins et al, 1979, 1980). As an example, Myers and Bird (1978) discussed image scale factors in relation to the detection of crown dieback in Australian forests. They found that it was possible to classify individual diseased trees at 1:2000, 1:3000 and 1:4000 but not at 1:6000.

The ultimate choice of the type of remotely-sensed imagery to be used is determined by the size of the communities and the aims of the project. However, in biogeographical work, studies have been traditionally of the smaller-scale, vegetation dynamics; plant ecologists are more concerned with plants at the society or species level. Therefore, for biogeographical studies, the most appropriate forms of imagery are the satellite systems and smaller-scale air photographs. Satellite imagery also has the advantage that repetitive coverage is available on a regular basis so that vegetation changes over time can be regularly monitored. Since the first Landsat satellite was launched in 1972, there should, theoretically, be an accumulated archive of imagery for most parts of the world. However, as the Landsat MSS operates in the visible and near infrared regions of the spectrum, its data acquisition is hampered by cloud cover, the proportion of cloud cover on a scene being inversely proportional to the amount of terrain information. Again, although repetitive coverage is frequent for many parts of the world, particularly in the humid tropical and humid temperate regions, the land is frequently obliterated by cloud. This has led some countries, for example Nigeria, Guatemala and Brazil, to fly radar imagery for vegetation inventories. However, SLAR imagery has a coarser resolution than low or medium-altitude photography so only vegetation formations and associations can be mapped. Nevertheless, in areas with persistent cloud, radar imagery is invaluable (Larin-Alabi, 1978), although Disperati and Keech (1978) found that SLAR imagery was less sensitive for vegetation boundary definition than Landsat MSS imagery or photographs of the Amazon Basin.

THE ROLE OF REMOTE SENSING IN THE DETECTION OF GEOBOTANICAL AND BIOGEOCHEMICAL ANOMALIES

Remote sensing has many different roles in biogeographical studies. In this section and the next its specific use as a mineral exploration tool in vegetated terrain will be analysed. The use of remote sensing in this context is a recent addition to the sciences of geobotany and biogeochemistry. Geobotany is concerned with the affinity of particular plants, plant associations or plant morphological aberrations to soils containing higher than normal contents of heavy metals, such as copper, lead, zinc or tin. Biogeochemistry enables the analysis of plant tissue to assess its heavy metal content. Geobotanical indicator species, plant association changes and biogeochemical anomalies can all be distinguished from the synoptic view of a remote sensing platform: so, during the last fifteen years, remote sensing, as an exploration tool, has become increasingly popular. This development has grown from, and has complemented, the use of satellite imagery and air photography for locating lineaments and faults with which ores are often associated (Abel-Gawad and Tubbesing, 1975; Hodgson, 1975; Liggett and Childs, 1975; Salas, 1975; Smith, 1977; Norman, 1980). Early remote sensing work in this field utilised various forms of air photography. The work of Cole and her colleagues in South Africa and Western Queensland, Australia, is a good example of the experimentation phase of exploration remote sensing, where geobotanical anomalies were successfully located on photographic imagery (Cole et al, 1974). Her photography of the Witvlei area of South Africa highlighted anomalous communities of Helichrysum leptolepis, which led to the discovery of copper-bearing sedimentary rocks. Multispectral photography to detect geochemically-induced reflectance changes were also conducted by Canney (1975) in the Philippines, Thailand, Brazil and the USA, and Røshold (1977) used aerial photography to locate geobotanical anomalies associated with copper in Norway. On black and white photographs the hill-top sites were revealed, while false colour infrared photographs enhanced all of the vegetation anomalies associated with soil copper levels greater than 1000 ppm.

This early remote sensing work, based on aerial photography, precluded the statistical analysis of the waveforms of plant reflectance but some digital classification and manipulation of the data were possible, provided the photography was digitised to a reference grey scale using a densitometer. Various computer package programs could then be used to manipulate the data and to extract information from it (Owen-Jones and Custance, 1974; Owen-Jones and Chandler, 1977). Custance (1974), for instance, adopted the densitometer method for the study of Australian photography to classify geobotanical anomalies and other vegetation associations.

After the launch of Landsat-1 in 1972, satellite imagery became an obvious additional source of data for geobotanical and biogeochemical mineral exploration. In the early years, straightforward image interpretation was employed to locate reflectances associated with ore bodies (Cole and Owen-Jones, 1977) and later, with the advent of easy access to digital image processors, sophisticated digital image analyses of Landsat data were used (Lyon, 1975; Bølviken et al, 1977; Lefevre, 1980; Darch, 1982; Darch and Barber, 1983).

The introduction of high-resolution spectroradiometers to detect plant spectra associated with heavy metals has also yielded results of great significance to mineral exploration in vegetated terrain. Healthy green leaves have a low visible and a high near infrared reflectance (Figure 10.1) while physiological disturbances, such as excesses of heavy metals, cause changes in leaf pigmentation,

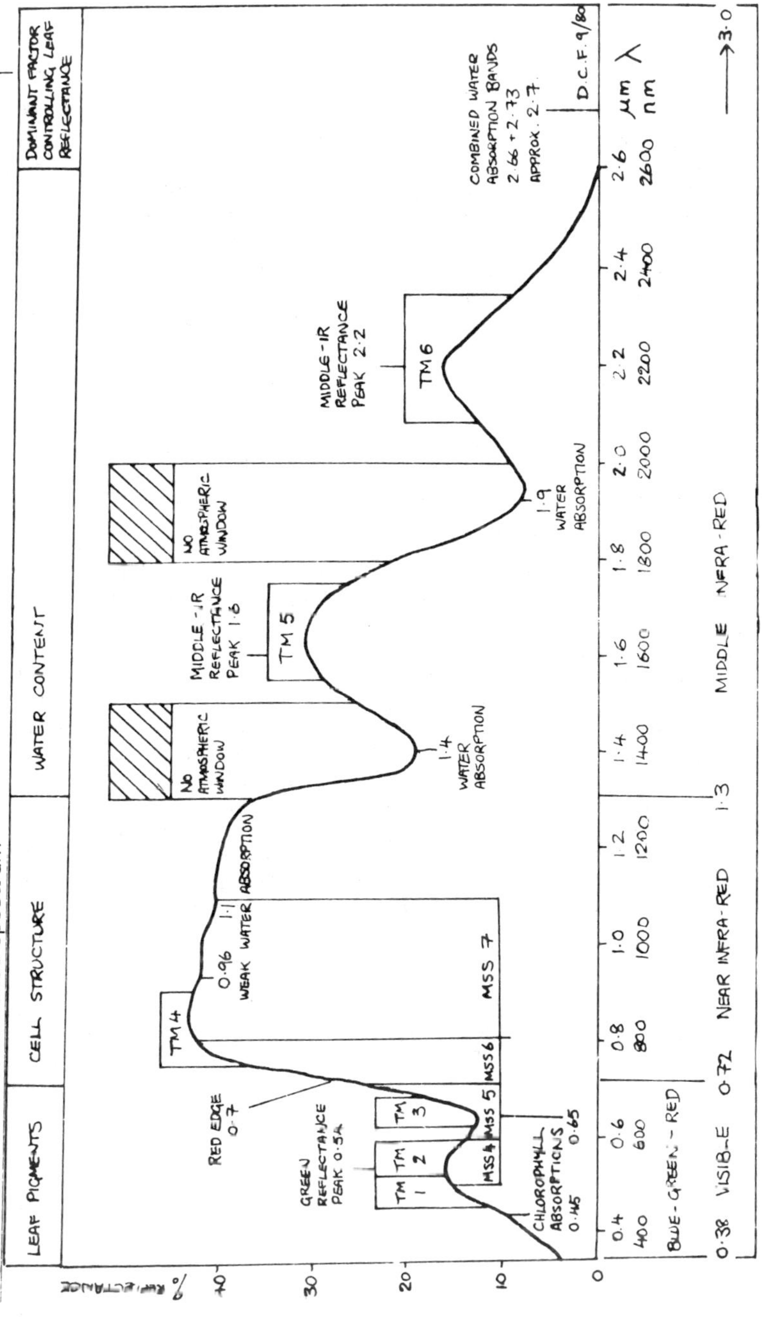

Figure 10.1: Plant Reflectance Spectrum

Table 10.4: The effects of heavy metals on plant reflectance, as measured in the field with a spectroradiometer

Author	Mineral	Species	Decreased Visible	Increased Visible	Decreased Near IR	Increased Near IR
Birnie & Dykstra, 1968	Cu & Mo	*Pinus contorta*		X	No Data	
Canney, Yost & Wenderoth, 1971	Cu & Mo	*Picea balsamea*		X	X	
	Cu & Mo	*Juniperus* sp.		X		
Yost & Wenderoth, 1971	Cu & Mo	*Abies balsamea*		X		X
	Cu & Mo	*Picea rubens*	X		X	
Press, 1974	Pb & Zn	*Quercus* sp.		X	X	
Lyon, 1975	Mo	*Pinus ponderosa*	No Data		X	
	Mo	*Juniperus utahensis*	No Data			X
Yost, 1975	Cu	*Quercus* sp.	X		X	
	Cu	*Pinus ponderosa*	X		X	
	Cu	*Juniperus* sp.	X		X	
Collins & Chiu, 1979	Cu, Pb & Zn	Conifers (species not stated)		X		No Data
Howard *et al*, 1971	Cu	*Pinus ponderosa*	No Effect			X

Table 10.5: The effects of heavy metals on plant reflectance, as measured in the laboratory with a spectrophotometer

Author	Mineral	Species	Decreased Visible	Increased Visible	Decreased Near IR	Increased Near IR
Press, 1974	Pb, Zn	Beans		X		No Change
Horler *et al*, 1980	Cd, Cu	Peas		X	X	
	Zn	Soybean	No Data		X	
Chang & Collins, 1980	Cu, Zn, Ni	Sorghum		X	X	
	Cu, Zn, Ni	Mustard		X	X	

structure and water content which leads to aberrant leaf reflectance. A number of studies have used spectroradiometers to record the changes in plant reflectance over natural geochemical anomalies (Table 10.4). Data collection in these studies has been achieved by elevating a spectroradiometer on a platform above the plant canopy. This position is near-vertical but is not entirely complementary to the true vertical angle achieved by a satellite or aircraft. Nevertheless, the results gained by these studies show a trend towards higher visible reflectance of the anomalous vegetation, together with a change in the near infrared reflectance; the results of Yost and Wenderoth (1971) and Yost (1975) are the exception.

The most successful detection of biogeochemical stress has been achieved with high spectral resolution spectroradiometers. Collins (1976) and Collins et al (1979, 1980) have developed such an instrument which scans in 500 narrow bands between 0.4 µm and 1.1 µm. In addition to allowing studies of reflectance changes in each narrow band, this instrument also allows for an interpretation of reflectance changes that occur on the red edge. It has been shown that the slope of the red edge is determined by chlorophyll concentrations and that chlorophyll concentration is affected by a plant's heavy metal content (Horler et al, 1980, 1983); in particular, an increase in the concentration of heavy metals in plant nutrient environments causes a reduction in chlorophyll production and a correlatable and progressive shift towards the blue part of the spectrum (Barber and Horler, 1980; Chang and Collins, 1980; Dockray, 1982). It follows that, for mineral exploration in vegetated terrain, the red edge is an important feature of plant reflectance, particularly as it is not prone to shifts due to background noise. Collins et al (1977) have already shown that conifers, growing over a metal sulphide-ore body, display a preliminary red edge shift to longer wavelengths, followed by a permanent shift to shorter wavelengths as chlorophyll production ceased. Laboratory experiments using spectrophotometers have also shown that it is possible optically to detect geochemically-stressed plants even when no visual signs of foliage damage are apparent (Table 10.5). But, as Knipling (1970) stated, although studies of single leaves are basic to the understanding of the reflectivity of an entire plant canopy, single leaf measurements are not directly applicable to the natural environment. However, in the studies of geochemical stress, the association of the two types of measurements are complementary.

CASE STUDY: REMOTE SENSING OF THE EFFECTS OF SEASONAL CHANGE IN GEOBOTANICAL AND BIOGEOCHEMICAL ANOMALIES, COED Y BRENIN, NORTH WALES, UK

Background

Many case studies, which have utilised remote sensing as an aid to mineral exploration in vegetated terrain, have adopted an insular approach in that they have made studies of only single sets of imagery rather than taking a multitemporal approach to the problem

Figure 10.2: The Harlech Dome, North Wales

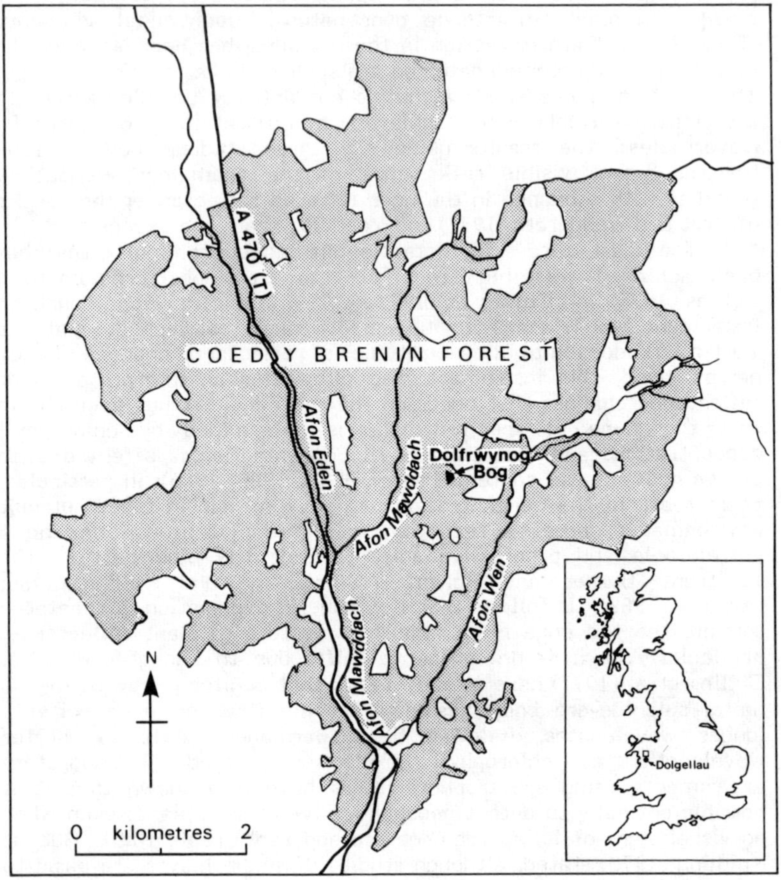

of anomaly detection. However, work by Levine (1975), Canney et al (1979), Lefevre (1980), Schwaller and Tkach (1980), Darch (1982) and Darch and Barber (1983) suggests that, as both the physical and biogeochemical state of plants changes during the year, some seasons may be better suited to the detection of vegetation anomalies by remote sensing than others. The study summarised here (for a more detailed account see Darch, 1982; Darch and Barber, 1983)) was conducted in the copper-mineralised Harlech Dome area of North

Wales (Figure 10.2) where both biogeochemical and geobotanical anomalies occur. The area was covered by four relatively cloud-free Landsat scenes and these were used to monitor changing plant reflectance with time over the main anomaly.

Initial interest in heavy metal cycling in plants and its affects on plant reflectance stemmed from work on cation cycling in various plants, conducted by Guhu and Mitchell (1966); Ruhling and Tyler (1970); Tyler (1971, 1976) and reviewed by Lepp (1979). Earlier work on seasonal copper cycles in pasture grasses was also significant (Beeson and MacDonald, 1951; Thomas, 1952; Wells, 1956; Fleming, 1965). All of these studies showed that copper reached peak concentrations in plant tissue at the beginning of the growing season, and then decreased in concentration with advancing plant maturity. Other heavy metals also showed distinct, but not necessarily similar, cycles. It seems that heavy metal cycling is related to plant phenology. In the case of copper, concentrations are highest just before flowering and then they decline. These temporal changes in plant biogeochemistry and structure may be successfully exploited to locate ore bodies masked by vegetation. The aim of the study in Wales was to test the hypothesis that vegetation anomalies are more easily detectable at particular times of the year, and that detectability is related to the biogeochemical status of the vegetation.

Field Site

Within the Harlech Dome, copper veins are densest in the south-eastern region, and the highest soil copper values have been recorded at the Dolfrwynog Bog (Figure 10.2). Here, total copper values of over 20,000 ppm have been recorded, and available copper levels of 16,000 ppm occur (Smith, 1978; Darch, 1982). The copper indicator, Armeria maritima, occupies the four-hectare site of the bog, but areas of lesser copper toxicity, where soils contain up to 500 ppm total copper, are vegetated with coniferous species or permanent pasture. Detailed descriptions of the site are given by Mehrtens et al (1972), Rice and Sharp (1976), Smith (1978), and Darch (1982), while documentation of the copper anomaly dates is from Davies (1813).

Procedure

The experimental methods adopted bifurcated towards (a) the biogeochemical monitoring of the copper cycles in twelve of the dominant species and (b) the multitemporal remote sensing study. The copper cycles were sampled at bimonthly intervals over the 1981-1982 calendar year, starting in May 1981. Six plant collections were made in that time and both whole plants and their constituent organs were assayed for copper immediately after each collection. The remote sensing part of the study comprised the processing of four Landsat images: 24 April 1975, 28 May 1977, 13 September 1977 and 17 April 1978, by various algorithms, including colour composite analysis, multispectral band ratioing, principal components analysis, density slicing and supervised classification. It was unfortunate that the Landsat images were not all from the same year but, as North Wales lies in a part of the world where cloudy conditions are a recurrent problem in visible and near infrared remote sensing, cloud-free imagery is a rarity.

Results

Each of the twelve dominant plants at the Dolfrwynog Bog displayed distinct copper cycles both on a whole plant basis and organ by organ. The most pronounced cycles were in Armeria maritima, Festuca ovina and Molinia caerulea (Figure 10.3).

Figure 10.3: Copper Cycles in Armeria maritima, Molinia caerulea and Festuca ovina

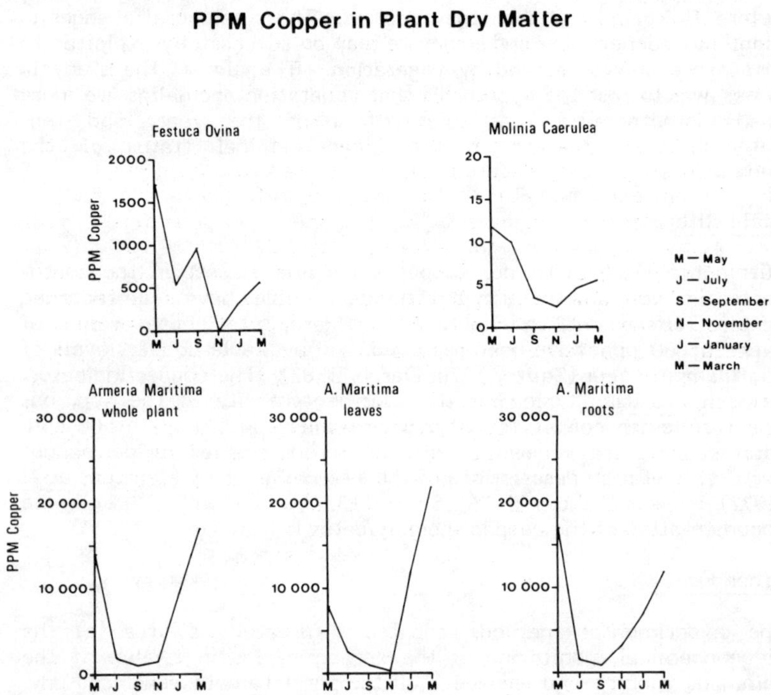

Their copper peaks occurred in May prior to flowering, after which copper levels generally declined. Similar cycles occurred in the Ericaceae but both their copper peaks and times of flowering were a

month later, in June. Mosses and lichens also underwent copper cycling, three mosses and a lichen reaching copper peaks in September and the other moss in January. The plants that peaked in May are of greatest significance because, combined, these three species, A. maritima, F. ovina and M. caerulea, cover 62 per cent of the bog surface and would therefore dominate the average reflectance of the site recorded by a Landsat multispectral scanner.

The results of the analysis of the Landsat colour composites showed that very different amounts of biogeographical information were extractable, depending on the time of year. The Dolfrwynog Bog copper anomaly was most easily distinguished from the rest of the plant associations in the scene in April and May, when the copper levels were highest in the dominant vegetation. In April, the bog was enhanced in a very distinct blue, but on the late May and September imagery, it was not immediately recognisable. These results are encouraging because they indicate that seasonal criteria are important in anomaly location. The results should be interpreted guardedly, however, because the distinct reflectance may be just as attributable to plant phenology as to enhanced copper concentrations. Nevertheless, seasonality is shown to be an important parameter in biogeochemical and geobotanical mineral exploration. This case study illustrates how remote sensing can be successfully integrated with field work to investigate plant/environmental relationships.

In an extension to the study, the reflectance of biogeochemically-stressed permanent pasture and coniferous forest sites were compared with control plots by using the digital reflectance data which were extracted from the Landsat computer compatible tapes. In these studies, too, the greatest contrast in the reflectance occurred in May, although the 'high' copper sites always had a lower reflectance than the control sites, proven at the 95 per cent level of confidence in many cases. The sites used were deliberately small (1000 m^2) so as to avoid changes in relief and aspect and edge pixels. The results, therefore, suggest the greater potential of vegetation anomaly detection that will be possible with finer resolution imagery. However, even Landsat multispectral scanner imagery has potential, although it was not designed for microscale studies, its purpose being originally for large-area small-scale studies.

CONCLUSIONS

The case study examined has illustrated only one of the many applications of remote sensing in biogeography, but it shows that, with improving spatial resolution, biogeographers are tending to move from macroscale to mesoscale and microscale remote sensing as a basis for their studies. This, in the opinion of the author, is because, immediately after the launch of Landsat-1, a trend to use imagery developed, and many researchers thought, optimistically, that remote sensing held the answers to all of their mapping problems. At that time, emphasis was placed on large area mapping of vegetation by the recognition of 'unique spectral signatures' often generated from a system of Landsat diapositives and filters on an additive viewer. Since then we have progressed from a subjective approach towards a pure science methodology where controlled laboratory studies are first used to deduce the reflectance properties of targets before detailed analysis of the reflectance of large areas is attempted. Image interpretation is no longer based on visual analysis alone, but includes the quantitative analysis of the digital reflectance data recorded by the sensor. This advance is still in progress but, for many, it has reached the pure science approach, aided by the increasing ease of access that remote sensers have recently gained to digital image-analysers. In biogeography, remote sensing has, in addition, moved from purely academic to practical applications, such as crop yield monitoring, forest and rangeland inventories, mapping the spread of plant and crop diseases, acting as an early warning system against disease, pest and drought, or the mapping of the spread of algal blooms in lakes and rivers, to mention but a few. Therefore, twelve years after the launch of the first Landsat satellite, remote sensing possesses a hierarchy of sensors that allow a full range of scales of investigation to be undertaken from the coarse resolution of 1.1 km on the AVHRR to the 1 cm resolution of laboratory-based spectrophotometers. Most developments in vegetation remote sensing have concentrated on the visible and near infrared parts of the electromagnetic spectrum, between 0.4 and 2.5 μm, because this region is sensitive to leaf chlorophyll, moisture and structural signals that indicate plant health, biomass and phenology, and which can therefore be used to predict crop yield, spread of disease and plant cover. The use of radar imagery for vegetation studies is not as advanced, even though, in many parts of the world, multitemporal monitoring of vegetation by radar is much more feasible than by visible or near infrared signals. Radar can penetrate cloud and haze, whereas the ground is blotted out for much of the time on visible and near infrared imagery. However, much less is known at present about the interactions of microwave energy and ground targets than about visible and infrared energy. The visible and near infrared are the most appropriate parts of the spectrum to use because of the physiological dependency that plants have on solar energy, so cloudy conditions have to be accommodated when projects are being planned and sensors are being chosen.

REFERENCES

Abdel-Gawad, M. and Tubbesing, L. (1975), 'Mineral target areas in Nevada from geological analysis of Landsat-1 Imagery', NASA Earth Resources Survey Symposium, Houston, Texas, June 1975, Vol. 1-13, pp. 1059-78.

American Society of Photogrammetry, Reeves, R. (ed.) (1983), Manual of Remote Sensing, 2nd edition.

Anon (1982), Landsat-4 Special Article, Remote Sensing Society, Newsletter No. 34, 10-16.

Beeson, K.C. and MacDonald, H.A. (1951), 'Absorption of mineral elements by forage plants: III The relationship of stage of growth to the micronutrient element content of Timothy and some legumes', Agron. Journ., **43**, 589-93.

Birnie, R.W. and Dykstra, J.D. (1978), 'Application of remote sensing to reconnaissance geologic mapping and mineral exploration', Proc. 12th Int. Symp. on Remote Sensing Environment, 20-26 April 1978, Manila, pp. 795-804.

Bølviken, B., Honey, F., Lyon, R. and Prelat, A. (1977), 'Detection of naturally heavy metal poisoned areas by Landsat-1 digital data', J. Geochem. Explor., **8**, 457-71.

Budd, J.T.C. and Milton, E.J. (1982), 'Remote sensing of salt marsh vegetation in the first four proposed thematic mapper bands', Int. J. Remote Sensing, **3** (2), 147-61.

Canney, F.C. (1975), 'Development and application of remote sensing techniques in the search for deposits of copper and other metals in heavily vegetated areas', Status Report, June 1975, US Dept. of Interior, Geol. Survey Project Report (IR NC-48, National Center Investigations, Sioux Falls, South Dakota, USA).

Canney, F.C., Cannon, H.L., Cathrall, J.B. and Robinson, R. (1979), 'Autumn colors, plant diseases and prospecting', Econ. Geol., **74**, 1673-6.

Canney, F.C., Wenderoth, S. and Yost, E. (1971), 'Relationship between vegetation reflectance spectra and soil geochemistry: new data from Cathart Mountain, Maine', 3rd Ann. Earth Resources Prog. Rev., Vol. 1, Geol. and Geogr., Houston, Texas, NASA Johnson Space Flight Center, 18.1-18.9.

Chang, S.H. and Collins, W. (1980), 'Toxic effects of heavy metals on plants', Abstract of 6th Ann. Pecora Symp., Sioux Falls, South Dakota, USA, April 1980, 122.

Chium, H. and Collins, W. (1978), 'A spectroradiometer for airborne remote sensing', Photogram. Eng. and Remote Sensing, **44** (40), 507-17.

Cole, M.M. and Owen-Jones, E.S. (1977), 'The use of Landsat imagery in relation to air survey imagery for terrain analysis in northwest Queensland, Australia', ERTS follow-on programme: study No. 2692D (29650), Final Report, Vols. 1, 2, 3.

Cole, M.M., Owen-Jones, E.S., Beaumont, T.E. and Custance, N.D.E. (1974), 'Recognition and interpretation of spectral signatures of vegetation from aircraft and satellite imagery in western Queensland, Australia', In (1974) European Earth Resources Satellite

Experiments, Proc. Symp., Frascati, Italy, European Space Research Organisation, pp. 243-87.

Cole, M.M., Owen-Jones, E.S. and Custance, N.D.E. (1974), 'Remote sensing in mineral exploration', in Barrett, E.C. and Curtis, L., (eds.), Environmental Remote Sensing - Applications and Achievements, Bristol Symp., Edward Arnold, London, pp. 51-66.

Collins, W. (1976), Spectroradiometric Detection and Mapping of areas enriched in Ferric Iron Minerals using Airborne and Orbiting Instruments, Ph.D Dissertation, Columbia University, USA.

Collins, W., Chang, S.H., Raines, G.L., Canney, F.C. and Ashley, R. (1980), 'Airborne biogeochemical survey test case results', Abstracts of 6th Annual Pecora Symp., Sioux Falls, South Dakota, USA, April 1980, pp. 71-3.

Collins, W. and Chiu, H.Y. (1979), 'Signature evaluation of natural targets using high spectral resolution techniques', Proc. 13th Int. Symp. on Remote Sensing Environment, University of Michigan, USA, 1, pp. 567-74.

Curran, P.J. (1979), Remote Sensing of Vegetation and Soil, Ph.D thesis, University of Bristol, UK.

Curran, P.J. and Milton, E.J. (1983), 'The relationship between the chlorophyll concentration, LAI and reflections of a simple vegetation canopy', Int. J. Remote Sensing, **4** (2), 247-55.

Custance, N.D.E. (1974), The Application of Terrain Classification Techniques to Visible, near Infrared and Thermal Infrared, Ph.D thesis, University of London, UK.

Darch, J.P. (1982), A Remote Sensing Study in Mineralized Vegetated Terrain, Ph.D thesis, University of London, UK.

Darch, J.P. and Barber, J. (1983), 'Multitemporal remote sensing of a geobotanical anomaly', Econ. Geol., **78**, 770-82.

Davies, W. (1813), A General View of the Agricultural and Domestic Economy of North Wales (publisher unknown), London.

Disperati, A.A. and Keech, M.A. (1978), 'The value of using SLAR satellite imagery and aerial photography for a forest survey in the Amazon Basin', in Collins, W.G. and van Genderen, J.L. (eds.) (1978), Remote Sensing Applications in Developing Countries, Remote Sensing Society, pp. 49-55.

Dockray, M. (1981), Verification of a New Method for Determining Chlorophyll Concentrations in Plants by Remote Sensing, M.Sc dissertation, University of London, UK.

Fleming, G.A. (1965), 'Trace elements in plants with particular reference to pasture species', Outlook on Agric., **4**, 270-85.

Gates, D.M. (1970), 'Physical and physiological properties of plants', in (1970) Remote Sensing with Special Reference to Agriculture and Forestry, Nat. Acad. Sci., Washington DC, USA, pp. 224-52.

Gausman, H.W. (1974), 'Leaf reflectance of near infrared', Photogram. Eng., **40**, 183-91.

Gausman, H.W., Escobar, D.E. and Rodriguez, R.R. (1978), 'Effects of stress and pubescence on plant leaf and canopy reflectance', in Hilderbrandt, G. and Boehnel, H.J. (1978), Proc. Int. Sym. Remote Sensing for Observation and Inventory of Earth Resources and the Endangered Environment, July 1978, Freiburg, West Germany, pp. 719-49.

Guha, M.M. and Mitchell, R.L. (1966), 'Trace and major element composition of the leaves of some deciduous trees: II seasonal changes', Plant and Soil, **24**, 90-112.

Hodgson, R.A. (1975), 'Regional linear analysis as a guide to mineral resource exploration using Landsat data', USGS Prof. Paper, **1015**, 155-72.

Horler, D.N.H., Barber, J. and Barringer, A.R. (1980), 'Effects of heavy metals on the absorbance and reflectance spectra of plants', Int. J. Remote Sensing, **1** (2), 121-36.

Horler, D.N.H., Dockray, M. and Barber, J. (1983), 'The red edge of plant leaf reflectance', Int. J. Remote Sensing, **4** (2), 273-88.

Howard, J.A. (1970), Air Photo-Ecology, Faber, London, UK.

Howard, J.A., Watson, R.D. and Hessin, T.D. (1971), 'Spectral reflectance properties of Pinus ponderosa in relation to the copper content of the soil, Malachite Mine, Colorado', Proc. 7th Symp. on Remote Sensing Environment, University of Michigan, USA, pp. 285-96.

Jackson, R.D., Pinter Jr, P.J., Reginato, R.J. and Idso, S.B. (1980), 'Hand-held radiometry, notes for use at the SEA/AR workshop on hand-held radiometry', Phoenix, Arizona, USA, February 1980.

Kimes, D.S. and Kirchner, J.A. (1983), 'Diurnal variations of vegetation and canopy structure', Int. J. Remote Sensing, **4** (2), 257-71.

Knipling, E.B. (1970), 'Physical and physiological basis for the reflectance of visible and near infrared radiation for vegetation', Remote Sensing of Environment, **6**, 155-9.

Larin-Alabi, F.B. (1978), 'Problems of remote sensing in the tropics: an appraisal of the Nigerian situation with regard to forest resources', in Collins, W.G. and van Genderen, J.L. (eds.) (1978), Remote Sensing Applications in Developing Countries, Remote Sensing Society, pp. 57-61.

Lefevre, M.J. (1980), 'Teledetection et anomalies geocliniques a échassières (Allier, France)', Chronique de la Récherché Minière, **453**, 41-63.

Lepp, N.W. (1979), 'Cycling of copper in woodland ecosystems', in Nriagu, J.O. (ed.) (1979), Copper in the Environment, John Wiley, London, Chapter 10, pp. 289-323.

Levine, S. (1975), 'Correlation of ERTS spectra with rock/soil types in California grassland areas', Proc. 10th Int. Symp. on Remote Sensing Environment, University of Michigan, USA, pp. 975-84.

Liggett, M.A. and Childs, J.F. (1975), 'An application of satellite imagery to mineral exploration', USGS Prof. Paper, **1015**, 253-70.

Lillesand, T.M. and Kiefer, R.W. (1979), Remote Sensing and Image Interpretation, John Wiley, New York, USA.

Lyon, R.J.P. (1975), 'Correlation between ground metal analysis, vegetation reflectance and ERTS brightness over a molybdenum skarm deposit, Pine Nut Mountains, Western Nevada', Proc. 10th Int. Symp. on Remote Sensing Environment, University of Michigan, USA, pp. 1031-44.

Mehrtens, M.B., Tooms, J.S. and Troup, A.G. (1972), 'Some aspects of geochemical dispersion from base-metal mineralisation within glaciated terrain in North Wales and British Columbia, Canada', in Jones M.J. (ed.) (1972), Geochemical Exploration, Inst. Min. Metall., London, UK, pp. 105-15.

Milton, E.J. (1980), 'A portable multiband radiometer for ground data collection in remote sensing', Int. J. Remote Sensing, **1**, 153.

Myers, V.I. (1970), 'Soil, water and plant relations', in (1970) Remote Sensing with Special Reference to Agriculture and Forestry, Nat. Acad. Sci., Washington, USA, pp. 253-97.

Myers, B.J. and Bird, T. (1978), 'Detection of a crown dieback in Australian eucalypt forests on large scale aerial photographs', in (1978) Symp. on Remote Sensing for Vegetation Damage Assessment, Amer. Soc. Photogramm., pp. 291-97.

Norman, J.W. (1980), 'Causes of some old crustal failure zones interpreted from Landsat images and their significance in regional mineral exploration', Trans. Inst. Min. Metall., Sect. B, **89**, 63-72.

Owen-Jones, E.S. and Chandler, B.J. (1977), 'The use of photographic imagery in earth resources studies', Journ. Brit. Interplan. Soc., **30** (5), 163-7.

Owen-Jones, E.S. and Custance, N.D.E. (1974), 'Digitized analysis of Skylark rocket imagery', Journ. Brit. Interplan. Soc., **27** (1), 18-22.

Press, N.P. (1974), 'Detecting the toxic effects of metals on vegetation from earth observation satellites', Journ. Brit. Interplan. Soc., **27** (5), 373-84.

Puckett, K.J., Niebor, E., Gorzynski, M.J. and Richardson, D.H.S. (1973), 'The uptake of metal ions by lichens: A modified ion-exchange process', New Phytol., **72**, 329-42.

Rice, R. and Sharp, G.J. (1976), 'Copper mineralization in the forest of Coed y Brenin, North Wales', Trans. Inst. Min. Metall., Sect. B, **85**, 1-13.

Røshold, B. (1977), 'Case history of copper mineralization within naturally copper poisoned areas at Raitevarre, Karasjok, Finnmark County, Norway', in (1977) Prospecting in Areas of Glaciated Terrain, Institute of Mining Metallurgy, London, UK, pp. 138-39.

Ruhling, A. and Tyler, G. (1970), 'Sorption and retention of heavy metals in the woodland moss Hylocomium spendens (Hedw) Br. et Sch.', Oikos, **21**, 92-7.

Salas, G.P. (1975), 'Relationship of mineral resources to linear features in Mexico as determined from Landsat data', USGS Prof. Paper, **1015**, 61-74.

Schwaller, M.R. and Tkach, S.J. (1980), 'Premature leaf senescence as an indicator for geobotanical prospecting with remote sensing techniques', Proc. 14th Int. Confr. Remote Sensing Environment, Costa Rica, 347-58.

Smith, R.F. (1978), Geographical Factors influencing the Distribution of Heavy Metal Tolerant Indicator Species in Parts of the United Kingdom and Europe, Ph.D thesis, University of London, UK.

Smith, W.L. (1977), 'Remote sensing applications for mineral resources', in Smith, W.L. (ed.) (1977), Remote Sensing Applications for Mineral Exploration, Dowden, Hutchinson and Ross Inc., Stroudsburg, Pennsylvania, USA, pp. 73-98.

Strahler, A.H. (1981), 'Stratification of natural vegetation for forest and rangeland inventory using Landsat digital imagery and collateral data', Int. J. Remote Sensing, **2** (1), 15-41.

Thomas, B., Thompson, A., Oyenugr, V.A. and Armstrong, A.H. (1952), 'The ash constituents of some herbage plants at different stages of maturity', Emp. J. Exp. Agric., **20**, 10-22.

Townshend, J.R.G. and Tucker, C.J. (1981), 'The utility of AVHRR of NOAA 6 and 7 for vegetation mapping. Matching remote sensing technologies and their applications', Proc. 9th Ann. Confr. Remote Sensing Society, London, UK, pp. 97-109.

Traizet, M. (1981), 'SPOT links with remote sensing users. Matching remote sensing technologies and their application', Proc. 9th Ann. Confr. Remote Sensing Society, London, UK, pp. 77-80.

Tucker, C.J. (1978), 'A comparison of satellite sensor bands for vegetation monitoring', Photogramm. Eng. and Remote Sensing, **44** (1), 1369-80.

Tucker, C.J., Jones, W.H., Kley, W.A. and Sundstrφm, G.J. (1981), 'A three-band hand-held radiometer for field use', Science, **211** (4479), 281-3.

Tyler, G. (1971), 'Distribution and turnover of organic matter and minerals in a shore meadow ecosystem: Studies in the ecology of Baltic seashore meadows - IV, Oikos, **22**, 265-91.

Tyler, G. (1976), 'Soil factors controlling metal ion absorption in the wood anemone, Anemone nemorosa', Oikos, **27**, 71-80.

Wells, N. (1956), 'Soil studies using sweet vernal grass to assess element availability, Pt. 1, Preliminary investigations', New Zealand Journ. Sci. Tech. B, **37**, 473.

Yost, E. (1975), Multispectral Color Photography for Mineral Exploration by Remote Sensing of Biogeochemical Anomalies, NASA-CR-144811, N77-10606, Washington DC, USA.

Yost, E. and Wenderoth, S. (1971), 'The reflectance spectra of mineralized trees', Proc. 7th Int. Symp. on Remote Sensing Environment, University of Michigan, USA, pp. 269-84.

FURTHER READING

Barrett, E.C. and Curtis, L.F. (1982), Introduction to Environmental Remote Sensing, 2nd edition, Associated Book Publishers Ltd., London.

Brooks, R.R. (1972), Geobotany and Biogeochemistry in Mineral Exploration, Harper and Row, New York, USA.

Curran, P. (1980), 'Multispectral remote sensing of vegetation amount', Progress in Physical Geography, **2** (3), 315-41.

Duggin, M.J. (1980), 'The field measurement of reflectance factors', Photogrammetric Engineering and Remote Sensing, **46** (5), 643-7.

Estes, J.E. and Senger, L.W. (1974), Remote Sensing - Techniques for Environmental Analysis, Hamilton Publishing Co., Santa Barbara, California, USA.

Fuller, R.M. (ed.) (1983), Ecological Mapping from Ground, Air and Space, ITE Symposium No. 10, Natural Environmental Research Council, London.

Gausman, H.W. (1977), 'Reflectance of leaf components', Remote Sensing Environment, **6** (1), 1-9.

Holz, R.K. (1973), The Surveillant Science - Remote Sensing of Environment, Houghton Mifflen Co., Boston, Mass., USA.

Lintz, J. and Simmonett, D.S. (1976), Remote Sensing of Environment, Addison-Wesley Publishing Co., Reading, Mass., USA.

Sabins, F.F. (1978), Remote Sensing - Principles and Interpretation, W.H. Freeman and Co., San Francisco, USA.

Swain, P.H. and David, S.M. (eds.) (1978), Remote Sensing: The Quantitative Approach, McGraw-Hill, International Book Company, New York, USA.

Chapter 11

BIOGEOGRAPHY: HERITAGE AND CHALLENGE

J. A. Taylor

INTERDISCIPLINARY SUB-FIELDS

Biogeography is associated with both the geographical (Simmons, 1980) and the biological (Vuilleumier, 1977) sciences which continue to nourish its development. It has also acquired a separate identity on its own terms, not only within geography but also within biology. Thus, as a two-way, interdisciplinary sub-field, it has been able to develop reciprocally within the orbits of the parent sciences with which it shares some common, if constantly changing, ground. There is nothing unique in these interdisciplinary borderlands which are essentially a means of expansion or contraction of adjacent sciences. At its positive best, such adjacency sparks off innovation and creates new hybrid sciences. At its negative worst, such proximity may, through academic power politics or the ill matching of the protagonists, generate controversy, prejudice, stagnation and even the extinction of the weaker science or selected sub-fields within it.

The splitting of the atom owed much to the combined research of physicists and, later on, chemists. The injection of chemistry into biology has created the biochemical sciences and new penetrations in the field of microbiology to interpret relationships between environmental signals and physiological responses in plants and animals. Geological research has been transformed by the introduction of geochemical and geophysical techniques. Civil engineering and geomorphology have together advanced hydrological research and teaching, not least in geography departments (Ward, 1980) almost the extinction of the flame of traditional Wooldridgian geomorphology. Geomorphologists have emigrated in large numbers from the Tertiary to the Quaternary scale of reference where a veritable multitude of palaeoenvironmental scientists (including a large contingent of British biogeographers (Simmons, 1980)) congregate and successfully promulgate the multidisciplinary Quaternary sciences (Chorley, 1971; Brown, 1975; Clayton, 1980).

In contrast, climatology continues to live dangerously, perhaps sinfully, with meteorology. Atmospheric physicists have achieved more rapport with biologists, agonomists and environmental scientists than with climatologists, with the exception of a very important few (Atkinson, 1980). Consequently, the burgeoning and critical field of climate change (Parry, 1978; Munn and Machta, 1979; Smagorinsky, 1981), especially recent manifestations and trends, appears to engender more scepticism from the meteorologists than collaboration (Mason, 1976 and 1982). Pedology, too, exhibits continual turbulence as Soil Survey hovers between taking international taxonomy too far for field survey purposes, on the one hand, and mixing Rothschild-derived, force-fed research programmes, with the target of achieving a national soil survey coverage within a few years, on the other. Academic pedology has found diverse lodgings in departments of agriculture, biochemistry and geography but prospers best when a wide range of specialists, e.g. soil chemists, soil physicists, soil biologists, soil palynologists, etc. apply their expertise to the soil medium under one roof. But the fact remains that in geography departments at least, soils are best taught, understood and researched as integral parts of the physical and cultural landscapes with which they have evolved.

The buoyancy and longevity of an interdisciplinary sub-field is guaranteed if its practitioners are equipped, collectively as well as individually, with expertise in both adjacent sciences. It is over-qualification and overexperience in the one rather than the other which can create the undesirable ill balance and malformed growth of the sub-field. This may win scientific acceptability from the one science but invite professional hostility from the other.

Inevitably, interdisciplinary sub-fields tend to be more controversial and often charismatic than many long-established core disciplines, and their practitioners are heavily committed to the scientific demands of at least two major study areas. But Chaucer's oft-quoted _cri de coeur_: 'The lyf so short - the craft so long to lern' must not be used as an excuse for not attempting a definition of any busy sub-field, let alone biogeography itself.

THE IDENTITY OF BIOGEOGRAPHY

Subjects and their sub-fields evolve continuously over time and experience episodes of convergence or divergence and sequences of ever-changing technology and affiliation. At any time, the evolutionary stage attained is reflected in the research outputs and teaching profiles of the contemporary practitioners involved. For the achievements of British biogeography or, more properly, British biogeographers, Simmons (1980) has usefully compiled a comprehensive survey but is forced to conclude that there is no such thing as 'a distinctive British school of biogeography'. However, he adds that 'it is scarcely possible to conceive of our teaching without the material on natural ecosystems and biota, and man's impact upon them and our attitudes to man-biota relationships'. In fact, therein lie the _three_ basic ingredients for a definition of the scope of biogeography as practised by geographers: (i) natural ecosystems, (ii) how they have changed by man and (iii) the relationships between (i) and (ii). Simmons' two major texts (1974 and 1979) detail his central concept of biogeography which may be expressed simply as an evaluation of the modification of natural ecosystems by man's activities. Simmons' material is drawn extensively from the biological rather than geographical literature. The shortage of biogeographical textbooks, once bemoaned by Oldfield (1967), has since been partially corrected. Watts (1971) presented the first major comprehensive study (since Dansereau (1957) and Polunin (1960)) of the principles of biogeography as focussed in ecosystem mechanics. Pears (1977) presented a 'basic' volume derived very much from the experience of teaching biogeography to undergraduate geographers but previously Seddon (1971) and Cox et al (1976), and later Jones (1980), all produced texts more oriented to the style of biogeography teaching in biology departments. Jones' earlier chapters smack of the kind of traditional, global paleobotanical (and palaeozoological) approach which characterises the type of palaeoecology researched and taught by biologists (Gould, 1981). A recurrent theme within these numerous texts, since Eyre (1964) and Stoddart (1965, 1967) promoted its importation from ecology, is the ecosystem, first proposed by Tansley in 1935, and embodying, in one concept or model, the flora, fauna, man and environment and their collective interactions (see Chapter 4 by Moss herein). Kellman (1980) and Stott (1981) have reacted to this excessive regurgitation of the one idea. Stott elaborates his views in Chapter 1 of this volume, to which we must now refer to trace the roots of biogeographical studies from the last century through to understand the stage they have reached in the late twentieth century.

Stott concludes, as does Simmons (1980), in effect, 'that there can be no such thing as the 'definitive' approach or the 'final' biogeography, implying that an eventual explicit identity will remain illusory. In that context, the complexity of the history of biogeography reflects its current dilemmas and debates. Prior to the impact of Fleure (1951) and Newbigin (1936), both zoologists, on pre-World War II developments in British biogeography, the pioneering of the subject was the exclusive domain of biologists, including some with a remarkable talent for early theorising and synthesis and also with an acute environmental awareness. Alexander von Humboldt (1769-1859) is regarded as one of the founder fathers of geography, as an explicit study, and also of biogeography as a systematic study (Humboldt and Bonpland, 1805).

The impact of Darwinism on later geographical thought has been summarised by Stoddart (1967) in cementing ideas on evolutionary timescale changes, competition between species, and the coexistence of both association and randomness in nature. Haeckel's broad concept of an ecology involving interactive organisms and environment, anticipated the expanded scope of modern biogeography as it is studied in geography departments, where it was natural (de Candolle's (1882) visionary work apart) that, following de Martonne (1932), Fleure (1951), Sauer (1952), Morgan and Moss (1965), Eyre and Jones (1966), Harris (1969) and Taylor (ed.) (1980), for example, the study of man should be integrated with the study of plants and animals in their respective ecological settings.

The distinctive commitment to the taxonomy and functioning of the 'raw materials' of biogeography, i.e. the plants and animals themselves, has remained, and does remain, very much the prerogative of departments of botany and zoology. Brown (1975) compares the relatively easy and accessible vocabulary of geomorphology with the more demanding and detailed vocabularies of plant and animal taxonomy, including simple identification. Yet not all biogeographical research insists on that expertise, as will emerge later.

It is essential to point out that British university geography was introduced by a diverse range of non-geographical specialists from both the arts and sciences, including Fleure (zoologist), Roxby (historian), Stamp (geologist - originally!), Miller (geologist) etc. in the early decades of this century and at least three or four academic generations ago. These were the 'first-generation' geographers (Fisher, 1959). The general objective was to achieve an integrated study of man and environment as expressed in traditional regional studies. Again, although biogeography achieved some representation before 1939, it languished in the 1940s and 1950s, as geomorphology assumed its vast dominance of physical geography but became less and less adaptable as the 'physical basis' of the subject in its human context (Martin, 1951). Intense specialisation in aspects of physical and human geography was also rampant at this stage, thus creating a divided, compartmentalised subject, and much disillusionment with integrated regional approaches. Consequently, geography was in no condition to respond to, and capitalise on, the remarkable reawakening of environmental awareness dating from the mid-60s, an awareness which was to lead to the creation of Environmental Science departments at several, in particular the newer, universities. Here, potentially, was the perfect arena for biogeographical studies to prosper but it was mostly biologists, earth scientists and atmospheric scientists who foregathered at Lancaster, Norwich and Coleraine. However, internally, within established geography departments a minority of specialists, coming in, as Oldfield (1967) explained, from the 'pedological, geomorphological or climatological margins' (of biogeography), initiated and expanded biogeography options from about the mid-1960s. These 'first generation' biogeographers have now produced an increasing number of 'second generation' specialists, among whom it is obviously premature to look for a common philosophy (Simmons, 1980; Stott, Chapter 1 herein).

Nonetheless, recent (1976-81) surveys, reported in Simmons (1980), reveal that two major and related interests dominate contemporary British biogeography. These are (i) the study of palaeoenvironments and (ii) the ecological basis of land-use which, taken together, create a focus on a broad-based kind of environmental history and on the ecological condition of the derived, modern environment: in other words, environmental and man-induced changes over time, their successive cyclical stages, their impacts and their management. Palaeoecologists, in particular those trained in the Cambridge sub-department of Quaternary Studies, pioneered pollen-analytical research in Britain in the '40s and '50s, and it is significant that Godwin (1975) subtitles his classic work on the history of the British flora 'A factual basis for phytogeography'.

Biogeography then is more than the study of the distributions of flora and fauna. Yet the spatial dimension has varied remarkably in its importance to geographical, as distinct from biological, investigations. 'Geographical' to the non-geographer means spatial, fundamentally, and therefore distributional and, *ipso facto*, locational. But, to the geographer, the need to explain distributions involves the full range of environmental and cultural factors that may impinge on those distributions. To the geographer, then, 'geographical' is as muuch aspatial as spatial and is ultimately and embarrassingly encyclopedic in its terms of reference. To the biologist, 'geographical', traditionally and persistently, means spatial and often global, and his explanations of the development of distributions (the current Darwinist/cladist controversy aside (see Stott, Chapter 1 herein)) are given in evolutionary rather than ecological terms but, initially, with the exclusion of 'the complex subject of man' (Newbigin, 1936). It follows that biogeography for geographers is destined to be an expanded combination of plant ecology, animal ecology and human ecology, whilst biogeography for biologists is essentially a spatial platform for the debate on the manner in which modern plant and animal distributions are related to their antecedents, in both biological and geological terms.

Hare (1969) has remarked how developments in geography (and physical geography in particular) has been persistently out-of-step with scientific fashion and trend during this century. The geologists and historians, etc. who founded the first departments of geography in the inter-war period, promoted integrated, cross-Faculty studies but the nineteenth-century fixation for separating the arts from the sciences was generally to persist. Geographical studies were given a major fillip by the Second World War, not least because of the need for detailed, accurate and up-to-date geographical information for intelligence purposes. The new technique of air-photo interpretation was pioneered but it did not flourish in geography departments until the '50s. 'Second-generation' (Fisher, 1959) geographers became committed to increasingly intense specialisation, especially in geomorphology, historical and urban geography, and the stage was set for the decline, slow at first and very rapid later, of integrated, including regional, studies. In the meanwhile, the statistical revolution which encompassed the biological sciences in the '50s could not penetrate into geography until the '60s. Physical geography, compartmentalised, divided and no longer liaising satisfactorily with human geography, was ill-prepared for the resurgence of interest from the late '60s in integrated studies yet again in the guise of the newly dubbed 'environmental' sciences. Human geography, in the meantime, through the auspices of Harvey (1969), Haggett (1965) and Hall (1970) *et al*, went through a fundamental reappraisal in the late '60s and '70s and has emerged with a well developed, if also well debated, theoretical base (Johnston, 1983). Physical geography, including biogeography, still awaits this methodological and philosophical renaissance (Haines-

Young and Petch, 1980). Nonetheless, biogeography, by virtue of its biological data source, together with climatology and hydrology, has embraced the full range of statistical techniques now available, although the 'opportunist' or ad hoc nature of much geomorphological, palaeoenvironmental and pedological field data means that strict sampling procedures for site selection are of limited application, and data sets are often subjectively derived and not robust enough for statistical treatment. The analysis of pollen diagrams, relating to the prescribed and self-selecting locations of organic deposits, is a nice case in point. Techniques like cluster analysis may achieve unbiassed zonations, and computer programmes are available to take the labour out of pollen diagram construction, but it is a moot point whether 'eye-balled' interpretations are in any sense extended but merely confirmed.

Again, since the 1950s, physical geography (and human geography) has become obsessed with studies of process rather than form (the latter being the focus of the '50s), and a switch in emphasis to the time-scale from the space-scale was inevitable. Moreover, human geographers, looking for human explanations of human data, adopted the featureless isotropic plane as the theoretical platform on which their human distributions are displayed, thus symbolising the complete demise of the one-time sacred 'physical basis of geography'. In the meanwhile, economists and sociologists have tended to adopt their own spatial and regional approaches, especially when concerned with planning problems, and regional synthesis survives under the umbrella of new agglomerations styled as 'area studies' or 'development studies'. 'Regional Science' now applies mathematical measures to spatial distributions. Geography, thus, retains the doubtful privilege of taking the wrong direction at the wrong time.

Biogeography, however, has been fortunate in two respects. First, it reasserted itself in geography departments at the time of the environmental reawakening in the 1960s (but could certainly have contributed more than it did in the '70s to environmental research (Simmons, 1980)). Second, it has enjoyed continuity of development in biology departments, being the beneficiary not only of a prompt statistical transformation in the '50s but also of a theoretical renaissance in the '70s, albeit now clouded in fundamental controversy in the '80s (Stott, Chapter 1; Watts, Chapter 2, Flenley, Chapter 3, herein). At the same time it must be conceded that the kind of biogeography which has been nurtured in geography departments over the last two decades has been so obsessed with applications of the ecosystem concept, on the one hand, and palaeoenvironmental reconstructions, on the other, that the response to the 'new biogeography' being so fiercely debated by the biologists has been muted. As Edwards (1982) has observed, such selective, narrow and fragmentary specialisation cannot be maintained indefinitely.

THEORETICAL BIOGEOGRAPHY

Stoddart (1965, 1966, 1967, 1978, 1981 and 1982) has made regular appraisals of recent developments in biogeography with particular reference to theoretical advances. He quotes extensively from the biological literature, in particular the zoological sources from which the major innovatory thrust has recently come (May (ed.), 1981).

MacArthur and Wilson's (1967) pioneer development of the equilibrium theory of island biogeography, although criticised and heavily attacked in particular by Gilbert (1980), has been constructively reviewed by Diamond and May (1981) and Higgs (1981) with particular reference to the design of nature reserves, an application which Gilbert regards as premature. Islands are naturally well defined, biophysical units or prescribed theatres of biological operation with measurable and controllable dimensions, discrete locations and accessibilities, and less complex histories than is usual for less well circumscribed habitats. Moreover, the island concept is equally applicable to any geographically isolated area, e.g. a lake, a mountain peak, a peatbog or the detail of the type of mosaic landscapes developed, for example, in periglacial environments, deciduous forest clearings, or within the savannas, to name but three. The equilibrium theory of island biogeography is based on the postulate that a dynamic relationship exists between (1) the species arriving and colonising, which is a function of distance of the island from source, and (2) species becoming extinct, which is a function of island area. There are logistic problems in data assembly. Can one differentiate between resident, seasonal and chance species? Are simple positive sightings reliable population indicators? Are varying arithmetic totals of parallel varying ecological singificance? Extinction as such is a difficult stage to define. Again, the physiography and habitat potential of islands varies spatially and also eventually over time, e.g. if climate changes or ocean currents deviate.

But is not the modification and eventual replacement of hypotheses of the very essence of progress in the scientific method? Ball (1975) calls for this development by the introduction of hypothetico-deductive procedures <u>à la</u> Popper (1968) to supplement the empirical-descriptive (i.e. the establishment of distributions and especially discontinuities) and the so-called 'narrative' (explaining the distributions) methods already well practised. Cracraft (1975) makes a parallel three-fold recommendation. Much biogeographical work requires to progress from the inductive to the deductive and, ultimately, the predictive (Haines-Young and Petch, 1980). The current Darwinist/cladist debate is very intense and often <u>ad hominem</u> between the (a) <u>traditional school</u>, which assumes <u>centres-of-origin</u>, dispersal or migration tracks, and the availability and applicability of many different explanations, e.g. continental drift and <u>not</u> vicariance <u>alone</u>, and (b) the <u>new school</u> (Croizat, 1958, 1964, 1973, 1981; Nelson, 1973; Rosen, 1975; Nelson and Rosen, 1981), which assumes only the one explanation, i.e. the fragmentation of antecedent distributions (i.e. <u>vicariance</u>), generalised 'tracking' of species and continental movements. As Stoddart (1981) comments, biogeography would be 'weakened by neglecting either school of thought'.

Equally ripe for revision are the traditional ecological concepts of 'succession' and 'climax' (Flenley, Chapter 3 herein; Horn, 1981(a)). Clementsian succession is rare in the real world and palaeoecological work is revealing that what was thought to be the largest, established and most stable of the vegetation climax formations, the tropical rain forests, is vulnerable to climatological, pedological and physiological changes over the short-term and over relatively short distances (Taylor, 1975; Flenley, 1979; Newman (ed.), 1982). Whitmore _et al_ (1982) recommend that the tropics, rather than, conventionally, the temperate latitudes, might be considered as the _norm_ in both biogeographical and geomorphological studies. The uniformitarian-inspired, nineteenth century-style, latitudinal zonations of vegetation and soils may now melt away into more detailed calibrations over time and space of Quaternary environmental oscillations. Multivariate statistics enables the analysis and interpretation of massive if localised data sets. Horn (1981a) refers to the representation of succession as a Markovian replacement process. The probability of plants being replaced in a specific time period can be estimated and future community compositions can be predicted. He suggests that, where successional convergence occurs, it is a statistical phenomenon rather than a biological one. Again, as Matthews (1979) has shown, plant succession may be divergent. Linear models have proved most successful, especially for predicting forest successions, but many interesting non-linearities must occur due to varying sensitivities to disturbance, for example.

The climax theory, a symptom of the old, unicycle, ternary model, must also subside with the decline of the parallel and analogous concepts of 'grade' in river profile development, 'maturity' of soil profile development and long-term 'stability' of climate regimes. These concepts were developed and applied in the context of 'closed systems'. Following Stoddart (1965 and 1967), Chorley (1971) and Bennett and Chorley (1978), the 'dynamic equilibrium' concept may now be applied in the 'open-system' context, allowing the operation of continuous change in both process and form. Oldfield (1983) has proposed a 'steady state' model of ecosystem change as an additional alternative to the standard successional and cyclic models.

Biogeographers, very much a minority group in geography departments, have been concentrating on extending the research frontiers in the study of British (mainly but with some overseas investigation) vegetation and environmental history. The parallel reconstructions of edaphic, climatic, topographic and hydrological histories and related archaeological and historic successions _per se_ (e.g. Curtis and Simmons (eds.), 1976; Taylor (ed.), 1980) are testimony to the broader expertise in palaeoenvironmental science which physical geographers are usually more able to apply than palaeoecologists, among whom, nonetheless, is often displayed an equally penetrating environmental awareness (Dimbleby, 1962; Smith, A.G., 1970; Birks, 1977). As with the spatial dimension, geographers have no automatic case for monopoly or even priority in expertise, so also it is apparent that palaeoclimates, palaeosols, palaeohydrologies demand the combined interpretations of palaeoenvironmentalists of different training and backgrounds. But systematic training in a number of specialisms within physical geography is especially valuable in this context.

Ironically, the global space dimension, now adjustable in the light of plate tectonics, has provided the ideological battle ground for the zoologists. The new biogeography incorporates welcome contemporary theories on the short-term and episodic nature of climatic and geological changes, and updated Darwinism provides a temporal framework for theoretical comparison and context. Biogeographers in geography departments must work now towards their own hypothetico-deductive models to calibrate and predict the environmental changes with which they are so obsessed, and then tackle the even more difficult problem of environmental management both in theory and practice.

APPLIED BIOGEOGRAPHY: WITHIN PHYSICAL GEOGRAPHY

Biogeography, being concerned to study biological and man-derived processes operating in both natural and man-made environments, is bound to contribute to interpretation of pattern and change in both physical and cultural landscapes. It follows that it has applications (a) within physical geography as a whole, (b) within human geography as a whole, (c) in areas involving both physical and human geography. Along with climatology, pedology and hydrology, biogeography has comprehensive linkages throughout the length and breadth of the subject and especially at what has been allowed to become its central divide between the human and physical sectors within which vigorous sets of introverted specialisations are independently pursued, the one side becoming more oblivious to the other and producing the kind of one-sided stance that Wooldridge once described as 'an ugly monocular squint' (Wooldridge, 1945 and 1950). It is, however, biogeography, with pedology, that provides the biophysical arena where land-use systems convert natural energy for man's dietetic and technological consumption. The biosphere is the engine room where biogeochemical exchanges take place and where man's cultural adaptations are concentrated and focussed within physical geography.

The full interpretation of the physical landscape must include the delicate veneer of soil, vegetation and land use that covers its surface. In the micro-detail of this veneer are the residual effects of both long-term and short-term geomorphological, hydrological and climatological processes and the abundant evidence of recent environmental history. Terracettes, rotational slips, evidence of colluviation, indurated horizons, polycyclic profiles, stone pavements, charcoal bands, fossil pollen records: such is the range and detail of the harvest of evidence available. Gerrard (1981) has presented an integrated view of geomorphology and pedology and Dimbleby (1962) and Smith (1970) pioneered investigations in soil palynology. Equally, palaesol horizons containing residues of associated flora and fauna allow the identification of warmer interstadial or interglacial periods, and thus contribute to the reconstruction of environmental histories, especially of the late Quaternary where the major thrust of current geomorphological research is located (Bowen (ed.), 1977).

Lockwood (1983) has outlined the role of vegetation cover in the operation of the world's climatic systems. Currently, there is probably more agreement among climatologists and meteorologists on the impact of rapid deforestation (through changes in the carbon dioxide content of the atmosphere), on the direction (if not the scale) of climate trends, than on other factors affecting those trends (Smagorinsky, 1981; Clark, 1982). Fossil pollen (Godwin, 1975), snails (Evans, 1972) and beetle remains (Coope, 1977a and 1977b) etc., have been used to reconstruct past climates but the exercise is in some ways presumptuous and generates its own interpretational controversies (Pennington, 1977 and Coope, 1977b). No changes have to be assumed in the autecologies of the biota concerned and it is important to realise that the impact of a change in climate will register at different times and intensities on the ground, according to variations in (a) relief, soil conditions and associated ground hydrology, (b) the state, vigour and continuity of the vegetation cover, (c) above all, in the inherent sensitivity, including mobility in fauna, of the biota involved and (d) finally, the site-impact of prehistoric man (Taylor, 1975 and 1980 (ed.)). This hypothesis could be applied to explain some of the time discrepancy (having regard to different types of evidence) between the onset of warming and cooling phases in the Devensian Late-glacial. A fuller explanation, however, must lie in a broader biogeographical awareness of the varying rates of colonisation opportunities and migration rates from southerly sources between *circa* 14,500 and 10,000 BP for (a) grasses, (b) trees, (c) animals like the deer, (d) insects like the beetle. The summers and afternoons of British-style, relatively cold, periglacial phases (with the angle of the sun's rays as now (Williams, 1975)) could well have been microclimatically warm enough to provide very early, if patchy, habitats for grasses and insects well before the ungulates and, later, shrubs and trees, could respond to a fuller-scale climatic warming with tree-canopies enjoying the required levels of maximum temperatures. The fullest bioclimatic interpretations are possible when palaeoecosystems are reconstructed, an example of which is provided in Figure 11.1 (Taylor, 1980a). The collective interrelationships within and between environmental and cultural factors, in terms of change and response, is fully displayed, thus promoting equilibrium and accuracy of interpretation of the initial causes and consequences of change. However, the *inferential* nature of such models must be stressed since the basic biological, pedological and archaeological and radiometric evidence is adapted to reconstruct other components of the environment for which direct data are not as available, e.g. the climate, the hydrology and, to a great extent, the fauna. Such palaeoecosystematic models are as yet inductive and must survive the accumulation of further

Figure 11.1: Ecosystem model for the Middle and Late Bronze Age (after Taylor, 1980a)

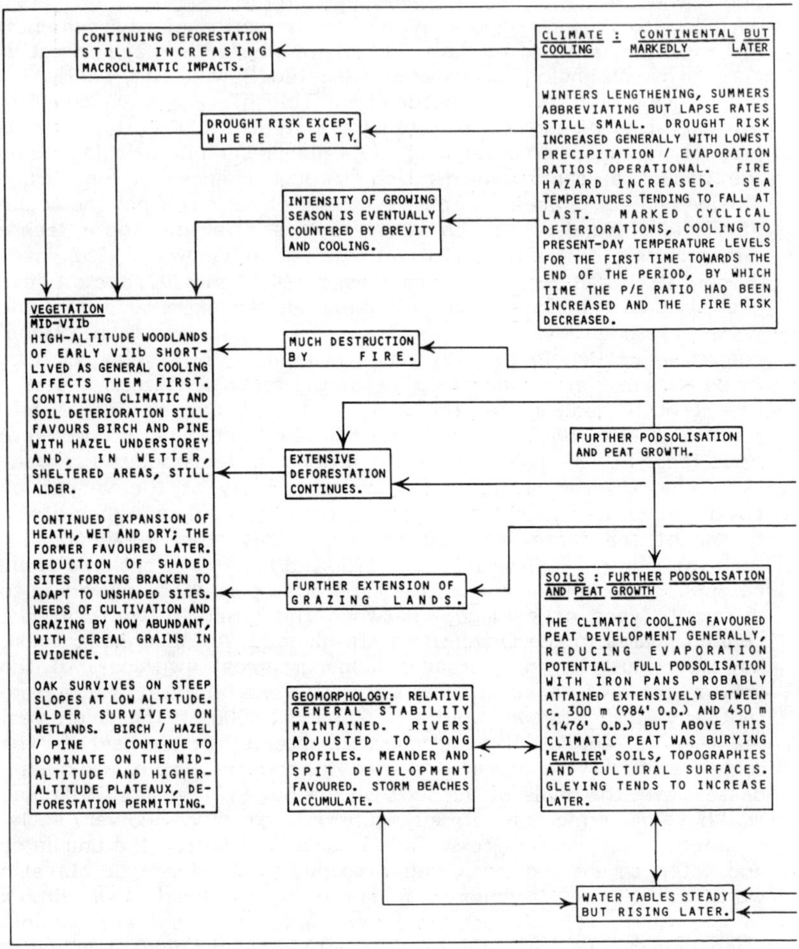

The model has been constructed on the basis of palaeoenvironmental and archaeological evidence available for sites in Wales and adjoining Welsh Borderland. It should be emphasised that much of the evidence for climatic, hydrological and geomorphological changes has been inferred from the more directly available evidence on the history of flora, fauna and soils.

Figure 11.1 cont'd.

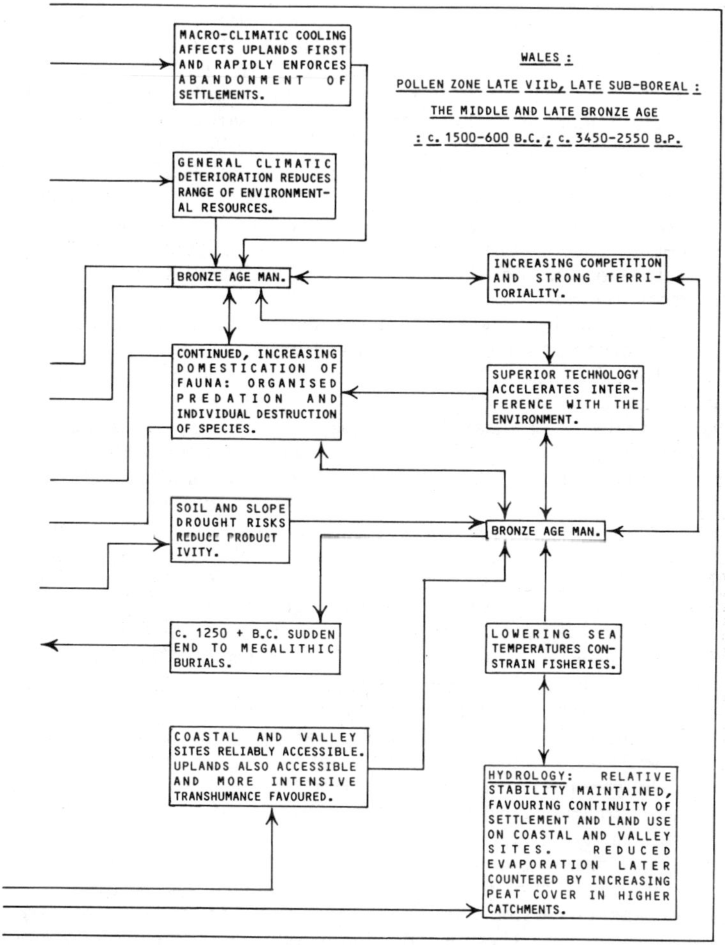

environmental and archaeological data and the development of new dating techniques in the future.

Smith and Taylor (1969) and Taylor and Smith (1972 and 1980) have shown how the analysis of vegetation history can aid the explanation of soil profile development, especially when organic layers are present either at the surface in the form of peat (at first, soil-induced, and termed pedogenic but, later, climatically controlled, and hence termed climatic peat), or within the profile, for example as the mor layer of a buried podzol. Biological processes, in collaboration with physical and chemical processes, continuously create (and successively destroy) the soil medium (Smith, Chapter 7 herein). Biological residues from these interactions provide clues of past stages and events and mirror the general ecological status of the soil as witnessed in the simple but often underrated technique of worm counts in modern soils.

The hydrological cycle involves losses, gains, surpluses and deficiencies by courtesy of the vegetation/land-use layer which lies at the interface with water-usage by plants, animals and man. Penman (1948) pioneered evaporation studies from different vegetation surfaces and Law (1956) was to discover that as much as 30 per cent of incident rainfall can be intercepted by a closed tree canopy. This statistic was derived for conifer plantation in the central Pennines. Later, Newson et al (1982) confirmed that the afforested upper Severn catchment, near the summit of Plynlimon in mid-Wales, also intercepted rainfall at comparably high rates which did not apply in the grasslands of the adjacent headwaters of the River Wye. Again, Taylor et al (1980) usefully combined data for soil moisture storage with data for potential soil moisture deficits for a 20 km^2 grid network covering Great Britain for the 1976 drought. This gives a more refined local measure of 'functional drought' as both soil and climate parameters are built into the distributions. The biospheric component of the hydrological cycle is the 'operations room' where both quantity and quality water resources are continuously required. Rapid vegetation change, whether it be in Amazonia or in the Southern Uplands of Scotland (Robinson, 1980), has immediate hydrological repercussions on land use, especially in the context of drainage costs and water quality. The afforestation of water-supplying catchments beyond, say, 50 per cent of their land area is increasingly being challenged by Water Authorities since the type of remote countryside and localised water supply systems involved normally allow very few alternative supply options in summer. Test cases are likely to increase as all budgets become tighter, with margins too narrow to allow more flexible land-use policies to operate. Clearly, the long-term planning of water resources over the coming decades should fully incorporate extrapolated land use and vegetation changes to reduce and perhaps avoid the recurrence of such friction (Pereira, 1973; Parker and Rowell, 1980).

WITHIN HUMAN GEOGRAPHY

The abandonment of any traditional 'physical background' in favour of isotropic surfaces may be excusable for theoretical studies in human geography. Otherwise what human geographers do, if they confine their objectives to explaining human distributions in exclusively human terms, might as well be located outside, as well as inside geography departments. The way forward is in a reassessment of, first, the way environment factors encourage or discourage the taking up of options and, second, the cost-benefit, operationally, socially, economically and technically, of accepting or refusing such options. Advanced and sophisticated, urbanised and industrialised cultures are not immune from such a reappraisal which may be deceptively more applicable to the rural and developing worlds.

It is patently obvious that the cruder physical underpinnings of the past which have been used to explain settlements (e.g. physical site and position, regionality, gaps, accessibility, defensibility, simple area and distances available, spring-lines, water and other natural resource availability, bridging and fording points, etc.) are basic geomorphological, hydrological and strategic factors which operated functionally in the dark and distant past in the foundation of pioneer settlements. Such associations have now been overprinted by new economic, social and political forces which have re-shaped settlement patterns and industrial locations. More pertinent, however, as argued by Taylor (Chapter 9 herein), is the pattern of evaluation over time of land (= biophysical) resources, which may present some degree of consistency along cultural successions from the most primitive and prehistoric to the most modern and intense. Again, non-rural landscapes and cities are equally interpretable as ecosystems (Figure 11.2) (Douglas, 1981 and 1983) with energy exchanges, biogeochemical cycles, balances and imbalances, storage requirements and pollution hazards.

Human systems involve the circulation of information as well as the fundamental cycling of energy in both is natural and converted forms. Birch (1972) has modelled the farm unit in this context where both energy transformation and information flows combine to produce outputs for transfer of marketing systems or for the negative or positive feedback into the original system (Figure 11.3). The use of the same concepts, models and methods enables a new liaison to be established with human geography.

Historical geography (Thomas, 1963; Darby, 1977), history and archaeology have long laboured without the benefit of a contemporary environmental settings to provide context when attempting to account for those peregrinations and predicaments of prehistoric and historic man which hint of environmental inducement, steering or constraint. Not that such environmental linkages are ever causal, consistent or punctual as Taylor (1968) commented:

> It is illogical to rank human motivations with environmental inducement. The former may or may not be conditioned by the latter. It will certainly predetermine the value and effect of

Figure 11.2: Phosphorous circulations in Sydney (A) and Hong Kong (B) in Kg day-1 The symbols used are those employed by Odum, H.T. (1971). The original data are from Aston et al (1972) and Newcombe (1977)

A SYDNEY

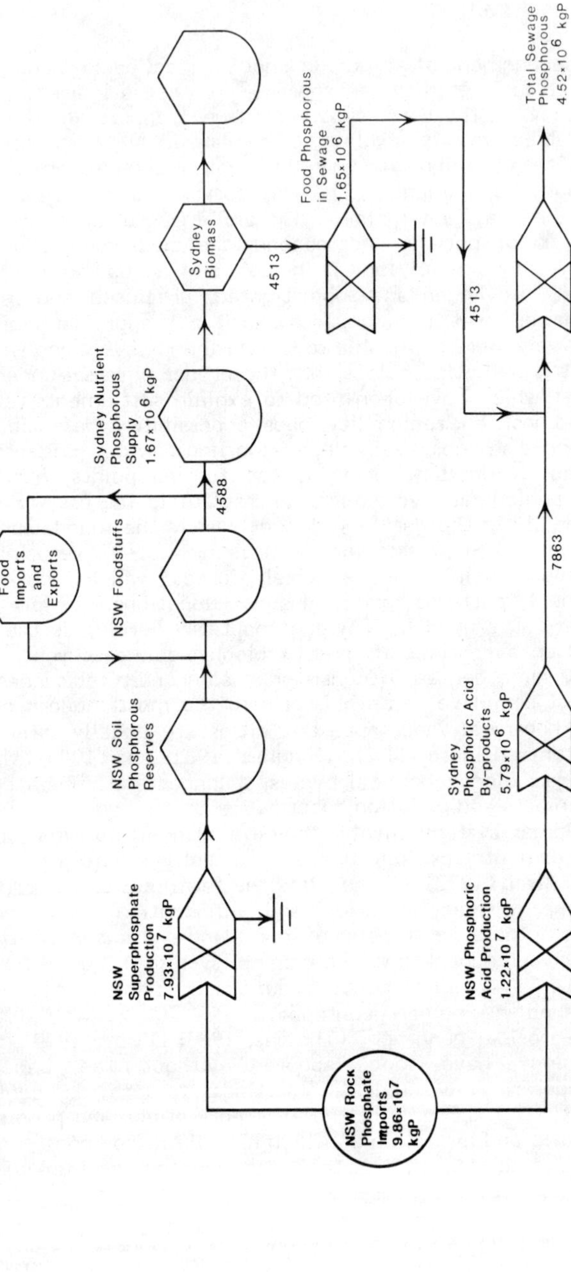

Figure 11.2 cont'd.

B HONG KONG

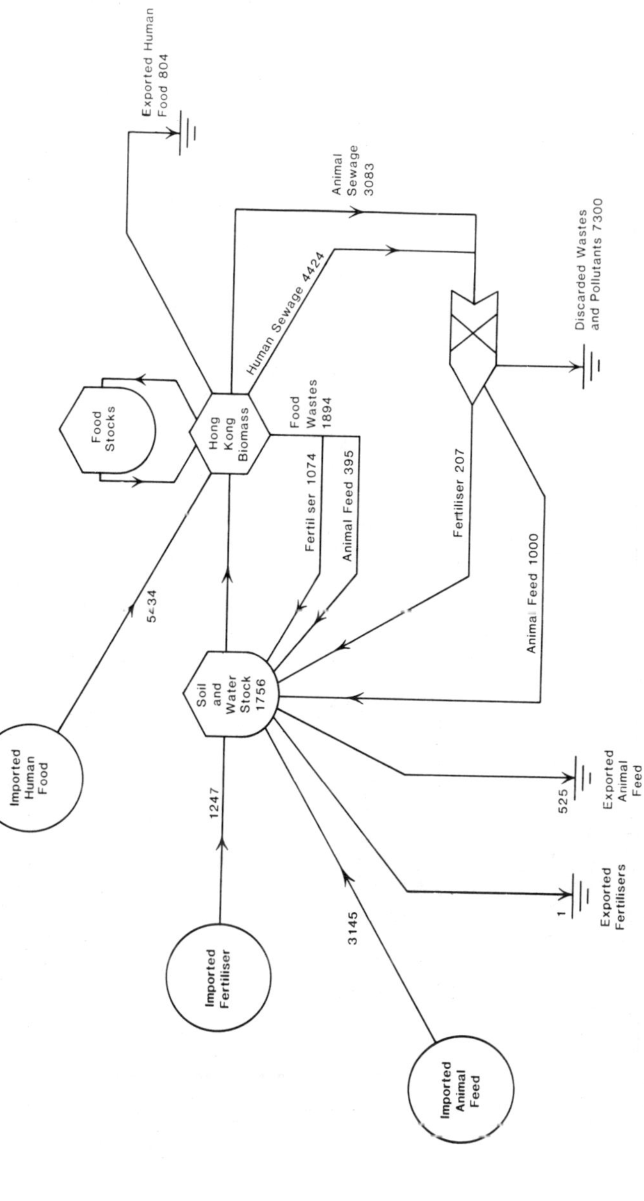

Figure 11.3: Model of a farm system, showing flows and exchanges of energy and information (after Birch, 1972)

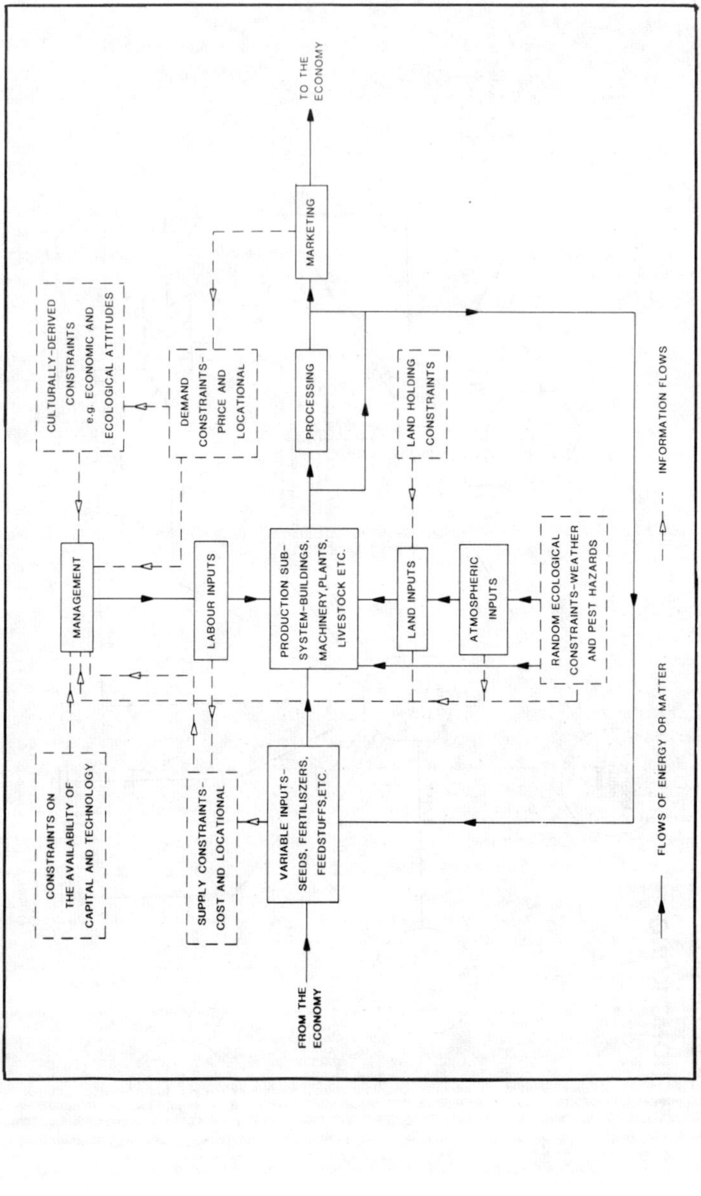

the latter. It may accommodate it or surrender to it but it is always independent of it *sensu stricto*. Again, a human decision is often short-term and sometimes instantaneous although it may have been incubating for a long time. On the other hand environmental factors often involve longer time-scales except for catastrophes, e.g. landslides, floods, etc.

Thus, the final burial of any vestiges of direct physical determinism may be followed by an historical application of behavioural geography in the context of the contemporary environmental conditioning. Lynch (1975) has discussed imaginatively the possible landscape appraisals attributable to Bronze Age man in his decision about the location of his settlements and burial places. Parry (1978) and Taylor (1980b) have examined behavioural attitudes to environmental changes, initially climatic but also eventually biogeographical, pedological and hydrological, in prehistoric and medieval times, respectively, in Wales and the Southern Uplands of Scotland, respectively.

At the same time, new direct environmental linkages are emerging from studies in medical geography, in particular the incidence of environmentally-related diseases and the occurrence of selected dietetic deficiencies (Howe, 1963, and Howe and Loraine (eds.), 1980). Recent research in Wales and Scotland is confirming that the bracken fern (*Pteridium aquilinum*) is a source of toxins, in particular the carcinogenic shikimic acid which, when transmitted via water or milk, is capable of inducing cancers in animals and possibly humans also (Evans, 1976; Evans *et al*, 1982; Royal Society of Edinburgh, 1982). Taylor (1978 and 1980c) has presented initial evidence that the bracken areas of Wales have doubled since the 1930s and the collation of local rates of expansion for a number of representative but well scattered sites in Britain indicates a national average rate of increase of 1.0 per cent, but on selected common lands it is in excess of 3.0 per cent. This rate of land-loss is all the more serious as bracken occurs mostly on well drained, sunny, sheltered slopes of high grazing potential. This land-use change will rank with the more familiar rural land-losses to afforestation and urbanisation, as a major twentieth-century event of both economic and ecological significance (Taylor, 1978). The problem is so interdisciplinary in nature as to require balanced and broad-based biogeographical solutions.

It follows that an alternative biophysical basis is available as a backcloth for selected studies in human geography where environmental constraints, inducements, resources or relationships are explicitly involved.

AS AN INTEGRATING FORCE BETWEEN PHYSICAL AND HUMAN GEOGRAPHY

The current fissiparous tendency throughout geographical studies has created an estrangement between physical and human geography that is unprecedented and ill-advised from the point of view of maintaining the intellectual identity and integrity of the subject which must include a commitment to study both man and his environment and ipso facto, their interrelationships as well. The integrative birthright could well be up for the sale for a mess of blinkered, systematic potage (Wooldridge, 1945, 1949 and 1950). Johnston (1983) has reviewed the possible role of resource geography in achieving some integration between physical and human geography, and concludes that such a role is both 'unlikely' and 'unnecessary' because, he argues, resource geography is basically 'a social science' which can import environmental considerations as and when required! He cites 58 references and concludes they are comprehensively ineffectual as integrators because they all fail to explain how either physical processes or human processes or both impinge on the resource problem they are analysing. He argues that both sets of processes should be fully and explicitly involved in such attempts at integrative study. This is very much the blinkered view of the systematist. Integrated studies involve much retailing as well as wholesaling, and much secondary evaluation of facts as well as primary establishment of facts. But they are also concerned with the operation of processes which involve both physical and human geography (Haggett and Chorley, 1969; Peel et al, (eds.), 1975) and with the quite different processes which occur at the interface between the two and nowhere else (Eyre and Jones (eds.), 1966; Tivy and O'Hare, 1980).

Resource management is not an exclusively social science and does possess an ecological basis (Taylor, 1974a; Eyre, 1978; Park, 1979). Natural resources have inherent qualities and rankings prior to assessment for usage by man. Usage in nature provides a pre-existing hierarchy, be it air, land, water, slope, soil, nutrients, flora, fauna, etc. Pre-use assessments of inherent productivity provide a base-reference for subsequent land-use performances and alternates (Taylor, Chapter 9 herein). Fully integrated studies must involve both the environmental base and the land-use activity performed on it. Such a perspective is eminently appropriate for biogeographical analysis, a point to which Johnston (1983) pays scarcely any attention.

The work of Isard (1969) and his co-workers (1967 and 1972) and of Clarke (1981) point the way to the successful integration of ecological and economic systems as an example of a man/land combination. Both systems may be conceived in budgetary terms, with inputs and outputs which are variably interchanged and converted, some within either system and a few in both. Using coefficients, the processes and commodities of both systems can be evaluated in comparative accounting terms. Thus the 'combined operation room' at the heart of the two systems is identified and penetrated. Isard's method is illustrated in Figure 11.4 and Tables 11.1 and 11.2 with supporting explanatory notes. Information flows (see Figure 11.3) interlock with converted energy flows in land-use systems and industries in general, and inputs and outputs can be expressed in the common language of either energy equivalent or monetary values and related productivities derived (Linton, 1965; Duckham and Masefield, 1970; Simmons, 1978).

Figure 11.4: Framework for the analysis of interregional economic-ecologic activity (after Isard et al, 1972)

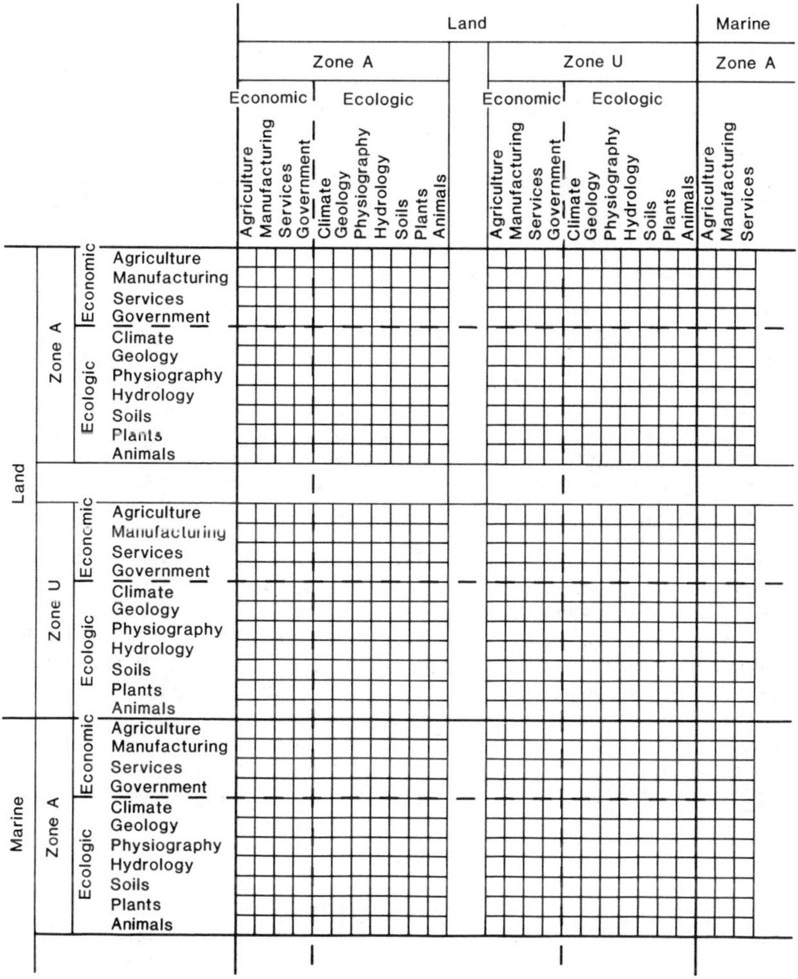

Isard establishes a numerical ecological classification system and derives ecological interrelation coefficients as applied to some more accessible case-studies such as selected fisheries.

Table 11.1: Input-output Coefficients: RIS sector 2911 Petroleum Refining

Sector			Coefficient
RIS 1311	Crude Petroleum and Natural Gas		-0.612006
1509	Construction, Maintenance and Repair		-0.001410
2652	Set-Up Paperboard Boxes		-0.008054
2655	Fibre Cans, Tubes, Drums and Similar Products		-0.001557
2812	Alkalis and Chlorine		-0.000935
2818	Industrial Organic Chemicals, n.e.c.		-0.027180
2819	Industrial Inorganic Chemicals, n.e.c.		-0.009190
2911	Petroleum Refining		+0.954950
2992	Lubricating Oils and Greases		-0.014954
3411	Metal Cans		-0.005294
4811	Telephone Communications		-0.001210
4890	Telegraph and Other Communications		-0.000100
4911	Electric Utilities		-0.007629
4920	Gas Companies and Systems		-0.012336
4941	Water Supply		-0.001082
4990	Sanitary and Other Systems		-0.000142
6020	Interest		-0.005088
6301	Insurance, Nonlife		-0.001242
6510	Real Estate Services		-0.007680
7301	Business Services, excluding Advertising		-0.017228
7310	Advertising		-0.002312
7400	Research and Development		-0.000390
7500	Automotive Repair		-0.001040
9000	Local and State Taxes		-0.008049
9100	Federal Income Tax		-0.014755
9826	Office Supplies		-0.000280
9842	Transportation Costs		-0.040831
9888	Wages and Salaries		-0.089378
9899	Residual		-0.063589
WPC 1001	Water Intake, Sanitary Use,	1,000 Gal/$ Output	-0.000455
1002	Water Intake, Production	"	-0.022827
1003	Water Intake, Cooling	"	-0.114861
1011	Water Discharged, Sanitary Use	"	+0.000323
1012	Water Discharged, Production	"	+0.070219
1013	Water Discharged, Cooling	"	+0.050100
1031	Biochemical Oxygen Demand BOD, 5-Day	1,000 Lb/$ Output "	+0.000065
1032	Ultimate Oxygen Demand, UOD	"	+0.000076
1033	Chemical Oxygen Demand, COD	"	+0.000169
1041	Suspended Solids	"	+0.000084
1042	Settleable Solids	"	+0.000124
1047	Turbidity	"	+0.000365
1051	Alkalinity	"	+0.000051*
1052	Acidity	"	+ n.a.
1061	Oils and Greases	"	+0.000012
10951	Phenols	"	+0.000003

* The pH factor for this industry is the range of 4.5-10.7, with 7.0-8.0 most likely.
Source: Isard *et al* (1967).

The upper part of Table 11.1 expresses typical inputs and outputs of economic commodities as coefficients. For example, the first coefficient listed, -0.612006, is a dollar's worth of *input* of crude petroleum or natural gas *per* dollar *output* of the industry. Similarly, the coefficient next to last, -0.89378, is the *input* of wages and salaries *per* dollar *output* of the industry. Concerning inputs/outputs other than economic ones, a classification for water intake from, and pollutants discharged via water to, the environment is presented in Table 11.2. For example, the fourth item is water intake for cooling purposes expressed in 1,000 gallons per dollar intake, and the penultimate item is radioactive waste. The set of non-economic

Table 11.2: Proposed Water Pollution Classification (WPC) Code

WPC Code	Water Related Item	Units
1000	Water Intake, Total*	1,000 Gal/$
1001	Sanitary Use	"
1002	Production Use	"
1003	Cooling	"
1004	Boiler Feed	"
1008	Irrigation	"
1009	Other, n.e.c.	"
1010	Water Discharge, Total	"
1011	Sanitary Use	"
1012	Production	"
1013	Cooling	"
1019	Other, n.e.c.	"
1020	Water Consumed	"
1031	Biochemical Oxygen Demand, BOD 5-Day	1,000 Lb/$
1032	Ultimate Oxygen Demand, UOD	"
1033	Chemical Oxygen Demand, COD	"
1040	Solids, Total+	"
1041	Suspended Solids	"
1042	Settleable Solids	"
1047	Turbidity	"
1048	Colour	"#
1051	Alkalinity	"
1052	Acidity	"
1061	Oils and Greases	"
1062	Surfactants	"
1070	Pathogenic (Disease Causing) Organisms	-x
1080	Temperature	-x
1090	Other Pollutants	"
1095	Toxic Material**	-x
1096	Radioactive Waste	-x
1099	Not Classified	"

* Cost of water intake is given by SIC code 4941 - water supply

\+ 1040 = 1041 + 1042 (i.e. WPC Codes)

\# In addition, colour should be specified by kind by its wave length

x No one satisfactory measure was decided upon, although thermal pollution may be specified most satisfactorily in terms of millions of BTUs per dollar output

** Phenols, which fall in this category, may be identified by a five-digit code such as WPC 10951

Source: Isard et al (1967).

commodities in Table 11.2 is listed in the lower half of Table 11.1 in the form of coefficients. These tabulations are not complete (e.g. air and air pollution inputs/outputs are excluded) but at least the first empirical linkages between economic and ecological systems are demonstrated.

Note: SIC = Standard Industrial Classification
WPC = Water Pollution Classification

RECIPROCATION WITH THE BIOLOGICAL SCIENCES

Whilst biogeography in geography departments has become the beneficiary of the statistical and theoretical advances achieved in biology in recent decades, a reciprocal contribution can be identified in the injection of a broader environmental expertise both in the field and the laboratory. This derives from a training in physical geography as a whole, including geomorphology (and geology), climatology, hydrology and pedology as well as biogeography and their individual technologies. In addition, traditional and new skills in survey, cartography and remote sensing have been brought to bear on the problems of representing and interpreting biological distributions.

Pollen-analytical studies of inorganic and organo-mineral sediments as distinct from organic ones have been expanded by geographers as well as ecologists (Dimbleby, 1962; Smith, 1970; Taylor, 1973; Maltby and Crabtree, 1976, for example). Many palaeoecologists have restricted their pollen sources to organic profiles. Yet mire sites, like lake basins, blanket peats or soil profiles, also exhibit their own 'site bias' in pollen records, and much more research is required into the variability of pollen records from different but proximate sites to calibrate such bias. Again, the additional flexibility of selection of field sites for pollen analysis in areas of upland peat or peaty soils will allow for much-needed experimentation in spatial sampling techniques to identify representative records on which vegetation distributions for particular stages and areas can be reconstructed. Birks *et al* (1975) has pioneered the way with a series of isopollen maps of Britain for *circa* 5,000 BP and an atlas containing similarly derived maps for north-west Europe has been compiled (Huntley and Birks, 1983). However, these studies adopt the large-scale, cartographic base beloved of the ecologist, with the background of the map consisting merely of coastlines (Dunnet, 1982). For such maps to be of biogeographical use in the fullest environmental sense, significant contours and other gradients (climatic, pedological for example) should be selectively shown, as indeed they would enhance the interpretation of much needed, small-scale, isopollen maps, e.g. of Wales or county size (Smith, 1970; Moore, 1971). The reconstruction of significant biogeographies of the past as local maps offers an eminently worthy challenge to the expertise of both biogeographer and ecologist, and would provide new spatial perspectives for archaeological research.

A similar technological overlap occurs in the methods now available to map modern vegetation distributions which may be recorded at *three* different levels. First, on the micro-scale, quadrats or *relevés* (Shimwell, Chapter 5 herein) may be taken according to a prescribed spatial sampling procedure (stratified random is usually recommended), the size, shape, intensity and frequency distribution of the quadrats being very much determined by the degree of floristic variability of the study area and the scale required (and utility) of any proposed, derived map. Second, on the meso-scale, reconnaissance mapping of visible vegetation distributions can be attempted in the field on, for example, the scale of 6" to the mile ($^1/10560$) provided the vegetation pattern is so standardised as to be legible in panoramic view (Taylor, 1968). Such descriptive recordings are liable to errors in field locations and cartographical transpositions, and should be underpinned by quadrats which could determine discontinuities and dominants and provide much needed quantification. Thirdly, using conventional air photography or satellite sources, the vegetation canopy can be remotely-sensed and checked for ground truth but the spatial continuity and contemporaneous, unbiassed quality of this source goes hand-in-hand with an unavoidable dependence of interpretation on the scale and standards of the initial photography (Darch, Chapter 10 herein). The bridging of the two extremes (that is, between intensely taken ground quadrats and very remotely sensed satellite sources with pixels of, say, 50 m^2, and even the bridging of quadrat and conventional air photographic sources), is not by any means a straightforward task (Davies, 1974). It is one where geographers could contribute professionally, especially from 1984 when the French satellite imagery source, SPOT, will be supplying new photography with a much finer resolution of 10 m^2 pixels.

It could be argued that bioclimatology as applied to both animals and plants is a sub-field shared between climatology and biogeography or biology (Putman, Chapter 6; Greenland, Chapter 8 herein). Ollerenshaw's successful work (1966) on role of weather factors in disease incidence and prediction for farm animals is applied bioclimatology at its most effective. In crop climatology a major contributor has been W.H. Hogg, as reported in Taylor (ed.) (1967). Without doubt, interdisciplinary qualifications for work in this area are a necessary prerequisite for its advancement.

IN SERVICE OF CONSERVATION

In contrast to conventional and systematic biological arguments for conservation (e.g. on the basis of individual species of plant or animal at risk), a broader-based biogeographical view could incorporate rock, soil, climate and total land-use, past and present, to derive a measure of the distinctiveness of the landscape or habitat concerned. Such a view would keep conservation policies in touch with the economic, social and technical realities of land usage and, in the end, protect heritage sites within the tempo of land-use change. No ecosystem can survive in any form of suspended animation. Change is a natural process within ecosystems as well as being imposed externally by man. Catering for both is preferable to the all too familiar friction between positive and negative policies when the forces of conservation and land-use are allowed to oppose each other. O'Riordan's (1982) bureaucratic prescription to reconcile the problem, as manifested in the British countryside, takes too little account of the institutional inertia (which already exists in the form of complex land ownership and tenure), open-market competition for land purchase, hierarchies of protected or conserved areas and sites, and incompatibility of local and national priorities. Real local conflicts between users, conservationists and planners in the British context (and probably elsewhere) are normally soluble only on a local scale, once full consultations have established what must be admitted are often fragile arbitration agreements. Nonetheless, O'Riordan certainly shapes the problem and its possible national solution, but inevitably from a conservational viewpoint, initially, rather than a developmental one.

On the theoretical side, both the theory of island biogeography and its application in conservation have been criticised (Gilbert, 1980). Wilson and Willis (1975) make four recommendations on the design of nature reserves (Figure 11.5). First, they argue that reserves should be as large as possible, initially for strategic reasons, to allow fall-back positions in relation to specific extinction rates. Second, they suggest that unique habitats are best protected in closely adjacent multi-reserves since extinction has a strong random component. Thirdly, because of additional vulnerability of peninsula shapes, reserves should be as round, regular-shaped and as continuous as possible. In Figure 11.5, the designs on the left in each case result in lower spontaneous extinction rates than their counterpart designs on the right. Fourthly, comprehensive surveys of both the least vulnerable and the minority species should be completed as soon as practicable as well as the urgent construction of extinction models for vulnerable and conspicuous organisms (Terbough, 1974 and Diamond, 1975). Diamond and May (1981) offer four qualifications. First, several smaller reserves in lieu of one large one may be preferable in that each smaller reserve may favour the survival of a different species. Higgs (1981) has exposed arguments for and against a single large reserve (SLR) (see Table 11.3). Second, many scattered small reserves are less susceptible to the spread of epidemic diseases or some analogous disaster, e.g. fire. Thirdly, 'edge' preferring species will thrive better in irregular-shaped and smaller reserves with a high perimeter-to-area ratio. Fourthly, non-random periodic changes in species population dynamics could help to determine the optimal size of reserves and the time-scales of management policies.

The continuing debate on the application of island biogeography theory is dramatically highlighted in two contemporary events. First, the well documented deforestation of Amazonia evokes the challenge as to best programme to be devised for the partial survival of the last major area of tropical rain forest on this planet. The survival plan would be well concentrated on the 'islands' of greatest ecological stability where the greatest continuity of forest cover occurs (Bromley and Bromley, 1982). The second refers to the annihilation or evacuation of the entire 17 million bird population of Christmas Island in the mid-Pacific which has reported early in 1983 by the Los Angeles County Natural History Museum. This is possibly the largest island extinction ever recorded. It may have been due to the aberrant timing of the warm surface El Niño ocean current which not only deprived the birds of access to food supplies in the immersed, cooler waters but also has been associated with climatic anomalies on the American Pacific seaboards and also possibly worldwide climatic repercussions in 1983. The events on Christmas Island make two

Figure 11.5: The design of nature reserves (after Wilson and Willis, 1975)

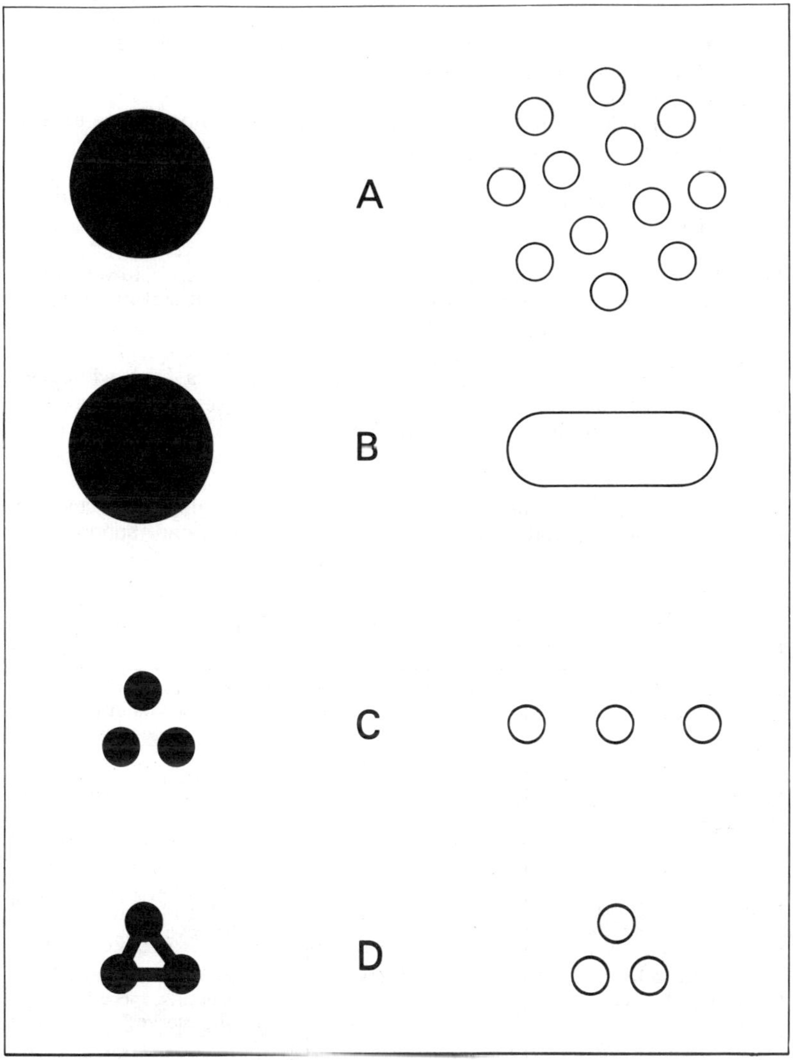

The above diagrams incorporate some geometrical rules for the design of nature reserves on the basis of some conventional biogeographic theory. The designs on the left (in black) all result in lower spontaneous extinction rates than the corresponding designs (outlined) on the right. Both the left-hand and the right-hand designs have the same total areas and assumed homogeneous environments.

A A continuous reserve is better than a fragmented one because of the effects of distance and area.

B A round design is the better one because of the peninsula effect (MacArthur and Wilson, 1967).

C A close grouping is better than a linear one because of distance effects.

D If the reserve must be divided, extinctions will be lower when the fragments are connected by corridors of natural habitat, no matter how thin the corridors.

Another principle, not incorporated in this figure, is that whatever the design of a given reserve, its extinction rates can be greatly reduced by the proximity nearby of similar reserves.

Table 11.3: The arguments for and against a single large reserve (SLR), indicating that theoretical considerations can support or oppose the setting up of a SLR (after Higgs, 1981)

For Reserve <1	Species-area curve	Against Reserve <1
Species lost by fragmentation	Irreversibility	Species lost from specialised habitats outside large reserve
(a) Better buffering	Catastrophes	'all eggs in one basket'
(b) Easier recolonisation		
Several, e.g. less interference, less edge per area	Management	Several, e.g. easier to set up, less likely to be developed
Proportionally more non-edge habitat	Edge Effects	If the edge species are interesting, more in small reserves
Fewer extinctions	Extinctions	Preservation of mutually exclusive groups of species

points most forcibly. One, the catastrophe theory, not only reinforces the short-term sensitivity of all ecosystems to sharp environmental changes or disasters but, second, it also puts island biogeography theory at risk because of the possible random or periodic abbreviation of the assumed time-scales of its application. What an immense opportunity for research is now afforded by the birdless Christmas Island.

NEW DIRECTIONS AND APPLICATIONS

Biogeography is in a critical re-emergent stage in both the geographical (Watts, 1978) and biological sciences (Taylor, 1984). Whilst major theoretical reappraisals in the latter group should evolve from current controversies over the origins and patternings of the distributions of biota over the earth, the continuing maturation and eventual convergence of the numerous thematic studies within the former group (symbolised in the chapter sequence in this volume), will enable the establishment of new integrated methodologies within physical geography, within biogeography itself and also within geography as a whole. The calibration of man/land linkages in mutually compatible terms will rebridge the current intellectual void (Johnston, 1983) between two major foci of the subject.

NEED FOR INNOVATION IN PALYNOLOGICAL RESEARCH

However, concerning the reality of the scope of biogeography as currently taught and researched in departments of geography, it was Hill (1975) who first argued that the then concentration on palynology and cultural biogeography had developed a historical perspective which served only to alienate biogeography from many of the contemporary trends in physical and in human geography. Later, Edwards (1982) constructively analysed what was by then an increasing dominance of palynological work in geography departments but deplored, with some justification, its lack of innovation in ideas, methods and techniques. Following the strictures of Haines-Young and Petch (1980) but responding to the earlier exhortations of Moss (1977), physical geography as a whole requires to develop deductive strategies which go beyond the observation, collation and analysis of field-based and laboratory-derived data. A palaeoecosystematic approach, as demonstrated by Taylor (1980a) (see Figure 11.1), is eminently biogeographical in concept and application but requires the cross-fertilisation of ideas on contemporary meso-climates, hydrologies and soils. A more balanced and mature perception of environmental relationships emerges as well as the place and response of prehistoric (and historic) man within them. Archaeological and historical correlations over space as well as time cannot but gain from such appraisals. As Edwards (1982) also argues, the tedium of pollen counting has other 'applied' rewards in that the calibration of rates of change in past ecosystems allows for the building of models to predict probably directions and rates of change in modern ecosystems (including reconstructed ones (Bradshaw,

1983)). This important principle and its application should become the conventional and repeatable basis for the determination of the conservation status of specific habitats and the time-scale of conservation periods, which should not necessarily be assumed to be, in any sense, permanent. In a constantly changing pattern of both natural and man-made land-usage, strategies of alternating conservation and usage may well be more sustainable and should not be dismissed as satisfying neither side of the argument.

INTEGRATIVE, BALANCED APPROACHES IN LAND-USE ECOLOGY

Provided it develops and attacks its own research frontiers, palynology will continue to dominate biogeographical programmes in geography departments in the immediate future. But the full development of biogeography as a competitive sub-field requires broader interests, including its current 'second string', i.e. land-use ecology (Simmons, 1980; Taylor, Chapter 9 herein). The injection of biogeographical approaches to land classification systems and to the measurement of causes-and-effects of land-use changes and environmental impacts enables (i) a necessary reconciliation of short-term and long-term, nature-derived and man-derived processes, and ecological and economic thresholds, (ii) a maximisation of productivity and conservation of ecological resources, and (iii) the final optimisation of land-use strategy and planning. This may sound perfectionist and unattainable but such reconciliations lie at the heart of the solution to land-use problems, no matter how pragmatic the targets and techniques adopted for standard planning procedures or *ad hoc* consultancy in what are often politically highly inflammable situations, involving complexities of ownership, tenure, responsibility and prejudice.

To quote a few examples, Cruickshank and Armstrong (1971) proved a functional relationship between sub-soil parameters and gross margins for a range of cropping enterprises on a sample of Londonderry farms in Northern Ireland (see Chapter 9 herein). Maxwell *et al* (1979) have devised an economic model, including the technique of discount rating, for the allocation of land as between agriculture and forestry in the context of the British uplands, but the elegance of the model is in contrast to the crudity of their classification of land inputs. Tubbs and Blackwood (1971) produced a workable classification of ecological habitats to be protected from development within the Hampshire Structure Plan but the problem of reconciling ecologically-superior, unploughed meadow with economically-superior intensive arable land is not resolved. The report of the MAFF (1966) on land classification techniques was correct in proposing that economic classifications per unit area or groups of holdings should be expressed in precise correlation, spatially and functionally, with the ecological classification of the land types and area available to a specific economic category of farm area, farm group or farm type. Future research should explore this methodologically difficult avenue: difficult because of the incomplete, fragmentary or unrepresentative nature of farm data and land data available (Taylor, 1974).

Simmons (1980) has provided details of recent contributions by geographers to the study of modern ecosystems and their management. Notable works are those of Harrison and Frenkel (1974) and Harrison (1975) on vegetation-analysis, Eden (1974) on selected savana ecosystems, Tivy (1972), Barkham (1973), Warren and Goldsmith (1974 and 1976), and Tivy and Rees (1978) on recreation ecology. With increasing land-use pressures and the imminent expansion of the international scale of tourism, environmental information is sorely needed both for conservation practice and also the successful operation of adjacent or on-site land-use systems, including tourism (e.g. Taylor (1979a), who presents a commissioned review of climate and weather factors affecting British tourism, and points the way to the need for similar regional, biogeographical surveys along the lines of those by Coppock and Duffield (1975) in Scotland and Warren and Goldsmith (1974) in the Scilly Islands, for example).

FUTURE REMOTE SENSING TECHNOLOGY

Darch (Chapter 10 herein) has demonstrated the present and future applications of remote sensing techniques in biogeographical research. Cole (1968) and Barrett and Curtis (1976) have made major pioneer contributions in this field. New remote-sensing platforms available for experimentation in the immediate future range from the French Satellite Source, SPOT (operational in 1984 and, offering a pixel resolution of only 10 m^2, of great potential for vegetation and land-use mapping) to the use of a microlight aircraft as pioneered by Francis, Taylor and Williams (1984) (which would appear to be the ideal platform for the photography and study of small and medium-sized (e.g. 10 m^2 to 1000 m^2) ecosystems). It has already proved superior for the study of erosion forms and processes in blanket peat in central upland Wales. It provides high resolution aerial photography at low cost. Again, low take-off and landing requirements and low flying speed, together with direct pilot and camera control, make the technique superior, in cost-benefit terms, to other available methods of remote-sensing, e.g. conventional aircraft, helicopter or balloon. It has great potential for biogeographical and other forms of field-based research.

CONSULTANCY

In addition to academic establishment and scientific acceptability, biogeography in its applied aspects has much to contribute in the fields of public and private consultancy. Cole (1968), Barrett and Curtis (1976) and others have been involved internationally, for example, with resource and land-use survey for agricultural and industrial purposes. Taylor (1976, 1977, 1979a and 1979b) has advised on peatland resources and development, internationally, agroclimatological problems, land-use strategies, and conflicts between land-use and conservation interests. Other consultancy activities include references quoted above concerning recreation ecology and ecosystem management, etc. (Warren and Harrison, 1970; Tivy, 1972; Warren and Goldsmith, 1974 and 1976; and Tivy and Rees, 1978; etc.).

These many and varied developments in applied biogeography complement the extension of academic biogeography in both physical and human studies and in the currently neglected borderland between the two. The re-emergent sub-field of biogeography appears well set to liaise reciprocally also with the biological and environmental sciences to the collective benefit of all practitioners, no matter what their first-degree subject was or what particular name their department carries. There are far too many scientific challenges and pressing problems of environmental monitoring and management to allow internecine quibbles on academic territoriality to interfere with progress in pure and applied research. The thrust must be interdisciplinary; the resultant advancements in biogeographical theory and practice will be of benefit to all biogeographers, regardless of their training or affiliation.

ACKNOWLEDGEMENTS

The Editor would like to thank the following for their generous assistance in the preparation of this Chapter: Philip Stott for useful comments on an early draft; Carol Parry and Linda James for typing the manuscripts; Michael Gelli Jones and Huw Hughes for drawing the illustrations and David Griffiths and Anthea Cull for photographing them.

REFERENCES

Aston, A.R., Millington, R.J. and Kalma, J.D. (1972), 'Nutrients in the Sydney area', Proc. Ecol. Soc. Australia, **7**, 177-91.

Atkinson, B.W. (1980), 'Climate', in Brown, E.H. (ed.), Geography Yesterday and Tomorrow, Oxford University Press, UK, pp. 114-29.

Ball, I.R. (1975), 'Nature and formulation of biogeographical hypotheses', Syst. Zoo., **24**, No. 4, 407-30.

Barkham, J.P. (1973), 'Recreational carrying capacity', Area, **5** (3), 218-22.

Barrett, E.C. and Curtis, L.F. (1976), Introduction to Environmental Remote Sensing, Chapman and Hall, London.

Bennett, R.J. and Chorley, R.J. (1978), Environmental Systems: Philosophy, Analysis and Control, Methuen, London.

Birch, J.W. (1972), 'Farming systems as resource systems', in Vanzetti, C. (ed.), Agricultural Typology and Land Utilisation, Centre of Agric. Geog., Verona, Italy, pp. 13-21.

Birks, J.H.B., Deacon, J. and Pegler, S. (1975), 'Pollen maps for the British Isles 5000 years ago', Proc. Roy. Soc. London, **B**, 189, 87-105.

Birks, J.H.B. (1977), 'The Flandrian forest history of Scotland: a preliminary synthesis', in Shotton, F.W. (ed.), British Quaternary Studies: Recent Advances, Clarendon Press, Oxford, pp. 119-35.

Bowen, D.Q. (ed.) (1977), 'X' INQUA Congress Excursion Guides (sixteen volumes), Geo Abstracts Ltd., Norwich, UK.

Bradshaw, A.D. (1983), 'The reconstruction of ecosystems', J. App. Ecol., **20** (1), 1-17.

Bromley, D.F. and Bromley, R. (1982), South American Development: a Geographical Introduction, Cambridge University Press, Cambridge, UK.

Brown, E.H. (1975), 'The content and relationships of physical geography', Geog. J., **141**, Part 1, 35-48.

Candolle, A. de (1882), Origine des plantes cultivees, Germer Bailliere, Paris.

Chorley, R.J. (1971), 'The role and relations of physical geography', Prog. in Geog., **3**, 87-109.

Chorley, R.J. and Kennedy, B.A. (1971), Physical Geography: a Systems Approach, Prentice Hall, Englewood Cliffs, NJ, USA.

Clark, C.W. (1981), 'Bioeconomics', in May, R.M. (ed.), Theoretical Ecology: Principles and Applications, Blackwells, Oxford, pp. 387-418.

Clarke, W.C. (1982), Carbon Dioxide Review, Oxford University Press, Oxford.

Clayton, K.M. (1980), 'Geomorphology', in Brown, E.H. (ed.), Geography Yesterday and Tomorrow, Oxford University Press, Oxford, UK, pp. 167-80.

Cole, M.M. (1968), 'Observations of the Earth's resources', Proc. Roy. Soc. A, **308**, 173-82.

Coope, G.R. (1977a), 'Quaternary Coleoptera as aids in the interpretation of environmental history', in Shotton, F.W. (ed.), British Quaternary Studies: Recent Advances, Clarendon Press, Oxford, UK, pp. 55-68.

Coope, G.R. (1977b), 'Fossil Coleopteran assemblages as sensitive indicators of climatic change during the Devensian (Last) Cold Stage', Phil. Trans. Roy. Soc. B, **280**, 313-40.

Coppock, J.T. and Duffield, B.S. (1975), Recreation in the Countryside: a Spatial Analysis, Macmillan, London.

Cox, C.B., Healey, I.N. and Moore, P.D. (1976), Biogeography: an Ecological Approach (second edition), Blackwell Scientific Publications, Oxford, UK.

Cracraft, J. (1975), 'Historical biogeography and earth history: perspectives on a future synthesis', Ann. Missouri Bot. Gdn., **62**, 227-50.

Croizat, L. (1958), Panbiogeography, Caracas, The Author.

Croizat, L. (1964), Space, Time, Form: the Biological Synthesis, Caracas, The Author.

Croizat, L. (1973), 'La Panbiogeographia in breve', Webbia, **28**, 189-226.

Croizat, L. (1981), 'Biogeography, past, present and future', in Nelson, G. and Rosen, G.N., Vicariance Biogeography, Columbia University Press, 501-23.

Cruickshank, J. and Armstrong, W.J. (1971), 'Soil and agricultural land classification in County Londonderry', Trans. Inst. Brit. Geogrs., **53**, 79-94.

Curtis, L.F. and Simmons, I.G. (eds.) (1976), Man's Impact on Past Environments, Thematic Number, Trans. Inst. Brit. Geogrs. NS1(3).

Dansereau, P. (1957), Biogeography, Ronald Press, New York, USA.

Darby, H.C. (1977), Domesday England, Cambridge University Press, Cambridge, UK.

Davies, L.M. (1974), A Model for Large-scale Vegetation Survey, unpublished Ph.D thesis, Biology Department, University of Southampton, UK.

Diamond, J.M. (1975), 'The island dilemma: lessons of modern biogeographic studies for the design of nature reserves', Biol. Conserv., **7**, 129-46.

Diamond, J.M. and May, R.M. (1981), 'Island biogeography and the design of nature reserves', in May, R.M. (ed.), Theoretical Ecology: Principles and Applications, Blackwell, Oxford, UK, pp. 228-52.

Dimbleby, G. (1962), The Development of the British Heathlands and their Soils, Oxford Forestry Memoirs, Clarendon Press, Oxford, UK.

Douglas, I. (1981), 'The city as an ecosystem', Progress in Phys. Geog., **5**, No. 3, 315-67.

Douglas, I. (1983), The Urban environment: a biophysical approach, Edward Arnold, London.

Duckham, A.N. and Masefield, G.B. (1970), Farming Systems of the World, Chatto and Windus, London.

Dunnet, G.M. (1982), 'Ecology and Everyman', J. Animal Ecol., **51**, 1-14.

Eden, M. (1974), 'The origin and status of savanna and grassland in southern Papua', Trans. Inst. Brit. Geogrs., **63**, 97-110.

Edwards, K.J. (1982), 'Palynology and biogeography', Area, **14**, 241-48.

Evans, I.A. (1976), 'Relationship between bracken and cancer', Bot. J. of Linnean Soc., **73** (1-3), 105-12.

Evans, I.A. et al (1982), 'The carcinogenic, mutagenic and teratogenic toxicity of bracken', in Bracken in Scotland, Proc. Roy. Soc. Edinburgh, **81B** (1-2), 65-77.

Evans, J.G. (1972), Land Snails in Archaeology with special reference to the British Isles, Seminar Press, London.

Eyre, S.R. (1964), 'Determinism and the ecological approach to geography', Geog., **49**, 369-76.

Eyre, S.R. (1978), The Real Wealth of Nations, Edward Arnold, London.

Eyre, S.R. and Jones, G.R.J. (eds.) (1966), Geography as Human Ecology, Edward Arnold, London.

Fisher, C.A. (1959), The Compleat Geographer, Inaugural Lecture (35), University of Sheffield, England.

Flenley, J.R. (1979), The Equatorial Rain Forest: a Geological History, Butterworths, London.

Fleure, H.J. (1951), A Natural History of Man in Britain, Collins, London.

Francis, I.S., Taylor, J.A. and Williams, T. (1984), 'The use of microlight aircraft in the remote sensing of peatlands', Proc. Inter. Peat Congress Symposium, Aberdeen, Scotland, pp. 23-42.

Gerrard, A.J. (1981), Soils and Landforms: an Integration of Geomorphology and Pedology, Geo Allen and Unwin, London.

Gilbert, F.S. (1980), 'The equilibrium theory of island biogeography: fact or fiction?', J. Biog., **7**, 209-35.

Godwin, H. (1975), A History of British Flora: a Factual Basis for Phytogeography, Cambridge University Press, Cambridge, UK.

Gould, S.J. (1981), 'Palaeontology plus Ecology as Palaeobiology', in May, R. (ed.), Theoretical Ecology: Principles and Applications, Blackwells, Oxford, UK, pp. 295-317.

Haggett, P. (1965), Locational Analysis in Human Geography, Edward Arnold, London.

Haggett, P. and Chorley, R.J. (1969), Network Analysis in Geography, Edward Arnold, London.

Haines-Young, R.W. and Petch, J.R. (1980), 'The challenge of critical rationalism for methodology in physical geography', Prog. in Phys. Geog., **4**, 63-78.

Hall, P. (1970), The Theory and Practice of Regional planning, Pemberton, London.

Hare, F.K. (1969), 'Environment: resuscitation of an idea', Area, **4**, 52-5.

Harris, D.R. (1969), 'Agricultural systems, ecosystems and the origins of agriculture', in Ucko, P.J. and Dimbleby, G.W. (eds.), The Domestication and Exploitation of Plants and Animals, Duckworth, London, pp. 4-15.

Harrison, C.M. (1975), 'The description and analysis of vegetation', in Chapman, S.B. (ed.), Methods in Plant Ecology, Blackwell, Oxford, UK, pp. 85-155.

Harrison, C.M. and Frenkel, R.E. (1974), 'An assessment of the usefulness of phytosociological and numerical classificatory methods for the community biogeographer', J. Biogeog., **1**, 27-56.

Harvey, D. (1969), Explanation in Geography, Edward Arnold, London.

Higgs, A.J. (1981), 'Island biogeography theory and nature reserve design', J. Biogeog., **8**, 117-24.

Hill, A.R. (1972), 'Ecosystem stability and man: a research focus in biogeography', in Adams, W.P. and Helleiner, F.M. (eds.), International Geography, University of Toronto Press, Toronto, pp. 255-7.

Hill, A.R. (1975), 'Biogeography as a sub-field of geography', Area, **7**, 156-61.

Horn, H.S. (1981a), 'Succession', in May, R.M. (ed.), Theoretical Ecology: Principles and Applications, Blackwells, Oxford, pp. 253-71.

Horn, H.S. (1981b), 'Sociobiology', in May, R.M. (ed.), Theoretical Ecology: Principles and Applications, Blackwells, Oxford, pp. 272-94.

Howe, G.M. (1963), National Atlas of Disease Mortality in the United Kingdom, Nelson, London.

Howe, G.M. and Loraine, J. (eds.) (1980), Environmental Medicine, Heinemann, London.

Humboldt, F.H.A. von and Bonpland, A.J.A. (1805), Éssai sur la géographie des plantes, accompagné d'un tableau physique des regions équinoxiales, Levrault, Schoell and Compagnie, Paris (German edition, 1807).

Huntley, B. and Birks, H.J.B. (1983), An Atlas of Past and Present Pollen Maps of Europe: 0-13,000 Years Ago, Cambridge University Press, Cambridge, UK.

Isard, W. (1969), 'Some notes on the linkage of ecologic and economic systems', Papers and Proceedings of the Regional Science Association, **22**, 85-96.

Isard, W. et al (1967), 'On the linkage of socio-economic and ecological systems', Papers of Regional Science Association, **21**, 79-99.

Isard, W. et al (1972), Ecologic-Economic Analysis for Regional Development, The Free Press, New York, USA.

Johnston, R.J. (1983), 'Resource analysis, resource management and the integration of physical and human geography', Progress in Phys. Geog., **7** (1), 127-46.

Jones, R.L. (1980), Biogeography, Hulton Educational Press, Amersham, UK.

Kellman, M.C. (1980), Plant Geography, Methuen, London.

Law, F. (1957), 'The effect of afforestation on the water yield of catchment areas', J. Inst. Water Engineers, **11**, 269-76.

Linton, D.L. (1965), 'The geography of energy', Geog., **50**, 197-227.

Lockwood, J.G. (1983), 'The influence of vegetation on the Earth's climate', Prog. in Phys. Geog., **7** (1), 81-89.

Lynch, F. (1975), 'The impact of the landscape on prehistoric man', in Evans, J.G. et al (eds.), The Effect of Man on the Landscape: the Highland Zone, CBA Res. Report, No. 11, London, pp. 124-7.

MacArthur, R.H. and Wilson, E.O. (1967), The Theory of Island Biogeography, Princeton University Press, Princeton, USA.

Maltby, E.M. and Crabtree, K. (1976), 'Soil organic matter and peat accumulation on Exmoor: a contemporary and palaeoenvironmental evaluation', Inst. Brit. Geogrs., New Series, **1** (3), 259-78.

Martin, A.F. (1951), 'The necessity for determinism: a metaphysical problem confronting geographers', Trans. Inst. Brit. Geogrs., **17**, 1-11.

de Martonne, E. (1932), Traité de Géographie Physique, T.III, Biogéographie, Paris.

Mason, B.J. (1976), 'Towards an understanding and prediction of climatic variations', Q. J. Roy. Met. Soc., **102** (433), 473-98.

Mason, B.J. (1982), Review of H.H. Lamb's Climate, History and the Modern World and M.I. Budyko's The Earth's Climate: Past and Future, Times Higher Educational Supplement, London, 24 December.

Matthews, J.A. (1979), 'Refutation of convergence in a vegetation succession', Naturwissenschaften, 66, (1), 47-9, Springer-Verlag, Berlin.

Maxwell, T.J., Sibbald, A.R. and Eadie, J. (1979), 'Integration of forestry and agriculture - a model', Agricultural Systems, Applied Science Publishers, England, pp. 161-88.

May, R.M. (ed.) (1981), Theoretical Ecology: Principles and Applications (2nd edition), Blackwells Scientific Publications, Oxford.

MAFF (1966), Agricultural Land Classification, Tech. Rep. No. 11.

Moore, P.D. (1971), 'Vegetation history', Cambria, **4** (1), 73-83.

Morgan, W.B. and Moss, R.P. (1965), 'Geography and ecology: the concept of the community and its relationship to environment', Ann. Ass. Amer. Geog., **55** (2), 339-50.

Moss, R.P. (1977), 'Deductive strategies in geographical generalisation', Prog. in Phys. Geog., **1**, 23-39.

Munn, R.E. and Machta, L. (1979), 'Human activities that affect climate', World Climate Conference Proceedings, WMO Geneva, Overview Paper 8, III, 101-23.

Nelson, G.J. (1973), 'Comments on Léon Croizat's Biogeography', Syst. Zoo., **22**, 312-20.

Nelson, G. and Rosen, G.N. (1981), Vicariance Biogeography, Columbia University Press, New York, USA.

Newbigin, M.I. (1937), Plant and Animal Geography, Methuen, London.

Newcombe, K. (1977), 'Nutrient flow in a major urban settlement: Hong Kong', Human Ecology, **5**, 179-208.

Newman, E.I. (ed.) (1982), The Plant Community as a Working Mechanism, Spec. Pub. No. 1, Brit. Ecol. Soc., London.

Newson, M.D., Calder, I.R. and Walsh, P.D. (1982), 'The application of catchment lysimeter and hydro-meteorological studies of coniferous afforestation in Britain to land use planning and water management', Crop Symp. Hydrological Research Basins, Sonderh. Landes. Hydrologie, Bern, Switzerland, pp. 853-63.

Odum, E.P. (1971), Fundamentals of Ecology, Saunders, Philadelphia, USA.

Oldfield, F. (1967), 'The linkage of ecological teaching with that in the earth sciences', in Lambert, J.M. (ed.), The Teaching of Ecology, Brit. Ecol. Soc. Symp., Blackwells, Oxford, UK, pp. 33-40.

Oldfield, F. (1977), 'Lakes and their drainage basins as units of sediment-based ecological study', Prog. in Phys. Geog., **1**, 460-504.

Oldfield, F. (1983), 'Man's impact on the environment: some recent perspectives, Geog., **68** (3), 245-56.

Ollerenshaw, C. (1966), 'An approach to forecasting the incidence of fasioliasis over England and Wales, 1958-62', Agr. Met., **3**, 35-53.

O'Riordan, T. (1982), Putting Trust in the Countryside. Earth's Survival: a Conservation and Development Programme for the UK, Report No. 7, Nature Conservancy Council, London.

Park, C.C. (1979), Ecology and Environmental Management - a geographical perspective, Dawson, Folkestone, England.

Parker, D.J. and Rowswell, E.C.P. (1980), Water Planning in Britain, Allen & Unwin, London.

Parry, M.L. (1978), Climatic Change, Agriculture and Settlement, Dawson, Folkestone, England.

Pears, N. (1977), Basic Biogeography, Longmans, London.

Peel, R.F., Chisholm, M. and Haggett, P. (eds.) (1975), Processes in Physical and Human Geography, Heinemann, London.

Penman, H.L. (1948), 'Natural evaporation from open water, bare soil and grass', Proc. Roy. Soc. (A) **193**, 120-48.

Pennington, W. (1977), 'The Late-Devensian flora and vegetation of Britain', Phil. Trans. Roy. Soc. B, **280**, 247-71.

Pereira, H.C. (1973), Land Use and Water Resources in Temperate and Tropical Climates, Cambridge University Press, Cambridge, UK.

Polunin, N. (1960), Introduction to Plant Geography and Some Related Sciences, Longmans, London.

Popper, K.R. (1959), The Logic of Scientific Discovery (trans. from the German), Hutchinson, London (originally as Logik der Forschung, Springer, Vienna, 1934).

Robinson, M. (1980), The effects of pre-afforestation ditching upon the water and sediment yields of a small upland catchment, Report No. 73, Institute of Hydrology, Wallingford, Berkshire, UK.

Rosen, D.E. (1975), 'A vicariance model of Caribbean Biogeography', Syst. Zoo., **24** (4), 431-64.

Royal Society of Edinburgh (1982), 'Bracken in Scotland', Proc. Roy. Soc. Edin., B, **81** (Parts 1/2).

Sauer, C.O. (1952), Agricultural Origins and Dispersals, American Geog. Soc., New York, USA.

Seddon, B. (1971), Introduction to Biogeography, Duckworth, London.

Simmons, I.G. (1978), 'Physical geography in environmental science', Geog., **63** (4), 314-23.

Simmons, I.G. (1979), Biogeography: Natural and Cultural, Edward Arnold, London.

Simmons, I.G. (1980), 'Biogeography', in Brown, E.H. (ed.), Geography Yesterday and Tomorrow, Oxford University Press, Oxford, UK, pp. 146-66.

Simmons, I.G. (1981), The Ecology of Natural Resources, Edward Arnold, London.

Smagorinsky, J. (1981), 'CO_2 and climate - a continuing story', in Berger, A. (ed.), Climatic Variations and Variability: Facts and Theories, NATO Adv. Study Instit. Series D., Reidel Publishing Company, Dordrecht, Holland, pp. 661-87.

Smith, A.G. (1970), 'The influence of Mesolithic and Neolithic man on British vegetation: a discussion', in Walker, D. and West, R.G. (eds.),

Studies in the Vegetational History of the British Isles, Cambridge University Press, Cambridge, UK, pp. 81-96.

Smith, R.T. (1970), Studies in the Post-glacial Soil and Vegetation History of the Aberystwyth Area, UCW, Aberystwyth, Wales, UK, unpublished Ph.D thesis.

Smith, R.T. and Taylor, J.A. (1969), 'The Post-glacial development of vegetation and soils in Northern Cardiganshire', Trans. Inst. Brit. Geogrs., No. **47**, 85-96.

Stoddart, D.R. (1965), 'Geography and the ecological approach: the ecosystem as a geographic principle and method', Geog., **50**, 242-51.

Stoddart, D.R. (1966), 'Darwin's impact on geography', Ann. Ass. Amer. Geogrs., **56**, 683-98.

Stoddart, D.R. (1967), 'Organism and ecosystem as geographic models', in Chorley, R.J. and Haggett, P. (eds.), Models in Geography, Methuen, London, pp. 511-48.

Stoddart, D.R. (1977), Progress report: biogeography, Prog. in Phys. Geog., **1**, 537-43.

Stoddart, D.R. (1978), Progress report: biogeography, Prog. in Phys. Geog., **2**, 514-28.

Stoddart, D.R. (1981), Biogeography: dispersal and drift, Prog. in Phys. Geog., **5**, 575-90.

Stott, P. (1981), Historical plant geography, George Allen and Unwin, London.

Tansley, A.G. (1935), 'The use and abuse of vegetational concepts and terms', J. Ecol., **16**, 284-307.

Taylor, J.A. (1968), 'Reconnaissance vegetation surveys and maps: including a Report on the Vegetation Survey of Wales 1961-66', in Bowen, E.G., Carter, H. and Taylor, J.A. (eds.), Geography at Aberystwyth, University of Wales Press, Cardiff, Wales, UK, pp. 87-110.

Taylor, J.A. (1973), 'Chronometers and chronicles: a study of palaeoenvironments in West Central Wales', Prog. in Geog., **5**, 247-334.

Taylor, J.A. (1974a), 'The ecological basis of resource management', Area, **6**, No. 2, 101-6.

Taylor, J.A. (1975), 'The role of climatic factors in environmental and cultural changes in prehistoric times', in Evans, J.A. et al (eds.), The

Effect of Man on the Landscape: the Highland Zone, Res. Report No. 11, Council Brit. Arch., London, pp. 6-19.

Taylor, J.A. (1976), Mossborough Hall Farm: Report on Soils, with special reference to the fields from which sandy subsoils have been extracted for glass manufacture, available via Hugh Owen Library, UCW, Aberystwyth, Wales, UK.

Taylor, J.A. (1977), 'The distribution and interpretation of the peat deposits of the British Isles', Fifth Int. Peat Congress, Poznan, Poland, Vol. IV, 228-43.

Taylor, J.A. (1978), 'British upland environment and its management', Geography, **63** (4), 338-53.

Taylor, J.A. (1979a), Recreation Weather and Climate: a State of the Art Review, SC/SSRC Panel, London.

Taylor, J.A. (1979b), A Report on Llanbrynmair Estate: An Assessment of its Land Uses and Development, available via Hugh Owen Library, UCW, Aberystwyth, Wales, UK.

Taylor, J.A. (1980a), 'Environmental changes in Wales during the Holocene period', in Taylor, J.A. (ed.), Culture and Environment in Prehistoric Wales, Brit. Arch. Report, Oxford, UK, pp. 101-30.

Taylor, J.A. (1980b), 'Man-environment relationships', in Taylor, J.A. (ed.), Culture and Environment in Prehistoric Wales, Brit. Arch. Reports, Oxford, UK, pp. 311-36.

Taylor, J.A. (1980c), 'Bracken: an increasing problem and a threat to health', Outlook on Agriculture, **10** (6), 298-304.

Taylor, J.A. (1984), 'Established and new growth points in biogeography', Prog. in Phys. Geog., **8**, No. 1, Edward Arnold, London, pp. 94-101.

Taylor, J.A. (ed.) (1967), Weather and Agriculture, Pergamon Press, Oxford, UK.

Taylor, J.A. (ed.) (1980), Culture and Environment in Prehistoric Wales, Brit. Arch. Reports, Oxford, UK.

Taylor, J.A. and Smith, R.T. (1972), 'Climatic peat - a misnomer?', Proc. 4th IPC Conference, Helsinki, Vol. **1**, 471-84.

Taylor, J.A. and Smith, R.T. (1980), 'The role of pedogenic factors in the initiation of peat formation and the classification of mires', Proc. Sixth Int. Peat Congress, Duluth, USA, pp. 109-18.

Taylor, J.A., Thomasson, A.J. and Wales-Smith, B.G. (1980), 'Potential moisture deficit and soil water availability', in Gregory, K.J. and Doornkamp, J.C. (eds.), Atlas of the 1976 British Drought, Inst. Brit. Geogrs., London, 51-2.

Terborgh, J. (1974), 'Preservation of natural diversity: the problem of extinction-prone species', BioScience, **24**, 715-22.

Thomas, D. (1963), Agriculture in Wales during the Napoleonic Wars: a Study in the Geographical Interpretation of Historical Sources, University of Wales Press, Cardiff, Wales, UK.

Tivy, J. (1972), 'The concept of carrying capacity in relation to recreational land use in the USA', Occ. Paper No. 3, CCS, Perth VII-58, Oliver and Boyd, Edinburgh, Scotland, UK.

Tivy, J. and O'Hare, G. (1980), Man and the Ecosystem, Oliver and Boyd, Edinburgh, Scotland, UK.

Tivy, J. and Rees, J. (1978), 'Recreational impact on Scottish lochside wetlands', J. Biog., **5**, 93-108.

Tubbs, C.R. and Blackwood, J.W. (1971), 'Ecological evaluation of land for planning purposes', Biol. Conserv., **3** (3), 169-72.

Vuilleumier, F. (1977), 'Qu'est-ce que la biogéographie?', Cr. Soc. Bio., **475**, 41-66.

Ward, R.C. (1980), 'Water: a geographical issue', in Brown, E.H. (ed.), Geography Yesterday and Tomorrow, Oxford University Press, UK, pp. 130-145.

Warren, A. and Goldsmith, F.B. (1976), 'The impact of recreation on the ecology and amenity of semi-natural areas: methods of investigation used in the Isles of Scilly', Biol. J. Linn. Soc. Lond., **2**, 287-306.

Warren, A. and Goldsmith, F.B. (eds.) (1974), Conservation in Practice, Wiley, London.

Warren, A. and Harrison, C.M. (1970), 'Conservation, stability and management', Area, **2**, 26-32.

Watts, D. (1971), Principles of Biogeography: an Introduction into the Functional Mechanisms of Ecosystems, McGraw Hill, London.

Watts, D. (1978), 'The new biogeography and its niche in physical geography', Geography, **63** (4), 324-37.

Whitmore, T.C., Flenley, J.F. and Harris, D.R. (1982), 'The Tropics as the norm in biogeography', Geog. J., **148** (1), 8-21.

Williams, R.G.B. (1975), 'The British climate during the Last-Glaciation: an interpretation based on periglacial phenomena', Geol. J., Spec. Issue, **6**, 95-120.

Wilson, E.O. and Willis, E.O. (1975), 'Applied Biogeography', in Cody, M.L. and Diamond, J.M. (eds.), Ecology and Evolution of Communities, Harvard University Press, USA, pp. 522-34.

Wooldridge, S.W. (1945), The geographer as scientist, inaugural lecture delivered at Birkbeck College, 30 November 1945, in Wooldridge, S.W. (1956), The geographer as scientist, Nelson, London, pp. 7-25.

Wooldridge, S.W. (1949), 'On taking the ge- out of geography', Geog., **34**, 9-18.

Wooldridge, S.W. (1950), 'Reflections on regional geography in teaching and research', Trans. Inst. Brit. Geogrs., **16**, 1-11.

FURTHER READING

Barrett, E.C. and Curtis, L.F. (1976), Introduction to Environmental Remote Sensing, Chapman and Hall, London.

Bennett, R.J. and Chorley, R.J. (1978), Environmental Systems: Philosophy, Analysis and Control, Methuen, London.

Black, J. (1970), The Dominion of Man: the Search for Ecological Responsibility, Edinburgh University Press, Edinburgh, Scotland, and New York, USA.

Chorley, R.J. (1973), 'Geography as human ecology', in Chorley, R.J. (ed.), Directions in Geography, Methuen, London.

Chorley, R.J. and Kennedy, B.A. (1971), Physical Geography: a Systems Approach, Prentice Hall, Englewood Cliffs, NJ, USA.

Collinson, A.S. (1977), Introduction to World Vegetation, George Allen & Unwin, London.

Cox, C.B., Healey, I.N. and Moore, P.D. (1976), Biogeography: an Ecological Approach (second edition), Blackwell Scientific Publications, Oxford, UK.

Curtis, L.F. and Simmons, I.G. (eds.) (1976), Man's Impact on Past Environments, Thematic Number, Trans. Inst. Brit. Geogrs. NS1(3).

Dansereau, P. (1957), Biogeography, Ronald Press, New York, USA.

Darlington, P.J. (1957), Zoogeography: the Geographic Distribution of Animals, Wiley, New York, USA.

Douglas, I. (1983), The Urban environment: a biophysical approach, Edward Arnold, London.

Duckham, A.N. and Masefield, G.B. (1970), Farming Systems of the World, Chatto and Windus, London.

Eyre, S.R. (1978), The Real Wealth of Nations, Edward Arnold, London.

Flenley, J.R. (1979), The Equatorial Rain Forest: a Geological History, Butterworths, London.

Gerrard, A.J. (1981), Soils and Landforms: an Integration of Geomorphology and Pedology, Geo Allen and Unwin, London.

Glacken, C.J. (1967), Traces on the Rhodian Shore, University of California Press, USA.

Godwin, H. (1975), A History of British Flora: a Factual Basis for Phytogeography, Cambridge University Press, Cambridge, UK.

Haggett, P. and Chorley, R.J. (1969), Network Analysis in Geography, Edward Arnold, London.

Harvey, D. (1969), Explanation in Geography, Edward Arnold, London.

Illies, J. (1974), Introduction to Zoogeography, Macmillan, London.

Isard, W. et al (1967), 'On the linkage of socio-economic and ecological systems', Papers of Regional Science Association, **21**, 79-99.

Isard, W. et al (1972), Ecologic-Economic Analysis for Regional Development, The Free Press, New York, USA.

Jones, R.L. (1980), Biogeography, Hulton Educational Press, Amersham, UK.

Kellman, M.C. (1980), Plant Geography, Methuen, London.

MacArthur, R.H. (1972), Geographical Ecology, Harper and Row, New York, USA.

MacArthur, R.H. and Wilson, E.O. (1967), The Theory of Island Biogeography, Princeton University Press, Princeton, USA.

de Martonne, E. (1932), Traité de Géographie Physique, T.III, Biogéographie, Paris.

Maxwell, T.J., Sibbald, A.R. and Eadie, J. (1979), 'Integration of forestry and agriculture - a model', Agricultural Systems, Applied Science Publishers, England, pp. 161-88.

May, R.M. (ed.) (1981), Theoretical Ecology: Principles and Applications (2nd edition), Blackwells Scientific Publications, Oxford.

Nelson, G. and Rosen, G.N. (1981), Vicariance Biogeography, Columbia University Press, New York, USA.

Newbigin, M.I. (1937), Plant and Animal Geography, Methuen, London.

Newman, E.I. (ed.) (1982), The Plant Community as a Working Mechanism, Spec. Pub. No. 1, Brit. Ecol. Soc., London.

O'Riordan, T. (1982), Putting Trust in the Countryside. Earth's Survival: a Conservation and Development Programme for the UK, Report No. 7, Nature Conservancy Council, London.

Parry, M.L. (1978), Climatic Change, Agriculture and Settlement, Dawson, Folkestone, England.

Polunin, N. (1960), Introduction to Plant Geography and Some Related Sciences, Longmans, London.

Simmons, I.G. (1979), Biogeography: Natural and Cultural, Edward Arnold, London.

Simmons, I.G. (1980), 'Biogeography', in Brown, E.H. (ed.), Geography Yesterday and Tomorrow, Oxford University Press, Oxford, UK, pp. 146-66.

Simmons, I.G. (1981), The Ecology of Natural Resources, Edward Arnold, London.

Stoddart, D.R. (1981), Biogeography: dispersal and drift, Prog. in Phys. Geog., **5**, 575-90.

Stott, P. (1981), Historical plant geography, George Allen and Unwin, London.

Taylor, J.A. (1973), 'Chronometers and chronicles: a study of palaeoenvironments in West Central Wales', Prog. in Geog., **5**, 247-334.

Taylor, J.A. (1976), Mossborough Hall Farm: Report on Soils, with special reference to the fields from which sandy subsoils have been extracted for glass manufacture, available via Hugh Owen Library, UCW, Aberystwyth, Wales, UK.

Taylor, J.A. (1979), Recreation Weather and Climate: a State of the Art Review, SC/SSRC Panel, London.

Taylor, J.A. (1984), 'Established and new growth points in biogeography', Prog. in Phys. Geog., **8**, No. 1, Edward Arnold, London, pp. 94-101.

Taylor, J.A. (ed.) (1980), Culture and Environment in Prehistoric Wales, Brit. Arch. Reports, Oxford, UK.

Tivy, J. and O'Hare, G. (1980), Man and the Ecosystem, Oliver and Boyd, Edinburgh, Scotland, UK.

Vuilleumier, F. (1977), 'Qu'est-ce que la biogeographie?', Cr. Soc. Bio., **475**, 41-66.

Watts, D. (1971), Principles of Biogeography: an Introduction into the Functional Mechanisms of Ecosystems, McGraw Hill, London.

Watts, D. (1978), 'The new biogeography and its niche in physical geography', Geography, **63** (4), 324-37.

Wratten, S.D. and Fry, G.L.A. (1980), Field and Laboratory Exercises in Ecology, Edward Arnold, London.

GLOSSARY

Abiotic	The non-living part of an ecosystem and any influences derived from its physical environment.
Adaptation	The process, physiological or genetical, by which an organism becomes adjusted to its environment.
Adaptive Radiation	The development of a new species from one ancestor species into a range of closely-related but adaptively very different species.
Allelopathy	The phenomenon whereby an organism selectively prejudices the environment against its own offspring, e.g. root exudates which prevent germination of the same species.
Allogenic Succession	Succession brought about by external changes in the environment, such as climatic change.
Allopatry	The total physical separation of ranges of similar species.
Autecology	The study of the interrelationships between an individual organism and its environment, including other organisms.
Autogenic Succession	Succession brought about by the action of the organisms themselves, e.g. a hydrosere, brought about by the reduction of water depth consequent upon the accumulation of organic debris.
Autotrophs	Organisms able to manufacture their own food requirement from inorganic substances.
AVHRR	Advanced, very high resolution radiometer carried on board the NOAA 6 and 7 and the TIROS-N satellites.
Bidirectional Reflectance	Reflectance whose angularity is determined by the sun and sensor angles.

Biocoenosis	Term first proposed by Möbius to describe the internal relationships of living plant or animal communities.
Biogeochemical Cycling	The circulations of chemical elements in the biosphere and through ecosystems from the biotic to the abiotic components.
Biogeochemistry	The chemical analysis of plant material to elucidate its chemical composition.
Biogeocoenosis	Term developed from Möbius' 'biocoenosis' to describe the sum total of ecological niches, plant and animal, with their environment.
Biogeography	The study and interpretation of distributions of plants, animals and man over space and time in relation to their environments.
Biomass	The weight of living material in an organism or ecosystem, usually expressed as dry matter per unit area, e.g. Kg/ha or g/m^2.
Biome	A generalised concept for a major community of plants and/or animals associated with an extensive area or zone of relatively uniform climate and soil types.
Biosphere	That part of the earth's terrestrial aquatic and atmospheric environments containing living organisms. The marginal zones of the biosphere where the presence of living organisms is temporary or occasional is termed the parabiosphere.
Biota	Groups of plants and/or animals located together in space.
Biotic	The living part of an ecosystem and any influences derived from the activity of living organisms.
Biotic Factors	Influences that derive from the activity of living organisms.
Blanket Peat	A term proposed by Tansley for the widespread shallow peat cover of 'the

British uplands, initiated mainly by increasing ground wetness consequent upon deforestation but accumulated mainly under the control of climatic factors, in particular a high precipitation/evaporation ratio. In western Ireland an analogous formation occurs at low altitudes.

<u>BP</u> — An absolute age of a fossil organic deposit, given in radiocarbon years <u>before present</u>, i.e. before 1950 AD.

<u>Buffering Capacity</u> — The capacity of a soil to resist temporal changes in its properties. The term is used to describe, for example, a soil's ability to resist leaching, counteract the effects of large applications of lime, or to maintain its supply of nutrients.

<u>Calcicole</u> — A plant tolerant of, and which grows best on, base-rich (calcareous) soils.

<u>Calcifuge</u> — A plant which is tolerant of, and grows best on, base-poor (acidic) soils.

<u>Catastrophe Theory</u> — A mathematical theory postulated by Rene Thom to explain how a system may be subject to gradual change, due to external variables, but attains a stage when its reaction to external change is discontinuous, i.e. a catastrophe occurs.

<u>Cation Exchange Capacity</u> — The capacity of clays and humus to adsorb cations (positively charged ions) from the soil solution.

<u>Characteristic Species</u> — Species which are present in two of Wallace's regions (q.v.).

<u>Chlorophyll</u> — The green pigment found in most plants which traps the energy of sunlight and uses it to produce organic compounds via photosynthesis.

<u>Cladistic Relationship</u> — A component of phytogenetic relationship referring to the pathways of ancestry, i.e. the branching patterns by which evolutionary lineages arose and diverged from one another. This is a method of numerical taxonomy which gives support to the idea of sudden change in evolution.

Cladogram	An expression of a cladistic relationship in the form of a tree-diagram in which the vertical scale is taken to indicate time or evolutionary advancement.
Clarke Profile Index	A formula proposed by G.R. Clarke which combines evaluations of soil texture and soil drainage to derive classifications of soil quality.
Climatic Peat	Peat which has accumulated under the predominant control of climatic factors.
Climax: Climax Community	The final stage (climax) of a successional sequence when the (climax) community has attained equilibrium with, and stability within, the prevailing abiotic environment.
Closed System	A system with no transfers of matter to and/or from its surroundings.
Colour Composite	A colour picture which comprises multi-spectral imagery where one colour is assigned to each spectral band used.
Community (Plant or Animal)	An organised aggregation of plants or animals of characteristic composition and form which has acquired a recognisable identity over time and in space.
Continental Drift	The theory that the continents have changed in position over geological time in response to the movement of lithospheric plates (see Plate Tectonics).
Cosmopolitan Species	Species which are present in at least five of Wallace's regions (q.v.).
Cultural Biogeography	The study of man's changing relationships with the natural environment over time.
Cultural Materialism	A view taken by some archaeologists that the nature, range and accessibility of materials available to a culture is determined by environmental controls.
Darwinism	The idea of evolution by natural selection of small heritable characters, as postulated by Charles Darwin.

Determinism — The philosophical doctrine that human action is not free but necessarily determined by motives which are regarded as external forces acting upon the will (see Physical Determinism and Possibilism).

Devensian Glaciation — The last glacial period in Britain based on a type-locality near Chester (Deva).

Dynamic Equilibrium — A state in a system where rates of change operate in relative balance to maintain that state.

Ecology — The study of the interrelationships of plants and animals to each other and to their abiotic environments.

Ecosystem — Term first proposed by Tansley for a community of organisms linked to each other and to their abiotic environments by the energy flow through trophic levels (food chains) and the cyclic interchange of materials.

Ecotone — The transitional zone between two ecosystems.

Ecotope — Following Tansley, that particular portion of the physical world forming a 'home' for the organisms which inhabit it (see Ecosystem).

Edge Pixels — Pixels which contain a mixed reflectance from two or more targets.

Electromagnetic Spectrum — The range of electromagnetic radiation that moves with the velocity of light and is characterised by wavelength and frequency. For remote sensing of vegetation the most useful part of the EM lies between 0.4 and 2.5 µm.

Endemic — Species which are confined to a particular region.

Environmental Archaeology — That branch of archaeology which studies the nature and impact of environmental factors on the prehistoric and historic cultural successions.

Environmental Determinism See Physical Determinism.

Environmental Science The comprehensive and integrated study of the natural environment and man's adaptations of it.

Eurytopic Tolerant of a wide ecological range, i.e. the opposite of Stenotopic.

Evapotranspiration The amount of water evaporated from plants, soils and other moist surfaces, i.e. the transport of water from the earth back to the atmosphere.

Feedback; Feedback Loop The interactive process whereby part of the output of an ecosystem is returned to become an input. Negative feedback loops encourage stability in ecosystems. Positive feedback loops promote change in ecosystems.

Flandrian Period (See Holocene.) Established term for the Post-glacial (post-10,000 BP) period in North-West Europe on the basis of primary evidence from sites in the Flanders area of the Low Countries.

Geobotany The visual detection of mineralization by means of peculiar plant distributions, indicator plants or the occurrence of plant morphology and changes induced by excess heavy metals.

Gleying The process whereby intermittent or permanent waterlogging creates alternating conditions of oxidation and reduction or a state of permanent reduction, respectively, in a soil profile.

Gradualism The kind of evolution originally postulated by Charles Darwin, i.e. by the natural selection of small heritable characters.

Gross Primary Productivity (GPP) Gross production in plants or ecosystems, or the total energy assimilated by the organism in a given time and over a given area.

Hemispherical Reflectance Reflectance with no angular dependence.

Heterotrophs — Organisms unable to manufacture their own food which must obtain their food requirement from the tissues of other organisms, living or dead.

Holism, Holistic — Holism is the theory concerning the natural tendency of groups of units to form themselves into wholes. Hence, a holistic approach is as widely based and as comprehensive as possible.

Holocene Period — A geological and international term for the present interglacial period which began in many parts of the world approximately 10,000 years ago (see Flandrian).

Homeostatis — The process whereby a change in the external environment stimulates a response to maintain equilibrium within an organism or ecosystem.

Humus; Humic Material — The organic content of the soil that has been decomposed and has lost its original structure. Organic additives to the soil are included in the definition.

Hydrophyte — A semi-aquatic or aquatic plant.

Hydroponics — The cultivation of plants in a solution of nutrient salts.

Infrared Film — Film which is sensitive to the 0.4 to 0.9 um region. It is generally used with a minus blue (yellow) filter so that, effectively, the film is sensitive to the 0.5 to 0.9 um region. The film can be either black and white or colour.

Inherent Productivity — The net radiant energy available at a given location area over unit time as conditioned by such variables at ground as slope, soil, sub-soil hydrology, land use history, etc. This concept underpins and is complementary to the concepts of Land Productivity and Net Primary Productivity.

IBP: International Biological Programme — Instituted during the International Geophysical Year (1958) as a multinational co-operative programme to monitor the major ecosystems of the earth.

Island Biogeography, Equilibrium Theory of	The theory postulated by MacArthur and Wilson (1967) that a dynamic relationship exists for a given island between (1) the colonisation rates of arriving species (which is a function of the distance of the island from source) and (2) the extinction rates of species on the island (which is a function of island area).
Isotropic Plane	An assumed regular planar surface against which spatial patterns of human geography may be interpreted in non-environmental terms.
K-Selection	Selection for a slow rate of population growth, and for a more efficient use of available resources.
Land Productivity; Land Potential	The ranking of topography, soil, sub-soil, hydrology, topoclimate and land use history as conditioning the biological and productivity of a given site/area over unit time.
Land Evaluation	An omnibus term for the several methods of evaluating land, including environmental, social, economic, and technical criteria.
Land Use Ecology	The study of the role and impact of ecological (or environmental) factors on land use patterns, productivities and management.
Land Use Optimisation	The adoption of forms of land use best suited to the environmental and managerial resources available at a given location and time.
Landnam	A term introduced by Scandinavian palaeoecologists, notably J. Iversen, for the very first interferences in, but not the first removals of, the mid-Flandrian woodland cover, usually by Mesolithic or Neolithic peoples, both directly and indirectly via hunted or partially herded wild animals.
Leaching	The gravitational translocation of soluble or particulate material from the upper horizons of the soil to deeper ones and, eventually, in part to subsoils and drainage waters.

<u>Leaf Area Index (LAI)</u> — The total area of leaves growing above a given area of ground divided by the area of the ground itself.

<u>Mesophyte</u> — A plant exhibiting moderate moisture requirements.

<u>MJ</u> — An abbreviation for MegaJoule or a million Joules. A Joule is a unit of work defined as that work performed when a force of one Newton moves a body one metre in the direction of the force. Both Joules and Watts are standard units of measurement of energy in the <u>System International</u> (SI) measurement system.

<u>Monophylesis</u> — The origin of a taxonomic group at one point in space and time, by evolution.

<u>Multispectral Imagery</u> — Remote sensing in two or more spectral bands, such as the green and the red visible light bands.

<u>MSS</u> — Multispectral scanner carried on board the Landsat satellites.

<u>Neo-Darwinism</u> — The evolutionary theory which combines Darwinism with modern genetics. It regards the gene-pool of a population as the fundamental unit of evolution, and takes into account larger mutations as well as the small heritable variations postulated by Darwinism.

<u>Net Primary Production (NPP)</u> — The rate at which plants store energy as organic matter in excess of that used in respiration, expressed usually as dry weight of organic matter, per unit area, per unit time, e.g. $g/m^2/day$ or kg/ha/yr.

<u>Nutrient Budget</u> — The balance sheet of nutrient transfers within any defined system as related to inputs and outputs of nutrients. This may be a simple soil-plant relationship or an overall assessment of mass balance within a catchment or part (or the whole) of the biosphere.

<u>Open System</u> — A system where energy and matter are constantly transferred across the system boundaries.

Oxidation-Reducation Potential — Otherwise known as environmental redox (Eh), this is a measure of the oxidising or reducing tendency of given conditions (cf. pH scale of acidity/alkalinity). As the capacity of the environment to take up electrons (to facilitate oxidation) decreases so the oxidation-reduction potential falls. Redox is dependent on temperature, pH and the amount of organic matter present.

Palaeoecology — The study of the interrelationships of fossil plants and fossil animals to each other and to their contemporary abiotic environments.

Palaeoecosystem — A fossil ecosystem.

Palynology — The study of fossil pollen.

Panchromatic Film — Black and white film sensitive to the whole of the visible part of the spectrum.

Pandemic — Species which exist in more than one of Wallace's regions (q.v.).

Paradigm — A pattern, exemplar or example.

Parapatry — The condition of contiguity which occurs in the ranges of some similar species.

Peat — An accumulation of plant debris which has remained incompletely decomposed principally due to the lack of oxygen in a waterlogged environment.

Pedogenesis — The transformation of regolith (i.e. the weathered mantle) or rock to soil by the initial plant (or animal) colonisations.

Pedogenic Peat — Peat which has accumulated under the predominant control of soil processes.

Pedon — The smallest individual soil-landscape unit as defined by the Comprehensive Classification System (USDA). This equates with the three-dimensional body of soil surrounding a given soil profile, and accords with the arbitrary notion of soils as comprising a collection of individual unit soil profiles.

<u>Phenology</u> — The study of the times of recurrence of natural phenomena, e.g. in the growth cycle of plants.

<u>Photosynthesis</u> — The production of organic compounds by green plants from water and carbon dioxide using the energy from sunlight trapped by chlorophyll.

<u>Photosynthetically Active Radiation</u> — That part of solar radiation that is potentially usable by a plant for photosynthesis.

<u>Photosynthetic Efficiency</u> — The ratio of the amount of energy taken into the plant by photosynthesis (specified as gross or net) to the amount of radiation received at the site (specified as total global or photosynthetically active).

<u>pH</u> — The soil reaction in terms of acidity or alkalinity.

<u>Phylogenesis</u> — The origin of a taxonomic group by evolution.

<u>Phylogeny</u> — The study of the evolutionary history of groups of organisms; hence phylogenetic or evolutionary classifications.

<u>Physical Determinism</u>, or <u>Environmental Determinism</u> — The view that attempts to explain human activities in terms of physical (or environmental) causes and controls (see <u>Determinism</u> and <u>Possibilism</u>).

<u>Physiognomy</u> — Study of the external features of appearance of an object, e.g. a plant.

<u>Physiography</u> — Scientific description of the physical features of the earth.

<u>Pixel</u> — Picture element; the spatial and spectral resolution cell of an image.

<u>Plate Tectonics</u> — The theory that the lithosphere is composed of up to six major plates capable of independent movement. Plates are built at mid-oceanic ridges and destroyed at subduction zones (see Continental <u>Drift</u>).

<u>Podzolisation</u> — The process whereby a soil is depleted of bases to become acid and develop an

eluviated surface 'A' horizon of removal and illuviated 'B' horizon of accumulation.

Positivism — The philosophical system of Auguste Comte (1798-1857) which recognises only observed phenomena and rejects speculation or theory.

Possibilism — The philosophical doctrine which stresses the operation of Man's free will in selecting his activities. The view was urged by the French historian, Lucien Febvre, in opposition to environmental determinism. Its creed was that 'there are no necessities, only possibilities'.

Production Ecology — The study of the energy flow through ecosystems, involving studies in photosynthesis, uptake and dissipation of energy, accumulation of biomass and food relationships.

Productivity — The level of production of an ecosystem compared with its potential or with other ecosystems.

Propagules — The units which are the means of dispersal and propagation of a species: seeds, spores, gemmae, etc.

Psammophyte — A plant adapted to the colonisation of sandy areas, e.g. dune systems.

Quadrat — A selected spatial sample of a community designed to be representative of the community as a whole (see Relevé).

Recreation Ecology — The study of the role and impact of ecological (or environmental) factors on recreational activities.

Red Edge — The steep rise in plant reflectance in the vicinity of 0.75 um.

Reductionism — The reductionist view (as advocated initially by Gleason) is to see the whole as no more than the sum of its constituent parts, and to believe that complex data and phenomena can best be explained in simple and individual terms. In ecology, the Gleasonian view, that communities of organisms are simply random aggregates of individuals with common or similar environmental requirements, with no emergent properties, contrasts with the Clementsian concept of community as a 'superorganism', with emergent properties not present at the level of the individual.

Relevé (Aufnahmen) — A selected spatial sample of a community designed to be representative of the community as a whole (see Quadrat).

Relict — Spatially, a small, restricted population of a species that was formerly much more widespread.

Remote Sensing — The acquisition of information about an object by a recording device that is not in physical contact with the object.

Respiration — That part of the assimilated energy converted by the plant or by the ecosystem to heat or mechanical energy or used in life processes.

r-Selection — Selection for a high rate of population growth and high productivity.

Saltation — The theory which postulates the origin of major new taxonomic groups by the occurrence of single massive mutations.

SAR — Synthetic aperture radar.

Sclerophyllous — Hard-leaved, usually leathery.

Semi-cosmopolitan — Species which are present in three or four of Wallace's regions (q.v.) normally in either tropical or extra-tropical areas but not in both.

SLAR — Side-looking airborne radar.

Soil Structure — The arrangement of soil particles into aggregates, including their morphology and stability.

Soil Texture — The composition of a soil based on the proportions of mineral particles of different sizes.

Spatial Resolution	A measure of the ability of the sensor to distinguish between spatially separate objects. Resolving power can be expressed in lines per millimeter or in meters.
Speciation	The formation of a new species.
Species Population	The continuous flux of species within a community of plants or animals.
Spectrophotometer	Laboratory instrument to measure the hemispherical reflectance of small targets. Measurements are made in spectral form.
Spectroradiometer	Field instrument which uses the sun as its radiation source to measure the bidirectional reflectance of targets. Measurements are made in spectral form.
Steady State	A condition of equilibrium where, in theory, the net rate of change of a system is zero but where, in practice, rates of change are insignificant in relation to the time scales under consideration (quasi-equilibrium) or are collectively in balance or nearly so (dynamic equilibrium).
Stenotopic	Tolerant of only a narrow ecological range, i.e. the opposite of eurytopic.
Succession	The regular and progressive stages in the components of an ecosystem from the initial colonisation of an area to a stable state or climax.
Supervised Classification	A machine classification conducted on the basis of training area information which is defined by the user.
Synecology	The study of living groups or communities of plants or animals.
System International (SI)	The internationally agreed system of standard measurements.
Systematics	The scientific study of the diversity and differentiation of organisms and of the relationships between them.
Taxon	A classification group of any rank in the hierarchy, such as a species, genus, family, etc., of plants or animals; pl. taxa.

<u>Taxonomy</u> — The study of classification, including its principles and practices.

<u>Trophic Level</u> — A grouping of organisms according to their food sources.

<u>Turnover</u> — The process of extinction of certain species and their replacement by others.

<u>Uniformitarianism</u> — The belief that there has been essential uniformity of cause and effect throughout the physical history of the earth. The corollary is that the present is the key to the past (the opposite belief is termed <u>Catastrophism</u>).

<u>Van t'Hoff Rule</u> — This rule states that the rates of chemical reaction in a system have a general tendency to double for every rise of 10°C in temperature.

<u>Vegetation</u> — Plants considered collectively: plant growth in the mass.

<u>Vicariance</u> — The splitting of old, ancestor populations into a series of new, spatially distinct sub-populations.

<u>W</u> — An abbreviation for Watt, which is a unit of power defined as one Joule per second. Both work and power are related to energy by the work-energy theory of physics. Watts and Joules are standard units of measurement of energy in the System International (SI) measurement system.

<u>Wallace's Regions</u> — The major faunal regions of the world, as delimited by A.R. Wallace (1876).

<u>Weathering</u> — This term refers to all the processes whereby rocks are decomposed or dis-integrated because of exposure at or near the earth's surface.

<u>Xenophyte</u> — A plant which is adapted to withstand or to avoid moisture stress and very dry environmental conditions.

SUBJECT INDEX